AF356593

High Voltage Insulating Materials-Current State and Prospects

High Voltage Insulating Materials-Current State and Prospects

Editors

Pawel Rozga
Abderrahmane Beroual

MDPI • Basel • Beijing • Wuhan • Barcelona • Belgrade • Manchester • Tokyo • Cluj • Tianjin

Editors
Pawel Rozga
Institute of Electrical Power
Engineering
Lodz University of Technology
Lodz
Poland

Abderrahmane Beroual
Ecole Centrale de Lyon
University of Lyon
Ecully
France

Editorial Office
MDPI
St. Alban-Anlage 66
4052 Basel, Switzerland

This is a reprint of articles from the Special Issue published online in the open access journal *Energies* (ISSN 1996-1073) (available at: www.mdpi.com/journal/energies/special_issues/HV_IM).

For citation purposes, cite each article independently as indicated on the article page online and as indicated below:

LastName, A.A.; LastName, B.B.; LastName, C.C. Article Title. *Journal Name* **Year**, *Volume Number*, Page Range.

ISBN 978-3-0365-1638-7 (Hbk)
ISBN 978-3-0365-1637-0 (PDF)

Contents

About the Editors

Pawel Rozga

Pawel Rozga is an associate professor at the Institute of Electrical Power Engineering of Lodz University of Technology, Poland; is the vice director of development at this institute; was a visiting professor at Mississippi State University, USA (2010/11); was a visiting researcher at Lappeenranta University of Technology, Finland (2013); and was a lecturer and scientific consultant at Chongqing University, China (2017). Currently, he is developing his own research group at Lodz University of Technology, researching the pre-breakdown and breakdown phenomena in dielectric liquids including esters and the influence of impregnation parameters on the electrical strength of solid insulation materials. He is also a senior member of IEEE, a member of the IEEE Technical Committee on Liquid Dielectrics, an associate editor of IEEE TDEI, and a member of the Section Editorial Board of *Energies*. He has published more than 110 papers and 2 book chapters.

Abderrahmane Beroual

Abderrahmane Beroual is a distinguished professor at the Ecole Centrale de Lyon, University of Lyon, France; was a recipient of the 2016 IEEE T. Dakin Award; is an IEEE fellow; is an associate editor for IEEE TDEI; was the head of the High Voltage Group at AMPERE Lab—CNRS (1999–2020) France; was a scientific expert at the SuperGrid Institute (1999-2020), France; is a distinguished visiting professor at the UK Royal Academy of Engineering at Cardiff University and at King Saud University of Riyadh, Saudi Arabia; was a visiting professor at many other universities; and is a member of the International Advisory Committee of the MALET Centre of Excellence—Wroclaw Division, Poland. He chaired the International Study Group on Streamer Propagation in Liquids of the IEEE DEIS, and he is a member of many advisory/scientific committees of international conferences (IEEE ICDL, IEEE CEIDP, IEEE ICHVE, CMD, etc.). His main research interests include high-voltage insulation, outdoor insulation, dielectric materials, and long air-gap discharge. He supervised more than 45 PhD theses, and he has published more than 490 papers, 5 patents, 2 books, and 6 book chapters.

Preface to "High Voltage Insulating Materials-Current State and Prospects"

The book is focused on the presentation of new solutions in the field of high-voltage insulating materials: gaseous, liquids, and solid. Its aim is to show the current trends and prospects of research on the development of insulating materials of different types. This topic was taken up due to the need to ensure the proper operation of high-voltage devices where gaseous, liquid, or solid insulation is applied. These materials have to ensure the correct, continuous, uninterrupted, and safe operation of various devices. It is also extremely important to ensure an appropriate level of environmental safety with the knowledge that high-voltage devices and, therefore, insulating materials operate under various stresses, which can be critical during their operation. The in-service conditions require high-quality materials, the properties of which do not change or have small changes over time. Hence, it is very important to investigate the properties of materials used in high-voltage equipment from various perspectives. Thus, electrical (AC, DC, LI, or combined exposure), thermal (e.g., accelerated aging), or chemical (impact of environmental factors or material compatibility) based factors must be taken into account in the assessment of these materials. It is the optimal solution that is sought for in a given application. Furthermore, the development of HVDC technology, which involves specific stress conditions and thus specific physical phenomena associated with stress, has become extremely important in this respect. This area of research is of particular interest of scientists, practicioners, engineers, and others who work in the area of high-voltage devices and high-voltage insulation.

Pawel Rozga, Abderrahmane Beroual

Editors

Editorial

High Voltage Insulating Materials—Current State and Prospects

Pawel Rozga [1,*] and **Abderahhmane Beroual** [2]

1   Institute of Electrical Power Engineering, Lodz University of Technology, Stefanowskiego 18/22, 90-924 Lodz, Poland
2   Ecole Centrale de Lyon, University of Lyon, Ampere CNRS UMR 5005, 36 Avenue Guy de Collongue, 69134 Ecully, France; Abderrahmane.Beroual@ec-lyon.fr
*   Correspondence: pawel.rozga@p.lodz.pl

Citation: Rozga, P.; Beroual, A. High Voltage Insulating Materials—Current State and Prospects. *Energies* **2021**, *14*, 3799. https://doi.org/10.3390/en14133799

Received: 26 May 2021
Accepted: 21 June 2021
Published: 25 June 2021

Publisher's Note: MDPI stays neutral with regard to jurisdictional claims in published maps and institutional affiliations.

Progression in the field of insulating materials for power transformers and other high voltage devices is visible regardless of the type of insulation: solid, liquid, or gas. This progression resulted from the necessity of ensuring the proper operation of high-voltage devices. These materials have to ensure correct, continuous, uninterrupted, and safe operation of various devices. It is also extremely important to ensure an appropriate level of environmental safety with the knowledge that high voltage devices and, therefore, insulating materials operate under various stresses, which can be critical during their operation. The in-service conditions require high-quality materials, the properties of which will not change or have small changes over time. Hence, it is very important to investigate the properties of materials used in high-voltage equipment from various perspectives. Thus, electrical (AC, DC, LI, or combined exposure), thermal (e.g., accelerated aging), or chemical (impact of environmental factors or material compatibility) based factors must be taken into account in assessment of these materials. It is the optimal solution that is sought for a given application. Furthermore, the development of the HVDC technology, which involves specific stress conditions and thus specific physical phenomena associated with stress, has become extremely important in this respect.

Since insulating materials are constantly developing and new materials keep appearing in the market (e.g., biodegradable insulating liquids in the case of liquid dielectrics or nano-fluids), this Special Issue is focused mainly on new solutions for the use in high-voltage applications.

The Special Issue "High voltage insulating materials-current state and prospects" has received good responses. From the 21 submissions, 13 were accepted, which gives the rate of acceptance on the level of circa 62%. Among the accepted paper, one is of the review type and twelve are regular papers.

The review paper [1] covers the topic of synthetic ester liquids for transformer applications. It represents a kind of guide for researchers and industrials working with synthetic esters on a daily basis. This paper describes most of the problems concerning the use of the synthetic esters in transformers as an insulating and cooling liquid. The authors presented the fundamental chemical properties of synthetic esters. their AC and DC breakdown voltage; lightning impulse breakdown voltage and pre-breakdown phenomena; synthetic ester-based nanofluids; combined paper-synthetic ester-based insulating systems; application of synthetic esters for retro-filling and drying of mineral oil-immersed transformers; DGA-based diagnosis of synthetic ester-filled transformers; and static electrification of synthetic esters. State of the art analysis indicated the opportunities for synthetic esters. However, the authors also underlined some important challenges for the possible application of the synthetic esters on a larger scale not only in distribution transformers but also in the power transformers of a high level of nominal voltages.

The subject of synthetic esters was also discussed in [2] where the authors proposed a model relating to the drying efficiency of cellulose-based transformer insulation by solely using synthetic esters. The results of the laboratory tests performed by the authors indicated that the drying process is strongly influenced by temperature and time of drying. In turn, the initial level of moisture of the ester has a weak effect on the drying efficiency. The aim of this work was to find the optimal drying conditions on the basis of which a mobile system for the on-site drying of the transformer's insulation will be developed.

The electrostatic charging tendency (ECT) of synthetic ester has been analyzed in [3]. The author also tested the ECT of mixture of synthetic ester and mineral oil. The findings of the author is that the ECT of synthetic ester may be reduced when combined with a small amount (up to 20%) of fresh or aged mineral oil. This result is particularly beneficial when retro-filling power transformers and replacing mineral oil with a synthetic ester.

The ECT of natural ester-mineral oil mixtures is discussed in [4]. It was stated in this paper that there is no clear correlation between the composition of the mixture and the electrification current. Some advantages are visible in the mixture with small amounts of natural ester (up to 10%), which manifested in the reduction in ECT.

The subject of natural ester liquids is continued in the papers [5] and [6]. The first one presents the results of the studies on the electrical behavior of natural ester-based ZnO nanofluids at different concentrations in the range of 0.05–0.4 g/L. The AC voltage based tests indicate that the dielectric strength of natural ester-based ZnO nanofluids increases with the concentration of nanoparticles, with the exception of 0.05 and 0.4 g/L of ZnO. The optimum is reached, however, for 0.1 g/L (for 1% and 10% probability). The authors also conclude that increasing nanoparticles concentration beyond the value of 0.4 g/L reveals the implication of a tunnelling/bridging mechanism that may reduce the value of AC breakdown voltage. In turn, the authors of [6] analyzed the natural ester (Marula Oil) treated with Al_2O_3 with concentrations varying from 0.1 g/L to 2 g/L and 0.25 g/L. They use mathematical considerations to determine survival and hazard rate. Similarly, as in the case of [5], the authors noticed that Al_2O_3 in natural esters that are tested proved to be effective against electrical, physical, and thermal stress.

Different liquid dielectrics (synthetic ester, natural ester, and mineral oil) were considered in [7] in terms of energy distribution of optical radiation emitted by the surface discharges. The liquid's influence on the emitted radiation range, such as the clear recognition, may be observed between the discharges developed in a given liquid.

A similar range of research in [7] was performed in [8], but in relation to the electric arc in the air. In addition to the AC voltage of different frequencies, the authors also considered DC voltage stress. They found differences in the percentage share of optical radiation energy for the particular spectral ranges when the arc is generated at AC and DC voltage. The authors highlighted the advantages of the used measurement method that may be helpful in the design of some apparatus where the electric arc is an undesirable phenomenon.

The results of the studies on glass-reinforced epoxy (GRE) core rods used in AC composite insulators with silicone rubber housing are presented in [9]. The authors subjected the samples to a temperature of 50 °C for 6000 h and AC and DC voltage, respectively. The 3 point bending test, micro-hardness measurement, and microscopic analysis were used as the comparative testing methods for the two mentioned aging tests. The findings from the measurements were that the microstructure of the GRE material can withstand the tests at both AC and DC voltage without reduction in its mechanical parameters.

The development of solid materials for transformers is also visible on the market, especially in relation to materials that can be subjected to high temperatures. Two papers presented the considerations on aramid-based paper applied in transformer applications [10,11]. In [10], the authors compared the influence of moisture content on the thermal conductivity of Kraft paper and aramid paper impregnated with different dielectric liquids, namely mineral oil, synthetic ester, and natural ester. The higher increase in thermal conductivity caused by moisture was observed in the case of cellulose paper than in the case of aramid one. At the same time the samples impregnated with mineral oil indicated higher thermal conductivity at the same moisture level. On the basis of the obtained relationships, the authors of [10] elaborated the equations that may be applied for determining the temperature field of the transformers at the design and operation stage. In turn, the authors of [11] tested the cellulose paper enhanced with aramid in terms of the polarization and depolarization current analysis method (PDC Method). The samples were analyzed from the viewpoint of accelerated thermal degradation as well as weight-controlled dampening. On the basis of the performed tests, the authors proposed the regression functions for the activation energy and the dominant time constants dependent on the parameters of the experiment representing the degree of aging. The findings of the authors may be used in the diagnostics of cellulose–aramid insulation applied in power transformers using the PDC method.

The studies related to the transformer diagnostics continued in [12], where frequency response analysis (FRA) was considered as the method for the detection of mechanical faults or short-circuits in transformer windings. The transformer with deformations introduced into the winding was compared with a 10 MVA transformer working at industrial conditions. The clue of the paper was to perform the analysis for various window widths and for various extents of the deformation. The results show that the choice of the data window width influences the results of the analysis performed. In addition, the rules for element number selection differ for various numerical indices.

Finally, the partial discharge detection in a non-uniform field in mineral oil was analyzed in [13] using the registration of even harmonic components in the leakage current. It was found that the fifth harmonic is dominant when partial discharges develop independently of the setup configuration and independently of the material used as the high voltage electrode. In addition, harmonics are correlated with the intensity of partial discharges and level of non-uniformity.

As presented above, this special issue covered a wide area of research on insulating materials of different types together with different measurement methods applied to assess the condition of these materials in the laboratory as well as in service. As the guest editors of this Special Issue, we would like to thank all the researchers who made their contribution to this Special Issue, as well as the reviewers who were involved in the assessment process that wished the papers included will be widely read and cited.

Conflicts of Interest: The authors declare no conflict of interest.

References

1. Rozga, P.; Beroual, A.; Przybylek, P.; Jaroszewski, M.; Strzelecki, K. A Review on Synthetic Ester Liquids for Transformer Applications. *Energies* **2020**, *13*, 6429. [CrossRef]
2. Przybylek, P.; Moranda, H.; Moscicka-Grzesiak, H.; Cybulski, M. Laboratory Model Studies on the Drying Efficiency of Transformer Cellulose Insulation Using Synthetic Ester. *Energies* **2020**, *13*, 3467. [CrossRef]
3. Zdanowski, M. Streaming Electrification of Nycodiel 1255 Synthetic Ester and Trafo EN Mineral Oil Mixtures by Using Rotating Disc Method. *Energies* **2020**, *13*, 6159. [CrossRef]
4. Zdanowski, M. Electrostatic Charging Tendency Analysis Concerning Retrofilling Power Transformers with Envirotemp FR3 Natural Ester. *Energies* **2020**, *13*, 4420. [CrossRef]
5. Duzkaya, H.; Beroual, A. Statistical Analysis of AC Dielectric Strength of Natural Ester-Based ZnO Nanofluids. *Energies* **2021**, *14*, 99. [CrossRef]
6. Raj, R.A.; Samikannu, R.; Yahya, A.; Mosalaosi, M. Investigation of Survival/Hazard Rate of Natural Ester Treated with Al_2O_3 Nanoparticle for Power Transformer Liquid Dielectric. *Energies* **2021**, *14*, 1510. [CrossRef]
7. Kozioł, M. Energy Distribution of Optical Radiation Emitted by Electrical Discharges in Insulating Liquids. *Energies* **2020**, *13*, 2172. [CrossRef]
8. Nagi, Ł.; Kozioł, M.; Zygarlicki, J. Comparative Analysis of Optical Radiation Emitted by Electric Arc Generated at AC and DC Voltage. *Energies* **2020**, *13*, 5137. [CrossRef]
9. Wieczorek, K.; Ranachowski, P.; Ranachowski, Z.; Papliński, P. Ageing Tests of Samples of Glass-Epoxy Core Rods in Composite Insulators Subjected to High Direct Current (DC) Voltage in a Thermal Chamber. *Energies* **2020**, *13*, 6724. [CrossRef]
10. Dombek, G.; Nadolny, Z.; Przybylek, P.; Lopatkiewicz, R.; Marcinkowska, A.; Druzynski, L.; Boczar, T.; Tomczewski, A. Effect of Moisture on the Thermal Conductivity of Cellulose and Aramid Paper Impregnated with Various Dielectric Liquids. *Energies* **2020**, *13*, 4433. [CrossRef]
11. Wolny, S.; Krotowski, A. Analysis of Polarization and Depolarization Currents of Samples of NOMEX®910 Cellulose–Aramid Insulation Impregnated with Mineral Oil. *Energies* **2020**, *13*, 6075. [CrossRef]

12. Banaszak, S.; Kornatowski, E.; Szoka, W. The Influence of the Window Width on FRA Assessment with Numerical Indices. *Energies* **2021**, *14*, 362. [CrossRef]
13. Aydogan, A.; Atalar, F.; Ersoy Yilmaz, A.; Rozga, P. Using the Method of Harmonic Distortion Analysis in Partial Discharge Assessment in Mineral Oil in a Non-Uniform Electric Field. *Energies* **2020**, *13*, 4830. [CrossRef]

 *energies*

Review

A Review on Synthetic Ester Liquids for Transformer Applications

Pawel Rozga [1,*], Abderrahmane Beroual [2], Piotr Przybylek [3], Maciej Jaroszewski [4] and Konrad Strzelecki [1]

[1] Institute of Electrical Power Engineering, Lodz University of Technology, Stefanowskiego 18/22, 90-924 Lodz, Poland; konrad.strzelecki@dokt.p.lodz.pl

[2] Ecole Centrale de Lyon, CNRS, Ampère UMR5005, University of Lyon, 36 avenue Guy de Collongue, 69134 Ecully, France; Abderrahmane.Beroual@ec-lyon.fr

[3] Institute of Electric Power Engineering, Poznan University of Technology, Piotrowo 3A, 60-965 Poznan, Poland; piotr.przybylek@put.poznan.pl

[4] Department of Electrical Engineering Fundamentals, Wroclaw University of Science and Technology, Wybrzeze Wyspianskiego 27, 50-370 Wroclaw, Poland; maciej.jaroszewski@pwr.edu.pl

* Correspondence: pawel.rozga@p.lodz.pl; Tel.: +48-42-631-26-76

Received: 23 October 2020; Accepted: 1 December 2020; Published: 4 December 2020

Abstract: Synthetic esters have become more and more popular in last few decades, explaining the increasing number of units filled with this liquid year by year. They have been investigated under different aspects, both from the fundamental point of view and breakdown mechanisms, well as from the application point of view. However, their use in high voltage equipment is always a challenge and deeper knowledge of the various aspects that can be encountered in their exploitation is needed. The intent of this review paper is to present the recent research progress on synthetic ester liquid in relation to the selected issues, most important for ester development in the authors' opinion. The described issues are the breakdown performance of synthetic esters, lightning impulse strength and pre-breakdown phenomena of synthetic esters, synthetic esters-based nanofluids, combined paper-synthetic ester based insulating systems, application of synthetic ester for retro-filling and drying of mineral oil-immersed transformers, DGA(dissolved gas analysis)-based diagnosis of synthetic esters filled transformers as well as static electrification of synthetic esters. The different sections are based both on the data available in the literature, but above all on the authors' own experience from their research work on synthetic ester liquids for electrical application purposes.

Keywords: synthetic esters; power transformers; liquid insulation

---

1. Introduction

The search for new insulating liquids respectful of the environment as well as the improvement of existing liquids for high voltage applications is a permanent task; the goal being the improvement of the properties of insulating systems in which these liquids have to be used while keeping the electrical properties required during exploitation in a given electrical device. Significant progress has been made in this area in recent years. This is how a new generation of insulating liquids appeared, namely liquid esters. These represent a potential alternative to mineral oils traditionally used in power transformers [1–8]. Among the ester liquids, the natural esters, synthetic esters and blended esters may be mentioned. In terms of natural esters, many companies have offered their products for many years [9–16]. These products differ from each other to a small extent and the differences result mainly of the source from which the esters are obtained (refining from soya, sunflower, rapeseed etc.) In turn, the blended esters were proposed recently and have not been so popular as natural ones yet.

Their main advantage in relation to the other esters is a lower viscosity in comparison to classical esters, which are closer to those of mineral oils [17–19]. While the number of natural esters available on the transformer market is growing year by year, the market of synthetic esters for electrical purposes is quite niche and only individual products are available for common use [20,21]. However, the last few years seemed to change this tendency, providing some new products worthy of consideration.

The 1970s are considered to mark the beginning of work into the use of synthetic esters for electrical applications. The first transformer unit in which a synthetic ester was introduced, was put into operation in 1976. In subsequent years, the works devoted to synthetic ester liquids were intensified. At present, the synthetic esters have found an application mainly in distribution and special transformers of relatively low powers and voltages. This situation results from the good environmental properties of synthetic esters (biodegradability and high fire point), which have been allowed to easily meet the restrictive environmental requirements for specific application such as for example high speed railways, wind turbines and shopping centers [4,20–23]. However, within the synthetic ester-based transformer applications, a continuous increase in the value of rated power and nominal voltage is being seen. Currently, the highest nominal voltage of transformer units using synthetic esters has reached 400 kV. With the development of synthetic ester-based applications, the standards describing the requirements for fresh synthetic esters have been developed too. The IEC (International Electrotechnical Commission) 61099 Standard "Insulating liquids—Specifications for unused synthetic organic esters for electrical purposes" [24] has been modified for years and its current form is dated to the year 2011. In turn, the ASTM (American Society for Testing and Materials) has just started to work on its own document entitled "New standard specification for less-flammable synthetic ester liquids used in electrical equipment", which is being considered currently.

The present paper is a review of the present knowledge and data available in the literature and more particularly is based on the authors' own experience from their research work into synthetic ester liquids and their application in a high voltage area. The subjects that will be dealt with are:

- Chemical properties of synthetic esters;
- AC and DC breakdown voltage of synthetic esters;
- Lightning impulse breakdown voltage and pre-breakdown phenomena of synthetic esters;
- Synthetic ester-based nanofluids;
- Combined paper-synthetic ester-based insulating systems;
- Application of synthetic esters for retro-filling and drying of mineral oil-immersed transformers;
- DGA-based diagnosis of synthetic ester-filled transformers;
- Static electrification of synthetic esters.

2. Chemical Properties of Synthetic Esters

Synthetic esters are chemical compounds resulting from a direct reaction of an alcohol molecule and a carboxylic acid molecule, called an esterification reaction. Polyols, i.e., molecules that contain more than one hydroxyl group, are usually used in this reaction. Each of the hydroxyl groups can participate in the reaction with another carboxylic acid molecule. As a result, the carbon chain from a polyol molecule is the backbone to which the carbon chains derived from carboxylic acid molecules are attached by ester bonds. In the synthesis process, one can use saturated molecules (containing only single bonds between carbon atoms) or unsaturated molecules (where there are double or triple bonds between carbon atoms). Since the compounds containing unsaturated bonds show less stability, the use of saturated compounds is preferred. Figure 1 presents schematically the esterification process [4,25].

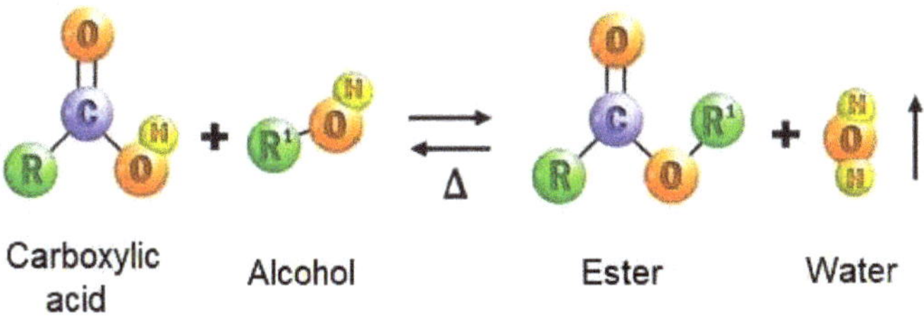

Figure 1. Graphical presentation of the esterification mechanism.

An example of a synthetic ester used for electrical purposes is a pentaerythritol ester. It consists of four ester groups, COOR, that are placed at the end of the cross structure of the main compound. The alkyl (organic) groups from R_1 to R_3 in this compound may be either the same or different and are typically from C_5H_{11} to C_9H_{19} (with saturated carbon chains). Such a chemical structure of synthetic esters makes it quite a stable compound [4,25,26]. Figure 2 presents the simplified structural formula of the compound of the mentioned type of ester.

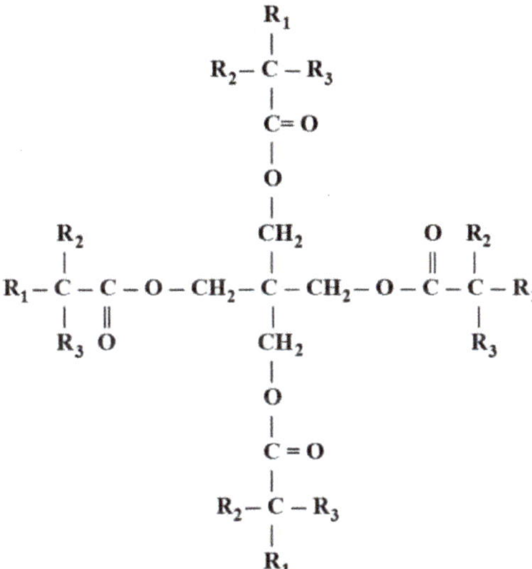

Figure 2. Simplified structural formula of pentaerythritol ester: R1, R2, R3—organic groups.

The requirements specified in the IEC 61099, which should be met by the fresh synthetic ester as received, are set in Table 1. For each parameter, the standard suggested for the determination of the distinctive properties is quoted. It is important to point out that all synthetic esters available on the transformer market fulfill these requirements [20,21].

Table 1. Requirements for fresh synthetic ester for electrical purposes as per IEC 61099.

Properties	Unit	Standard for Evaluation	Required Value
Density at 20 °C	kg/dm³	ISO 3675	Max. 1
Viscosity at 40 °C	mm²/s	ISO 3104	Max. 35
Flash Point	°C	ISO 2719	250
Fire Point	°C	ISO 2592	300
Pour Point	°C	ISO 3016	Max. −45
AC Breakdown Voltage	kV	IEC 60156	Min. 45
Dissipation Factor at 90 °C	-	IEC 60247	Max. 0.03
Resistivity at 90 °C	GΩm	IEC 60247	Min. 2

The density, but first of all the viscosity of a dielectric liquid, are the key parameters responsible for its cooling properties, which are important for the proper operation of the power transformer. It is well-known that viscosity of the liquid decreases with increasing temperature. The synthetic esters are more viscous than mineral oils over a wide range of temperatures. However, the differences are reduced when the temperature increases. For example, the viscosity of the synthetic ester Midel 7131 is 70 mm^2/s at 20 °C and 5.25 mm^2/s at 100 °C while mineral oil viscosity is 22 mm^2/s and 2.6 mm^2/s at 20 and 100 °C respectively. Hence it is clearly seen that the difference decreases from three to two times comparing 20 and 100 °C [1–4,20,21].

The synthetic ester liquids are eco-friendly products characterized by high level of biodegradability. Hence, in this field they provide a great alternative to other types of insulations used in power transformers such as mineral oils, silicone oils as well as to dry-type transformers. Since the synthetic esters are readily biodegradable, its usage in power transformers avoids environmental damage when leakage occurs as well as enabling reductions in containment measures. Typically, synthetic esters achieve 10% degradation by day 3 and on the 28th day they reach 89% degradation. With a high level of fire point, the synthetic esters significantly increase the fire safety of transformers and reduce the need for fire protection equipment. Synthetic esters are treated as non-flammable products of K3 class. Their flash and fire points reach temperatures higher than 250 and 300 °C respectively, which are much higher values than the corresponding temperatures characterizing mineral oils (150 and 170 °C). At the same time, synthetic esters show low smoke properties. Typically, mineral oils produce thick black smoke while synthetic esters are characterized by thin white smoke which is not as dense as the smoke generated by mineral oils. Through an extremely low pour point (−50 °C), transformers filled with synthetic esters become a highly effective solution in colder climates [4,20,21,27,28].

Synthetic esters have an exceptionally high moisture tolerance. The water content in the insulating liquid determines a number of its properties affecting the conditions and safety of the transformer operation. The water content in the insulating liquid is most often expressed in ppm by weight, and is measured using the standardized Karl Fischer titration method [29]. More and more often, the moisture of insulating liquids is described by means of relative saturation (RS), which is the ratio of the moisture content to the water saturation limit in the insulating liquid at a given temperature, and is usually expressed as a percentage. This parameter can be measured with a capacitive sensor installed in a transformer. In turn, the water saturation limit (water solubility) expresses the maximum concentration of dissolved water that can exist in insulating liquid at thermodynamic equilibrium at a specified temperature and pressure [30].

Synthetic esters are characterized by a much greater water solubility compared to other insulating liquids used in transformers. Different water solubility should be explained by the different polarities of particles of these liquids. Mineral oil is non-polar or is weakly polar. By contrast, ester linkages that are present in natural and synthetic esters make these liquids polar. Synthetic esters may have 2–4 ester linkages per molecule. Polar molecules tend to be most strongly attracted to other polar molecules like water. This allows for the creation of hydrogen bonds between the hydrogen atoms that belong to the water molecule and the oxygen atom which belongs to the ester molecule [4,31,32].

The water saturation limit in the insulating liquid can be calculated using the formula:

$$\log S = A - \frac{B}{T},$$
(1)

where:

A and *B*—water saturation coefficients characteristic for the given insulating liquid,
T—liquid temperature in Kelvin.

Table 2 summarizes the *A* and *B* coefficients for the new synthetic ester Midel 7131 taken from the literature, while Figure 3 shows the water saturation limit in the function of temperature determined on the basis of these coefficients and by means of the Equation (1).

Table 2. Water saturation coefficients for synthetic ester Midel 7131.

Water Saturation Coefficients		Water Saturation Limit, ppm		Reference
A	**B**	**20 °C**	**50 °C**	
5.42	629	1881	2975	[30]
5.320	608.28	1758	2739	[32]
5.4166	581.95	2700	4128	[33]
5.6614	695.74	1941	3224	[34]
7.1790	1191	1307	3115	[35]

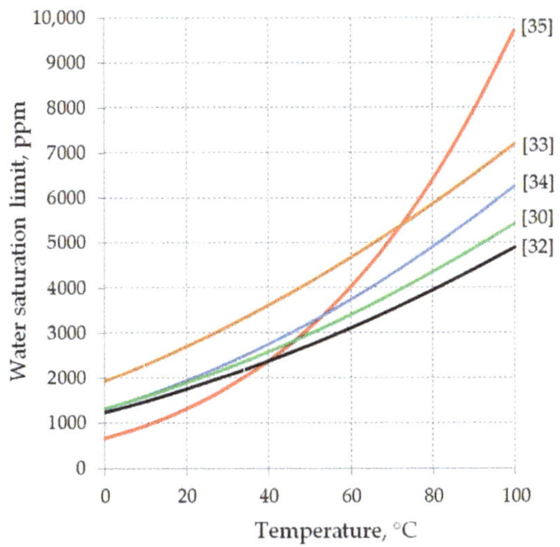

Figure 3. Water saturation limit in the function of temperature for synthetic ester Midel 7131 determined on the basis of water saturation coefficients taken from [30,32–35].

As shown in Figure 3, the water saturation limit, determined on the basis of water saturation coefficients given by various researchers, varies widely depending on the method that has been used to determine these coefficients. For example, for the temperature of 20 °C, the discrepancy in the results of the water saturation limit is as high as 1393 ppm. For further consideration, the water saturation coefficients provided in [30] was chosen. This choice is due to the fact that these coefficients were determined based on the results obtained during a Round Robin Test carried out by the Cigre D1.52 working group (WG). These factors represent an average value calculated basing on the results obtained from three methods in five different laboratories [30]. The results of the water saturation limit obtained on the basis of the A and B coefficients given by the Cigre WG are similar (especially in the temperature range from 0 °C to 50 °C) to the values presented in [32,34].

In accordance with [24,36], the water content in the synthetic ester (for untreated liquid as received) cannot exceed the value of 200 ppm, corresponding to a relative saturation equal to about 10% at 20 °C. In the case of such a high water content as in an ester, the drying process should be carried out before filling the transformer tank. In accordance with the standard [37], the water content in esters in service should not exceed the value of 400 ppm. This value is considered to be satisfactory for 35 kV or lower voltage equipment. Unfortunately, this standard does not specify the temperature for which the given water content limit applies. For the value of water content in the synthetic ester equal to 400 ppm, for 20 °C (S = 1881 ppm) and 50 °C (S = 2975 ppm), the relative saturation of the ester is equal to 21.3% and 13.4%, respectively.

The abovementioned acceptable high moisture content in synthetic esters is treated as an advantage from the viewpoint of interaction with solid insulation. The moisture from cellulose oxidation may be absorbed by the esters, which extends the lifetime of cellulose insulation by slowing down its aging [38,39]. In the same area, the synthetic ester dielectric strength is highly resistant against moisture content. The mentioned values of 200 or 400 ppm do not cause a reduction in AC breakdown voltage (AC-BDV) of a synthetic ester when it is measured in accordance with IEC 60156 Standard [4,20,21,40]. Reduction of this value is observed only at a moisture content of approximately 600 ppm, which is much higher than the 20 ppm which decreases mineral oil AC-BDV significantly.

Synthetic esters also have a higher electrical permittivity compared with mineral oils, which is an advantage for the distribution of electrical fields in the paper-dielectric liquid insulating systems of transformers [3,4,41,42]. The benefits of synthetic esters are indicated when adding nanoparticles, which at least similarly as in the case of mineral oils, improve the dielectric characteristics of esters [43–48]. The mentioned chemical properties of synthetic esters showed that they have found applications in the so-called retro-filling process. It involves the replacement of mineral oil in a transformer on ester liquid and thus reducing the potential environmental impact [49–53]. There are also some studies where the benefits of synthetic esters connected with its water solubility are used in the process of drying transformers [54–58]. The chemical structure of synthetic esters was found to be a reason for some disadvantages in the field of their behavior under lightning impulses (LIs): the dielectric performances of synthetic esters under these stresses are the worse ones [26,59–63]. Moreover, the lack of stability of the dielectric dissipation factor in the condition of thermal treatment, storage and transport is treated as an undesirable property of synthetic esters [64,65]. In addition, it has been found that due to the higher viscosity of ester liquids, impregnation of the solid insulation with synthetic esters is less effective under the same impregnation conditions than in the case of mineral oil [19,66]. Regarding static electrification, synthetic esters have an electrostatic charging tendency (ECT) stronger than mineral oils [67,68]. One of the open problems, which is influenced by the chemical properties of synthetic esters, is that of using the dissolved gas analysis (DGA) technique to assess the condition of in-service transformers filled with synthetic esters [69,70].

3. Breakdown Characteristics of Synthetic Esters

3.1. AC and DC Breakdown Voltage

AC breakdown voltage (AC-BDV) of a dielectric liquid is usually determined according to the standard [40]. As in the case of other dielectric liquids (mineral oils, natural esters), synthetic esters, in the form ready for filling transformers, are characterized by AC-BDV at least higher than 60 kV, but this value is often higher than 70 kV. However, some experimental work uses statistical approaches to analyze a wider range of data than that required in the mentioned standard. In [71], Martin and Wang analyzed 100 values of the AC-BDV of three dielectric liquids namely mineral oil, natural ester and synthetic ester using statistical approaches such as Gaussian and Weibull distributions to find correlations between the results obtained and to determine the withstand voltage defined as a voltage level for which the risk of failure is low at an acceptable level. Shortly, the results showed a strong similarity between the liquids such as all of them may be capable in acting as transformer liquid insulation. In turn, in [72], the authors tested the influence of conducting particles on the AC-BDV of ester liquids and mineral oil. They used two parallel plane electrodes with a 10 mm gap. They found that in all cases the conductive particles present in liquid volume significantly reduce the breakdown voltage of the liquids tested. However, this reduction is more visible in the case of mineral oil. It is because of the higher viscosities of both esters considered—the motions of the conductive particles in more viscous synthetic and natural esters are much slower than in mineral oil. In [73], the synthetic ester (tetraester) was tested together with three types of natural ester liquids and three types of naphtenic mineral oils. Similarly as in [71], the conformity of experimental data (32 values of AC-BDV in each case) with the Normal and Weibull distributions were checked and a comparison was made between the considered liquids. It was found from these studies that, in a quasi-uniform electric field (sphere electrodes with a 2.5 mm gap), the differences between liquids are minimal.

The synthetic ester tested had a AC-BDV intermediate between the natural esters and the mineral oils considered. Similar statistical behavior of synthetic esters with mineral oil and natural esters was noticed also in [74], but with some advantages (higher median of AC-BDV and lower standard deviations) for the natural ester liquids (vegetable oils). However, the obtained results in each case exceed 60 kV, which is enough of a value for the dielectric liquids to be applied in devices of nominal voltages equal to 69 kV or higher. The available data are limited in the case of DC stress-based tests of synthetic esters [75–77]. In [75,77] the authors compared simply the DC-BDV between selected types of dielectric liquids and their mixtures in a quasi-uniform electric field created by a sphere-to-sphere electrode system. When comparing liquids alone, there are no obvious differences between them. Statistically, the mean positive DC-BDV according to [75] is very similar. However, a decreased trend is noticed such as synthetic esters not having the highest value of DC-BDV and mineral oil having the lowest. Nothing new was noticed when comparing the probability distributions: the trend of mean values is valid whatever the considered quantile. Still, the synthetic ester seems to have the best properties under DC. In [77], Reffas et al. tested the liquids under DC voltage and found that olive oil (a new type of vegetable oil) seems to be the most resistant against DC stress for both polarities. However, a synthetic ester is comparable quantitively with mineral oil. Xiang et al. [76] reported totally different conclusions concerning the DC-BDV. These conclusions were based on measurements in a point-to-sphere electrode arrangement representing non-uniform electrical field distributions. Comparing synthetic esters and mineral oil, the authors showed a significant difference between the DC-BDV of both liquids. Mineral oil had higher DC-BDV for both polarities and independently of the considered electrodes gap (from 2 to 30 mm). The authors also investigated the streamers behavior in both liquids. They observed that in the condition of the experiment, there were no obvious differences in the shape of streamers recorded photographically corresponding with mineral oil and the synthetic ester. Similarly, the correlation between the stopping length of the streamers and their apparent charge were identical.

It appears from the above that it is difficult to obviously state whether synthetic esters behave similarly to mineral oil and other alternative liquids under AC and DC stresses. Based on a standard approach (2.5 mm gap and two sphere electrodes), there are no data indicating worse synthetic ester properties. They are rather treated as a good alternative for mineral oil similarly as natural esters. However, in a non-uniform electric field, the experimental data are not so optimistic and synthetic esters show some negative traits that are consistent with the observations under the lightning impulse voltages described later.

3.2. LI Breakdown Voltage and Associated Phenomena

Since synthetic esters started to be applied in power transformers, lightning impulse breakdown voltage (LI-BDV) as well as pre-breakdown and breakdown phenomena under this type of stress, have become issues for the special attention of researchers [25,26,59–63,74,78–86]. Impulse breakdown and pre-breakdown phenomena of synthetic ester liquids have been studied mainly at standard LI voltage 1.2/50 µs and at non-uniform electric field distributions.

A comparison of LI-BDV and switching impulse breakdown voltages (SI-BDV) of synthetic esters with Envirotemp FR3 natural ester and mineral oil Gemini X, have been presented in a comprehensive way in [26,79]. The authors performed a wide range of tests on these liquids using a sphere–sphere electrode system of 12.5 mm diameter and a spacing gap d of 3.8 mm. They analyzed the influence of voltage polarity and testing methods used for the determination of LI-BDV of the considered liquids. First, they found that mineral oil is characterized by slightly higher values of breakdown voltage (both lightning and switching), independently of the voltage waveform. The reduction of mean breakdown voltage of esters was in percentage 15% for the lightning impulse stress and 8% for the switching impulse stress. Second, the analysis of the voltage polarity show similar results: lower LI-BDV characterizes synthetic esters and natural esters and this is true for both polarities. On the other hand, four different testing methods were applied to compare the 50% LI-BDV: rising-voltage method when one shot per step was applied, rising-voltage method where three shots per step were applied, up-and-down method and finally multiple-level method. In each case, the best lightning performance characterized mineral oil. Synthetic and natural ester had clearly worse properties within the considered voltage stress. The results borrowed from [79] are summarized in Table 3.

Table 3. 50% LIBVs of the liquids tested with the use of various methods, d = 3.8 mm [79].

Method	Synthetic Ester (kV)	Natural Ester (kV)	Mineral Oil (kV)
Rising-voltage 1 shot/step	258.0	239.3	276.4
Rising-voltage 3 shots/step	205.0	200.4	251.9
Up-and-down	223.2	212.9	232.8
Multiple-level	248.9	230.8	270.0

However, the 1% breakdown probability (called a withstand voltage) determined from the two-parameter Weibull plots shows results more favorable for both esters. The 1% LI-BDVs for esters were 153.1 kV and 151.0 kV for synthetic esters and natural esters, respectively, which were only slightly lower than the corresponding withstand voltage of mineral oil (156.7 kV).

In contrast to the results presented in [79], Rozga et al. [62] proposed for the assessment of lightning performance of synthetic esters the approach based on the IEC 60897 Standard [87]. The 25 mm gap of point-to-sphere electrode arrangement and step method were applied for LI-BDV determination. The authors tested two synthetic esters, namely Midel 7131 and Envirotemp 200, which were then compared with Shell Diala naphtenic mineral oil. Independent of voltage polarity, the results obtained by the authors in the form of average values from 20 measurements indicate the similar values of LI-BDV for all three liquids tested with small deviation in favor of the esters at a positive polarity and in favor of the mineral oil at negative polarity (approximately 4–5 kV in relative values). This was confirmed by Rozga et al. [86] when testing a new synthetic ester liquid under a negative polarity. The authors obtained statistically identical negative LI-BDVs for synthetic esters as well as natural esters and mineral oil. However, opposite to the results presented in [62], positive LI-BDV was much higher for mineral oil than for both tested ester liquids.

In another work, Reffas et al. [74] compared the positive LI-BDVs of synthetic esters with olive oil, rapeseed oil and mineral oil. The tests were performed for a sphere-to-sphere electrode system with a 2.5 mm gap distance using the up-and-down method. The 1, 10, 50 and 90% breakdown probabilities indicate a lower lightning strength of the synthetic ester among the other considered liquids. For example, the median BV of synthetic esters was similar to the median of mineral oil and the mineral oil–olive oil mixture (20% to 80%), while in relation to the fresh olive oil and rapeseed oil, this value was definitely lower.

In turn, the paper [80] presents the results for three liquids, namely mineral oil, synthetic ester and natural ester, for the point-sphere electrode system and two spacing gaps of 5 and 10 cm. The tests were performed under an impulse voltage of 0.5 µs rise time and 1400 µs tail. For both gaps, similar results were obtained for ester liquids. For positive polarity, slightly higher values characterized the natural ester, while for negative polarity, the synthetic ester had a higher LI-BDV. Comparing the results concerning esters to those obtained for mineral oil makes both esters look worse. While for the 5 cm gap and positive polarity the difference is not high (around few kV), for 10 cm and a positive polarity and both gaps at negative polarity, the differences reached even 50 kV.

Wider study using an electrode system of non-uniform field distribution, was carried out by Liu and Wang [60]. The authors compared a synthetic ester with a natural ester and mineral oil at gap distances which varied from 25 to 100 mm for positive polarity and from 15 to 75 mm for negative ones. Except for 25 mm for positive polarity, where the LI-BDVs for all three liquids tested were almost identical, for the rest of the gaps and both polarities, ester liquids have lower BVs than mineral oil. In general, as the electrode gap increases, the differences between esters and mineral oil increases. However, comparing only ester liquids, a slightly higher LI-BDV characterized the tested synthetic ester.

In the work [85], the authors extended their analyses by placing the pressboard interface in parallel to the electrode system axis. Three gaps were considered: 25, 50 and 75 mm respectively; the liquids tested being the same as in [60]. The authors concluded that in terms of lightning impulse voltage, the pressboard did not influence the relationships between the liquids tested and noticed from the studies without the pressboard. For both voltage polarities, 50% breakdown voltage was lower in the case of esters. Note that it is well known that, in the case of a divergent electric field, the insertion of a pressboard barrier greatly improves the breakdown voltage whatever the type of oil, voltage waveform and polarity; the improvement is all the more important as the dimensions of the barrier are large.

Regarding the pre-breakdown phenomena in synthetic esters, the attention should be mainly focused on the parameters characterizing discharges (streamers) such as inception voltage, shape of discharge or stopping length of the streamers.

Concerning the inception voltage, the reported results indicate that, in a point-plane electrode system, the streamers in synthetic esters are initiated at similar values as in the case of mineral oil [59,63,81]. Furthermore, the impact of the radius of curvature of the high voltage electrode (point) on the streamer initiation in the synthetic ester is the same as for mineral oil. Figure 4 gives an example of comparison of the inception voltages of the three liquids tested taken from [81].

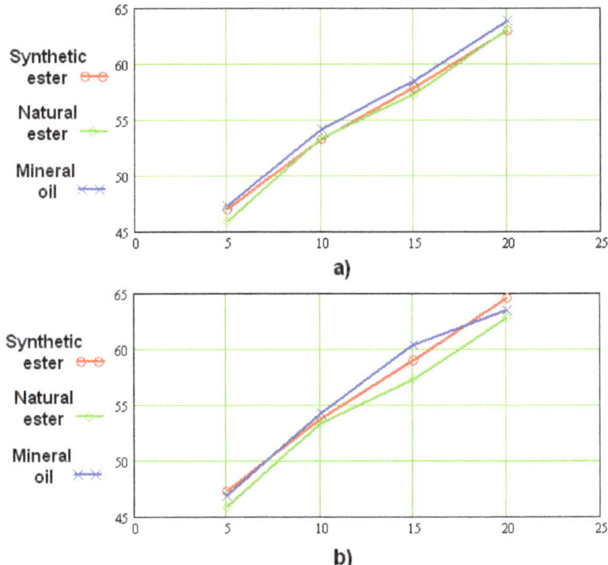

Figure 4. Inception voltage of streamers versus gap distance: (**a**) negative polarity of lightning impulse, (**b**) positive polarity of lightning impulse; *x*-axis represents gap distance in (mm), *y*-axis represents inception voltage in (kV) [81].

The authors of the mentioned works did not observe any difference in the spatial shapes of the streamers recorded photographically when they developed in esters and mineral oils at a voltage lower than the breakdown voltage. Figure 5 gives examples of streamers shapes recorded using a shadowgraph technique. The streamers at the inception voltage are characterized also by similar propagation velocities [59–63,78,80,81].

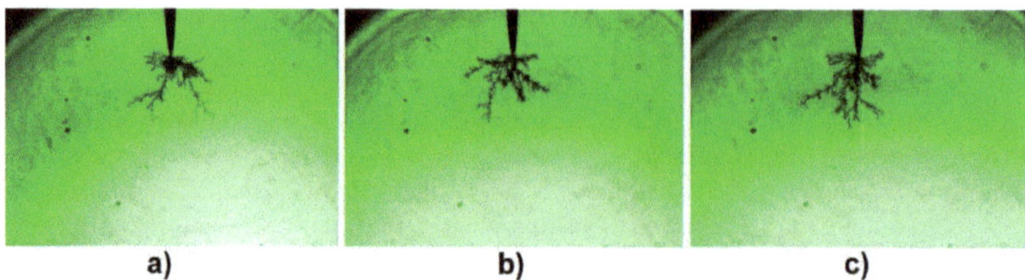

Figure 5. Photographs of negative streamers in: (**a**) synthetic ester, (**b**) natural ester, (**c**) mineral oil; d = 20 mm, V = 65kV.

A parameter often used when investigating the streamer development is their stopping length (L). Stopping length is the final length the streamer reaches during its propagation for a given voltage level. In general, the stopping length of streamers propagating in synthetic esters is longer than that of streamers in mineral oil, for the same experimental conditions. The well-known relationship that the stopping length increases with the applied voltage is valid also in the case of ester liquids for both voltage polarities [59,60,63,77,78,86].

A significant factor that describes liquid behavior under lightning impulse voltage is propagation velocity of the discharges (typically average value). That is the commonly considered quantity in the studies of streamers in liquids [15,18,59–63,80–86,88–92]. Typically, the propagation velocity increases with the increase in the testing voltage. In an assessment of dielectric liquids on the basis of propagation velocity, the so-called acceleration voltage has become a very important factor. It includes the voltage, which causes the sudden change of the propagation mode of the streamers from slow to fast. Concerning synthetic esters, for small gaps (up to 25–30 mm), the propagation velocity seems to be identical to that of streamers developing in mineral oil when considering the voltages equal to or below the breakdown voltage [60,62,74,83,84]. However, when increasing the testing voltage above the breakdown voltage, the differences for the mentioned small gaps appear between the streamers behavior in synthetic ester and mineral oil. The acceleration voltage of streamers in synthetic ester is lower, which means that they are more susceptible to the appearance of fast and energetic discharges. Figure 6 gives an example confirming this fact.

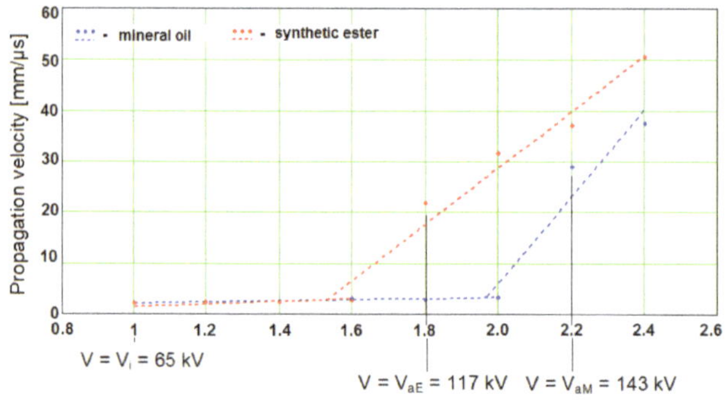

Figure 6. Relationship between the propagation velocity of positive streamers and multiples of inception voltage: d = 20 mm, V_i—inception voltage of the streamers, V_{aE}—acceleration voltage of streamers in synthetic ester liquid, V_{aM}—acceleration voltage of streamers in mineral oil, red dots—average propagation velocity of streamers in synthetic ester obtained from experiment, red dotted line—regression line for propagation velocity of streamers in synthetic ester, blue dots—average propagation velocity of streamers in mineral oil obtained from experiment, blue dotted line—regression line for propagation velocity of streamers in mineral oil [61].

For longer gaps (>30 mm), the differences between synthetic esters and mineral oils are greater. First, LI-BDV of synthetic esters, as reported in [60], is much lower than that of mineral oil. Additionally, a breakdown in synthetic ester is a result of fast developing streamers, while in mineral oil still slow streamers cause the breakdown independent of the gap distance. This means that the mentioned acceleration voltage is higher in mineral oil, which is its advantage—fast discharges in liquids are attributed to more intense ionization processes, so from the above, in mineral oil these processes appear at the over-voltages of higher values than in the case of synthetic esters. The worst properties of synthetic esters under LI stress are confirmed also by the fact that the differences in propagation velocities of streamers developing in esters and mineral oils increases with an increase in the gap distance.

Streamer characteristics may be also obtained from the analysis of streamer current, electrical charge as well as light emitted by the developing streamers. Streamer currents and emitted light are strongly associated with the streamer mode. The researches on light emission or streamer current in synthetic esters were presented in [60–63,74,78–86]. The main findings from these studies are that the intensity of light as well as the recorded current are always higher in the case of streamers developing in synthetic ester liquids. The light pulses in most cases are of higher peak values and simultaneously of higher frequencies, indicating that more intense ionization processes occur in esters and the branch number of streamers is more important, as shown in Figure 7.

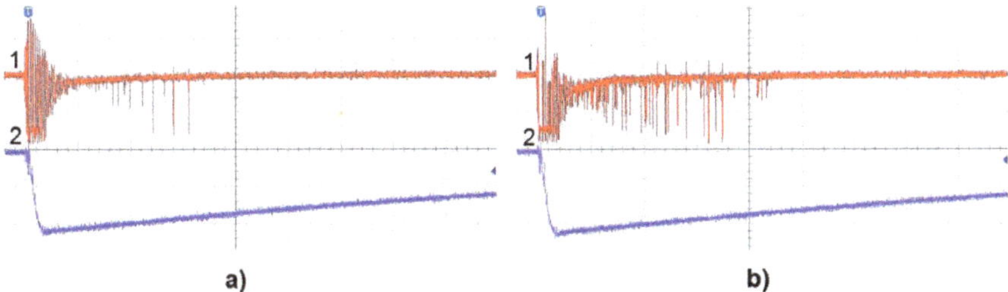

Figure 7. Light and voltage waveforms recorded at LI of negative polarity, t = 4 μs/div., V = 50 kV/div., 1—light, 2—voltage, V = 115 kV: (**a**) mineral oil, (**b**) synthetic ester.

In conclusion, the lightning performance of synthetic esters are in general of a similar nature when considering the small gaps. This would mean that the phenomena of initiation, development and breakdown of synthetic esters and mineral oils are qualitatively similar. However, when the gaps become longer, the differences are evident, indicating lower resistance of synthetic esters against lightning over-voltages. In general, the transition from slow to fast streamers occurs faster in synthetic esters than in mineral oil. So in terms of lightning performance of synthetic esters, it is a challenge for researchers to analyze the longer gaps, while this is especially important from the viewpoint of the usage of synthetic esters in power transformers of higher nominal voltages. Improving the lightning properties of synthetic esters so that they can be successfully used in high voltage applications is a task that researchers need to address in the near future.

4. Synthetic Ester-Based Nanofluids

As was mentioned above, despite the interesting properties of synthetic esters with regard to environmental protection, they present some disadvantages, the main ones of which are their viscosity and impulse breakdown voltage. The viscosity of esters is two to three times higher than that of mineral oils, which constitutes a concern for heat transfer from the transformer windings to the outside, even if the viscosity has a positive effect on the impregnation of insulant papers and pressboards. Thus, with an increase of electric energy demand and, consequently, the increase in energy density, which, in turn, causes an increase of the temperature of the transformer windings, the problem of transformer cooling has become even more crucial. Among others, it is this cooling problem which is at the origin of the introduction of nanoparticles (NPs) into insulating liquids and which has given rise to a new class of liquid insulators called nanofluids (NFs). Indeed, it has been well known for many decades that NFs are very efficient for cooling. They are used in many fields such as transportation (exemplary for cooling the systems of heavy power machines), heating buildings and in the reduction of pollution, the cooling of nuclear systems, space and defense (space stations and aircrafts) and electronics (chips, electronic circuitry components) as well as in solar absorption for heat-transfer performance. Their thermal characteristics are better than those of the base fluids. These strongly depend on the properties of added nanoparticles and especially their size, shape, volume fraction and type of material [93–98]. Rafiq et al. in [99] reported that some nanoparticles based on copper, silica and alumina improve thermal conductivity in a linear relationship with the mass fraction in the mixture. These properties, as well as the contact surface between the nanoparticles and the base liquid, are major parameters of influence both on the thermal properties and on the dielectric properties of the nanofluid. Recent work has shown that in addition to improving the heat transfer they provide, nanoparticles also improve the dielectric withstand of insulating liquids [100–109]. These two fundamental properties required for a possible use in high voltage applications, that is, good heat transfer and good dielectric strength make nanofluids possible alternatives for the base oils in oil-based devices. The most frequently used nanoparticles are TiO_2, Fe_3O_4, SiO_2, Al_2O_3 and ZnO. Other alternative NPs such as CuO, ZrO_2 and C60 are also used to increase the dielectric strength of oils [110].

4.1. Influence of Nanoparticles on the Breakdown Voltage of Synthetic Estes-Based Nanofluids

The electrical strength of ester-based nanofluids has been investigated at different electrical stresses, such as AC, DC and LI voltages. In general, it is mainly conducting nanoparticles and especially Fe_3O_4 NPs that gives the best improvement of BDV. However, the effect is more pronounced on mineral oil than on esters; the addition of Fe_3O_4 NPs to mineral oil can double the AC-BDV of the base mineral oil [111,112]. A similar effect has been also observed with semi-conducting and insulating NPs but with an improvement less important than that with Fe_3O_4 [19,20]. Unlike mineral oil, the effect of nanoparticles on the dielectric strength of esters and particularly natural esters, is not always so beneficial. Indeed, Peppas et al. [113,114] reported that the addition of Fe_3O_4 nanoparticles increases the AC BDV of natural ester oil Envirotemp FR3 up to an optimum value corresponding to a concentration of NPs of 0.008%; the improvement is of about 20% in relation to the base oil. Beyond this concentration, breakdown voltage drops radically below the BVs of the base oil. These authors also observed that with SiO_2 NPS, the AC-BDV is lower than that of the natural ester. Makmud et al. [115] reported that the addition of Fe_3O_4 (conductive) and TiO_2 (semi-conducting) NPs to a highly refined, bleached, deodorized palm oil (RBDPO) increases the AC-BDV of the base NE.

By considering Fe_3O_4, Al_2O_3 and SiO_2 nanoparticles, Usama and Beroual [43,112] reported that these NPs do not have a significant influence on the AC-BDV of natural esters (MIDEL 1204) as observed with mineral oil and synthetic esters (MIDEL 7131). Tables 4 and 5 give the mean AC-BDV of natural and synthetic ester-based nanofluids. The best improvement does not exceed 7%. Sometimes, the addition of NPs even decreases the electrical strength of natural esters, as is the case with SiO_2 at a concentration of 0.05 g/L, where the AC-BDV is reduced by 15% [112]. While it growths slightly by 5% at 0.2 g/L and 4% at 0.3 g/L, with Al_2O_3 NPS of a size of 50 nm and at 0.05 and 0.2 g/L concentration, the AC-BDVs are lower than the base oil while they are a little bit higher at a concentration of 0.3 g/L (by 6%); with 0.4%, the improvement is unimportant (1%). For the same kind of NPs of the size of 13 nm, there is a small improvement at 0.05% concentration that is at a percentage equal to 7%.

Table 4. Mean value of AC-BDV of different natural ester (MIDEL 1204)-based nanofluids [112].

	Natural Ester	Fe$_3$O$_4$ (50 nm)	Al$_2$O$_3$ (50 nm)	Al$_2$O$_3$ (13 nm)	SiO$_2$ (50 nm)
	Natural ester/0.05 (g/L) Nanofluids				
Breakdown voltage (kV)	68.77	57.53	61.53	73.93	58.20
	Natural ester/0.2 (g/L) Nanofluids				
Breakdown voltage (kV)	68.77	66.90	66.10	69.87	72.23
	Natural ester/0.3 (g/L) Nanofluids				
Breakdown voltage (kV)	68.77	69.53	73.00	67.33	72.73
	Natural ester/0.4 (g/L) Nanofluids				
Breakdown voltage (kV)	68.77	73.63	69.07	67.30	67.13

Table 5. Mean value of AC-BDV of different synthetic ester (MIDEL 7131)—based nanofluids [43].

	Synthetic Ester	Fe$_3$O$_4$ (50 nm)	Al$_2$O$_3$ (50 nm)	Al$_2$O$_3$ (13 nm)	SiO$_2$ (50 nm)
	Synthetic ester/0.05 (g/L) Nanofluids				
Breakdown voltage (kV)	60	56.57	66.3	80.83	71.5
	Synthetic ester/0.2 (g/L) Nanofluids				
Breakdown voltage (kV)	60	59.97	74.3	79.87	72.5
	Synthetic ester/0.3 (g/L) Nanofluids				
Breakdown voltage (kV)	60	70.07	75.27	72.3	74.6
	Synthetic ester/0.4 (g/L) Nanofluids				
Breakdown voltage (kV)	60	88.67	74.03	69.9	78.9

The adding of the same NPs to MIDEL 7131 improves its dielectric strength [43]. For a given type of nanoparticles and their sizes, an optimal concentration exists that gives the highest value of AC-BDV. The best improvement is obtained with Fe$_3$O$_4$ (50 nm) NPs. It is of 48% with Fe$_3$O$_4$ (50 nm), while it is of 35 and 25% with Al$_2$O$_3$ (13 and 50 nm in size respectively), and it is of 32% with SiO$_2$ (50 nm). Thus, synthetic ester-based NFs are found to be more beneficial than natural ester-based NFs.

Mohamad et al. [44] reported that the addition of 0.01 g/L of Fe$_3$O$_4$ NPs to Palm Fatty Acid Ester (PFAE) improves the AC-BDV compared to pure PFAE by 43%. Menlik et al. [26] also observed that the adding of silica-coated titanium (ST TiO$_2$) to ENVITRAFOL (a biodegradable oil) at a concentration of 0.25% increases the AC-BDV by about 33%.

As concerns synthetic esters, it has been reported that Fe$_3$O$_4$, Al$_2$O$_3$ and SiO$_2$ NPs improve the dielectric strength of MIDEL 7131. The improvement can reach 48% with 0.4 g/L of Fe$_3$O$_4$ (50 nm). While it is of 35 and 25% with Al$_2$O$_3$ for NPs of sizes 13 and 50 nm, respectively, and 32% with SiO$_2$ (50 nm).

Regarding the influence of NPs on the DC-BDV of esters, Beroual et al. [46,47] investigated three types of NPs, namely Fe$_3$O$_4$, Al$_2$O$_3$ and SiO$_2$. They observed that, for the 0.05 g/L concentration, the DC-BDV of natural ester MIDEL 1204 decreases despite the type of NP. The best improvements are for 0.2 g/L of Fe$_3$O$_4$ (50 nm) and for 0.3 g/L of Al$_2$O$_3$ (50 nm); these improvements are around

10.6% and 9%, respectively. The highest improvement of DC-BDV is noticed for Al_2O_3 NPs (13 nm), which is 7.6% for 0.4 g/L. However, the reduction of DC-BDV is observed independently of the concentration of SiO_2. The same authors also analyzed the effect of the same NPs on synthetic esters (MIDEL 7131) and observed the existence of an optimal concentration of NPs that gives the highest DC-BDV with the exception of SiO_2, which decreases the DC-BDV of pure synthetic esters, irrespective of the concentration. For the other types of NPs, the DC-BDV is enhanced by circa 25%, 13% and 10%, with Al_2O_3 (13 nm), Al_2O_3 (50 nm) and Fe_3O_4 (50 nm) NPs at the optimal concentration that is 0.05 g/L, respectively.

By adding silica nanoparticles dispersed in a synthetic ester, Mahidar et al. [48] observed an increase in the corona inception voltage and BDV of NP modified synthetic esters compared to pure synthetic esters; an improvement of approximately 20% in breakdown voltage is noticed with nano esters compared with pure esters.

As concerns the lightning impulse breakdown voltage (LI-BDV), it has been generally investigated under divergent electric fields and mainly in point-plane electrode configurations. The nanoparticles have generally had a beneficial effect on the LI-BDV of esters. So, Li et al. [116] reported that positive and negative LI-BDVs of vegetable oil (natural ester) based Fe_3O_4 nanofluids increase by 37% and 12%, respectively.

It was shown that in a quasi-uniform electrical field, the adding of Fe_3O_4, Al_2O_3 and SiO_2 NPs to natural (MIDEL 1204) and synthetic esters (MIDEL 7131) increased the negative LI-BDV. The best improvement of LI-BDV of synthetic esters is noticed with Fe_3O_4 [117,118]. The improvement is of about 25.6% with Fe_3O_4 at a concentration of 0.05 g/L; 22% with SiO_2 at a concentration of 0.3 g/L; and 18.3% with Al_2O_3 at a concentration of 0.3 g/L. For natural esters, the best improvements of the U50% value of LI-BDV are approximately 7.5% with Fe_3O_4 at a concentration of 0.2 g/L; 16.8% with Al_2O_3 at a concentration of 0.05 g/L; and 13% of SiO_2 at a concentration of 0.2 g/L [117,118].

Note that LI-BDV of pure synthetic esters and natural esters is more that about twice that of BDV at AC stress [117]. By measuring the AC as well as the positive and negative LI breakdown voltages of vegetable oil-based ZnO nanofluids at different particle concentrations, Chen et al. [119] observed improvements of BDV, these voltages are of 30.2%, 24.6% and 5.5%, respectively.

4.2. Processes Involved in Breakdown Voltage Mechanisms of Synthetic Esters-Based Nanofluids

Different mechanisms have been advanced to explain the processes by which NPs affect the breakdown voltage. According to some researchers, the improvement or degradation of the AC-BDV of ester-based nanofluids can be explained by the trapping of electrons by NPs or of slowing the propagation of the streamers [120]. However, there is one fact that seems to be unanimous, it is that of the influence of the dimensions of nanoparticles: for the same type of NPS and concentration, the smaller the nanoparticles, the higher the breakdown voltage of nanofluids is. On the one hand, this is due to the large volume fraction of interfaces in the material bulk and on the other hand the ensuing interactions between the nanomaterial surface charged and the liquid molecule.

According to Beroual and Khaled [46,117], the improvement in the BDV of NP-based esters may be due to (1) the electron trap sites which nanoparticles create around the electrodes by reducing the electron number and consequently by increasing the streamer initiation voltage, thus leading to an increase in the breakdown voltage; and (2) by reducing the number of electrons which move towards the opposite electrode, resulting in an accumulation of electrons at the nanoparticle-surrounding liquid interfaces, thus forming the so-called double layer, and thereby slowing down the propagation of the streamers leading to the breakdown. When the double layer is saturated, the excess electrons will no longer be trapped; they will contribute to the propagation/acceleration of the streamers and thus lead to the reduction of the BDV. This can explain the existence of optimal concentration, a threshold value of NPs beyond which the electrons are no longer trapped.

Some authors such as Du et al. [120], Hwang et al. [121], and Miao et al. [122] explain the improvement of BDV of esters-based nanofluids by the influence of NPs on the streamer development.

NPs, especially the conducting ones, catch the electrons in their movement and accumulate these electrons on the surfaces, creating a local electric field which will counteract to the external electric field. The NPs act as "electron traps", resulting in a slowdown of streamers, thus finally to increase the breakdown voltage. Such a theory has been developed for magnetic NPs in the presence of LI voltage. The interpretation advanced by Peppas et al. [114], and Makmud et al. [115] on the role of NPs goes in the same direction. Nanoparticles act as kind of electronic scavengers that reduce the charge density and hinder the streamer propagation in nanofluids, resulting in an increase of BDV.

The conductivity of NPs has also an influence on the BDV of NFs. The conductive NPs capture very rapidly fast moving electrons and convert them into slow negatively charged NPs resulting in the slowing streamer development and increasing the breakdown voltage [123]. This can explain why Fe_3O_4, conductive NPs, give better results than insulating ones.

According to Ibrahim et al. [124], the dielectric constant of NPs also has an effect on the BDV of NFs. They observed that the adding of iron nickel oxide (Fe_2NiO_4) NPs increases the DC-BDV of NFs. NPs of a higher dielectric constant tend to be positively charged while NPs particles of lower dielectric constants have a tendency to be polarized and negatively charged. Both positively and negatively charged NPs can trap streamer electrons. Therefore, both positive and negative DC-BDVs of NF are increased. As for Madawan et al. [125], the improvement of BDV by the addition of magnetic NPs is due to the relaxation time constant of NPs. The time constant value of Fe_3O_4 NPs is low in comparison to that of Al_2O_3 NPs. Hence Fe_3O_4-based NFs have greater BDV than NFs with other NPs. The NFs with Fe_3O_4 NPs have a higher rate of scavenging of charges from the streamer.

According to Hwang et al. [121], if the polarization-relaxation time of NPs is shorter than the streamer development time, the streamer is inhibited and the BDV increases. In turn, when the time constant is greater than the streamer development time, the streamer and the breakdown voltage are not affected. However, such an interpretation is in contradiction with the results concerning the influence of different types of electronic scavenger compounds (carbon tetrachloride, Iodobenzene or sulfur hexafluoride) [126,127]. Indeed, it is well-known that such additives accelerate the propagation of the streamers in liquids [128].

As concerns the influence of the voltage wave form (AC and DC), it can be explained by the injected space charge. Under DC voltage, the injected homo-charges make the liquid more conductive, leading thence to a drop in the breakdown voltage. While under AC voltage, the hetero-charges neutralize each other and their effect on liquid conduction and breakdown voltage is therefore less significant. The mechanisms making DC-BDV higher than AC-BDV in NFs would therefore be different. Thus the conductivity of the liquid at itself cannot explain this difference.

The increase in DC-BDV could be due to the fact that by accumulating on the surface of the nanoparticles, the number of homo-charges injected decreases. The electric field in the vicinity of the injector (that is to say of the HT electrode) and in the volume of the liquid decreases. This results in an increase in the initiation threshold voltage of the streamer and consequently an increase in the DC breakdown voltage.

5. Combined Paper-Synthetic Ester-Based Insulating Systems

5.1. Moisture Equilibrium Curves

The moisture equilibrium curves describe the dependence of water content in cellulose materials (expressed in percentage by weight) as a function of water content in electro-insulating liquid (expressed in ppm by weight) [30]. Figure 8 shows the equilibrium curves for cellulose insulation impregnated with a synthetic ester. In order to construct these curves, the method proposed by Oommen was used [129]. It consists of combining the curves of water content in electro-insulating liquid (Figure 9) and cellulose paper (Figure 10) as a function of air relative humidity. The curves from Figure 10 were obtained on the basis of water saturation coefficients for synthetic esters. Midel 7131 determined by Round Robin Test carried out by the Cigre D1.52 WG, while the water sorption isotherms for new cellulose paper (degree of polymerization equal to 1360) were determined on the basis of data from [32].

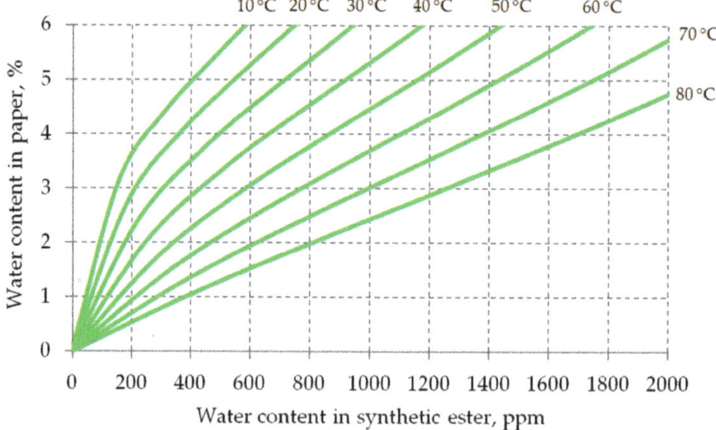

Figure 8. Moisture equilibrium curves for cellulose paper and synthetic ester Midel 7131.

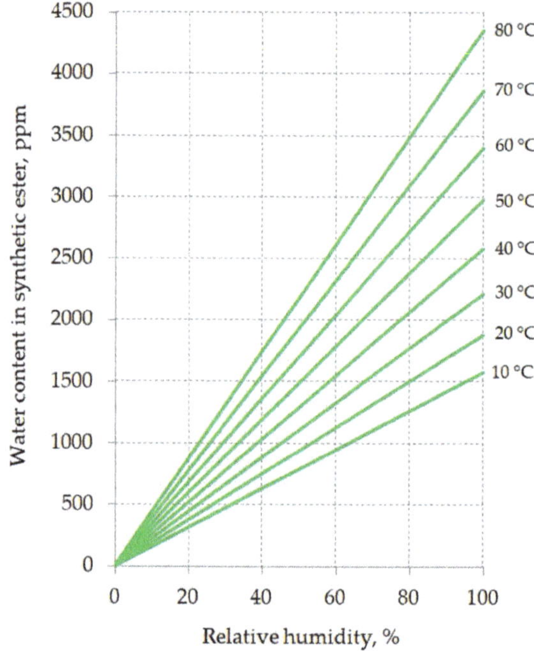

Figure 9. Water content in synthetic ester Midel 7131 of relative humidity, based on the data from [30].

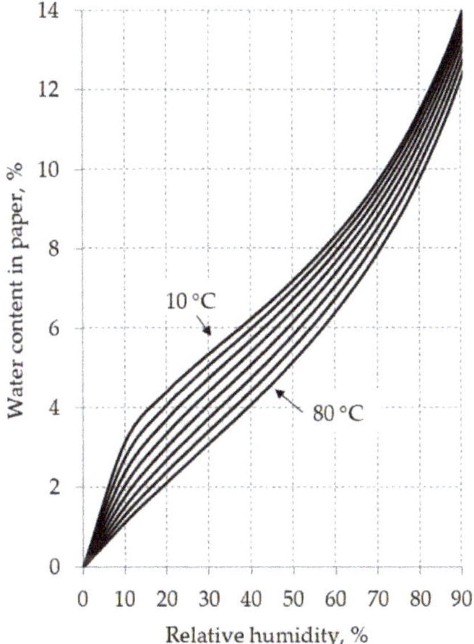

Figure 10. Water sorption isotherms of cellulose paper, based on the data from [32].

The moisture equilibrium curves from Figure 10 show the relationship between the amount of water accumulated in cellulose insulation and the amount of water dissolved in the synthetic ester. They allow, for example, to estimate how the water content in the insulating liquid will change, and thus the degree of ester saturation with water, as a result of changes in the insulation temperature. The moisture equilibrium curves also makes it possible to determine the water content in solid insulation impregnated with a synthetic ester on the basis of information on the water content in this liquid and the insulation temperature. However, it should be remembered that this method of assessing the moisture content of solid insulation has a number of limitations. To use this method, the cellulose-ester system must be in the moisture equilibrium or close to this state, which needs to maintain the temperature at a constant level for a long time. The higher temperature, the shorter time needed to achieve moisture equilibrium. A single measurement of water content in ester by means of the Karl Fischer titration method does not enable the assessment of the dynamics of water migration in the insulation system. An alternative for measuring the water content by the Karl Fischer method is a capacitive sensor, which, when installed on a transformer, gives a broader image of the insulation moisture. Moreover, in order to correctly apply moisture equilibrium curves, it is necessary to take into consideration the influence of the aging rate of both cellulose and liquid insulation [32,130,131].

Table 6 gives the guidelines for interpreting the condition of cellulose insulation on the basis of oil relative saturation. This interpretation was taken from the standard [132] prepared for mineral insulating oils. On the basis of the relative saturation values (Table 6) and the above-quoted water saturation coefficients (Table 2), it is possible to calculate the limit values of water content in synthetic esters for a given temperature, which corresponds to the different conditions of cellulose insulation. The limit values of water content calculated for temperature equal to 50 °C are shown in Table 6.

Table 6. The guidelines for interpreting the condition of cellulose insulation on the basis of water content in synthetic ester.

Relative Saturation, %	Limit Values of Water Content, ppm @ 50 °C	Condition of Cellulosic Insulation [132]
<5	<149	Dry insulation
5 to 20	149 to 595	Moderately wet, low values indicate fairly dry to moderate levels of moisture content in the insulation. The values toward the upper limit show moderately wet cellulose insulation
20 to 30	595 to 893	Wet insulation
>30	>893	Extremely wet insulation

5.2. Impregnation of Solid Insulation with Synthetic Esters

In order to obtain the optimal properties of a liquid–solid insulation structure, the impregnation process of the solid insulation by a given dielectric liquid should be thoroughly carried out. For this reason, it is particularly important to select proper parameters of drying and then impregnation processes, which will ensure a long lifetime of solid insulation and thus the transformer itself. In the case of mineral oil used as the dielectric liquid in power transformer, the manufacturers have possessed a great knowledge in selection of optimal parameters for impregnation process. This knowledge has resulted from the long presence of mineral oil in the transformer market. It is the opposite situation in the case of synthetic esters, which have only recently been used in the transformer applications; the practical knowledge of their behavior in different areas of exploitation is not yet sufficiently mastered.

The impregnation properties of synthetic esters were studied widely by Dai and Wang in [133], and also in a narrow range by Rozga [19] using the capillary effect. Although the capillary test is not a complicated, the results obtained with its use give the data on impregnation efficiency of a given liquid considered. In general, capillary tests concern the phenomenon of pulling up the liquid at the impact of capillary forces it and depends on the specific properties of a given material. The kinematic viscosity of the liquid, the number of open pores contained in the insulating material (pressboard), the diameters of the pores, as well as the pressure outside and inside of the material determine the rate of liquid penetration into the insulating material. In the tests described in [133], the 3 mm thick pressboard and two types of laminated blocks were investigated. The tests were carried out at selected temperatures of impregnation processes: 20, 40 and 60 °C. In the case of the pressboard, the authors indicated that the balance point at which the process of impregnation occurs in a similar way is 60 °C for synthetic esters and 20 °C for mineral oil. This finding was confirmed in the tests of laminated blocks where the synthetic ester at 60 °C impregnated the block identically to the mineral oil at 20 °C. Similar findings noticed in [19], where the considered temperatures were 20, 60, 80 and 100 °C and 2 mm thick pressboard samples were tested. Obtaining similar results of impregnation when using synthetic esters instead of mineral oil, was achieved neither by extending the impregnation time nor by increasing the temperature of the impregnation process. For example, saturation of the pressboard sample at the temperature 100 °C to the same height as in the case of mineral oil required approximately 1.5 longer time in the case of synthetic esters.

5.3. Breakdown Characteristics of Solid Insulation Components Impregnated with Synthetic Esters

The comparison of breakdown characteristics of pressboard samples impregnated with different liquid dielectrics under different drying and impregnation process conditions was done in [41]. In these studies, next to the full impregnation process applied to the 2 mm thick pressboard samples according to the IEC 60243-1 Standard [134], intermediate conditions were also considered. Thus, in the first cycle called "full impregnation" the samples, after drying under vacuum through 24 h in 100 °C, were flooded with a given dielectric liquid (mineral oil, natural ester, synthetic ester, respectively) and then impregnated in 100 °C through 24 h. In the second cycle, called "8 h, 60 °C", the samples, after drying identically to the first part, were impregnated at 60 °C for 8 h. In the cycle part called "8 h, 100 °C", the samples after drying were impregnated at a temperature of 100 °C for 8 h. The results obtained in the form of a median of 26 breakdowns are presented in Table 7.

Table 7. Median of breakdown voltages of pressboard impregnated in a different ways (kV).

Impregnation Cycle	Mineral Oil	Natural Ester	Synthetic Ester
8 h, 60 °C	38.85	38.10	37.55
8 h, 100 °C	45.50	41.80	44.00
Full, 24 h, 100 °C	81.95	83.40	88.60

The obtained results show some kind of benefit of ester liquids although capillary tests do not demonstrate it. In the case of all impregnation cycles applied, both esters impregnated the pressboard similarly as mineral oil assessing the process from the point of view of electrical strength of pressboard. Especially, full impregnation gave promising results for the synthetic ester because breakdown voltage was a few kV higher than in the case of the other liquids used for impregnation. As we assume that the first cycle ensures full impregnation of the pressboard samples, the highest electrical strength of the pressboards impregnated with synthetic ester is not a surprise: the synthetic ester has a much higher electrical permittivity (3.2) than mineral oil (2.2). This causes the pressboard to be impregnated with this liquid and also has a greater electrical permittivity, which leads to a decrease in the value of the electric field in the material, i.e., an increase in its electrical strength.

A similar conclusion was formed in [135]. A full impregnation procedure for 0.5 and 1 mm thick pressboard samples impregnated with esters and mineral oil gives a reflection of the statistically identical electrical strength of pressboards for a whole range of breakdown probabilities assessed using three-parameter Weibull distribution function.

However, breakdown tests of solid insulation showed some disadvantages of esters when observing the marks on the pressboard surfaces left by discharges leading to breakdown. These marks occurred denser in the case of pressboard impregnated with esters. It means that ester-impregnated pressboards are more susceptible to partial discharge influences which are not desirable in real insulating structures. Figure 11 depicts an example of pressboard surfaces after discharge leading to breakdown.

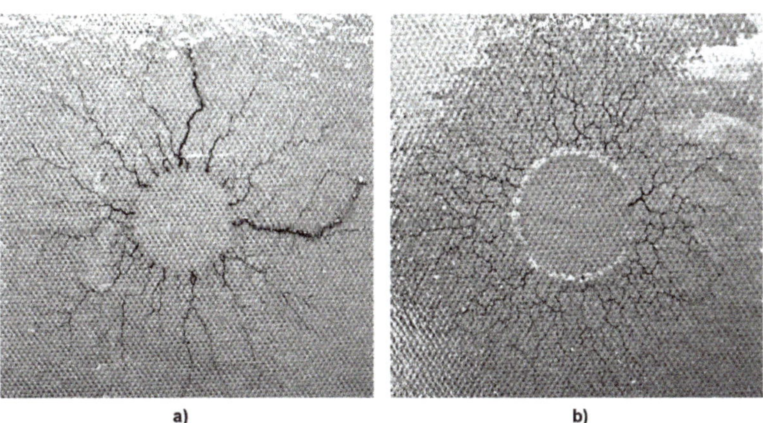

a)   b)

Figure 11. Differences in the shapes of marks due to discharges over the surface of pressboard samples: (**a**) mineral oil, (**b**) synthetic ester.

5.4. Electric Field Distribution in Insulating Systems with Synthetic Esters

Insulation structures based on cellulosic materials and synthetic esters have not been the subject of a wide range of studies. Rather, the natural esters are considered in such studies as representative of alternative dielectric liquids [136]. Thus, in [41], the studies in this field were achieved comparing synthetic esters and mineral oil in terms of electrical field distribution in simplified power transformer winding structures. Figure 12 depicts the electric field distribution of the assumed main insulation part isolated with synthetic esters and mineral oil, respectively. It is not easy from this figure to identify the differences. Thus, Figure 13 shows a selected magnified fragment from Figure 12.

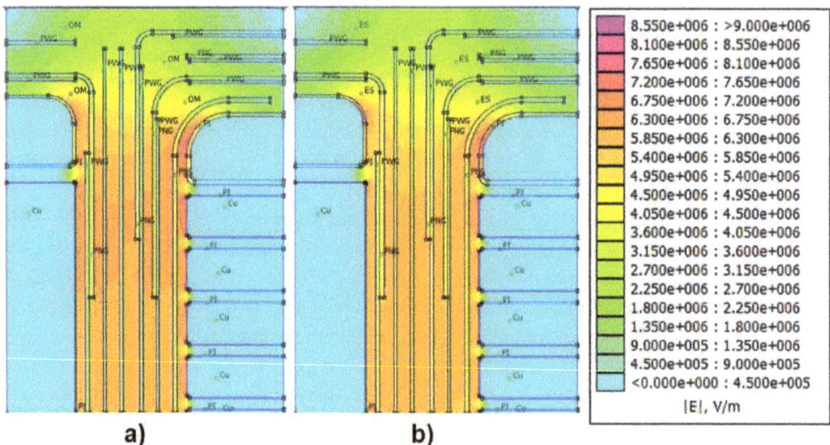

Figure 12. Electric field distribution in simplified insulation structure of power transformer: (**a**) mineral oil, (**b**) synthetic ester.

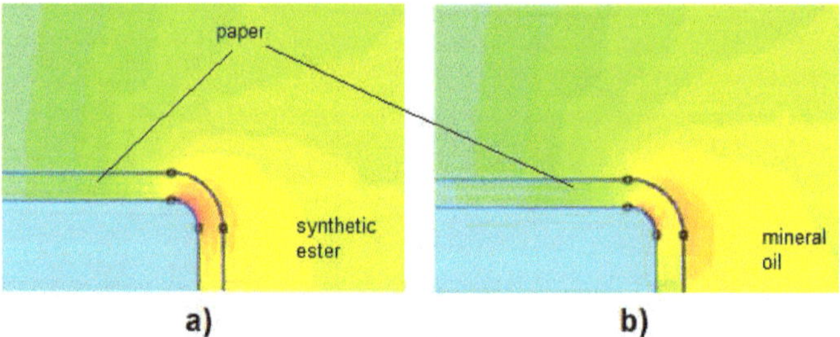

Figure 13. Magnified fragment of electric field distribution in transformer insulation: (**a**) synthetic ester, (**b**) mineral oil.

As can be seen, the intensity of electrical field is higher in paper when the synthetic ester is used as a dielectric liquid (concentrated region marked by deep red). However, in the "liquid part", the electric field stress decreases in the case of the synthetic ester where a light orange color is observed, while dark orange appears in the case of mineral oil. The obtained data are not, however, a surprise—higher electrical permittivity of synthetic esters in relation to the mineral oil causes a decrease in the electrical field in this medium at the expense of an increase in the electrical field stress in solid insulation. However, the differences between the distinctive regions are lower in the case of ester liquids, which is their benefit [41,136].

6. Application of Synthetic Esters for Retro-Filling and Drying of Mineral Oil-Immersed Transformers

6.1. Miscibility

In accordance with [4], synthetic esters are miscible with mineral oil and natural esters in all proportions. In turn, the synthetic ester is not miscible with silicone oil. There are some reports in the literature that even small concentrations of silicone oil in esters can cause foaming during the process of filling a transformer under a vacuum. The unlimited miscibility of synthetic esters with mineral oil enables the application of esters for retro-filling and drying mineral-oil-immersed transformers. In both situations, the mixture of synthetic esters with oil is formed whose properties differ from both base insulating liquids. This must be considered during the operation and diagnostics of transformers subjected to the retro-filling or drying process.

6.2. Retro-Filling

The retro-filling process should be understood as the replacement of insulating liquid in the transformer tank with another type of liquid. In the past, the retro-filling processes were performed mainly for environmental reasons. For example, this was the case with polychlorinated biphenyls (PCBs), which due to their good fire properties, were readily used in transformers between 1930 and the mid-seventies [49]. In the 1970s, it was recognized that PCBs pose a threat to the environment and therefore, the process of their removal began. Ultimately, the owners of PCB transformers had to make the decision either to "replace" the transformers with new non-PCB units or to "reclassify" them to non-PCB status using a retro-filling process [49]. Due to lower costs compared to the purchase of a new unit, the retro-filling process was eagerly chosen by transformer owners. Currently, the retro-filling process is most often associated with the replacement of mineral oil with synthetic or natural esters. The replacement of mineral oil with synthetic esters is performed mainly for two reasons. The first one is related to the better fire properties of esters compared to mineral oil. The second reason is related to synthetic esters being an environmentally friendly liquid which follows from the fact that it is readily biodegradable, unlike mineral oil which is both toxic and non-biodegradable [4,5,137].

The manufacturers of synthetic esters also point to other technical and non-technical benefits related to the replacement of mineral oil with synthetic esters, which include [52]:

- increasing the transformer life; which is related to the superior water tolerance properties of synthetic esters, which increase the life of solid insulation'
- lowering the insurance costs by reducing the likelihood of a serious fire;
- and demonstrating the social responsibility of the transformer owner.

All the technical benefits of the retro-filling mentioned above, especially fire safety, depend to a large extent on the quality of the liquid exchange treatment. The removal of as much of the mineral oil as possible from the transformer tank is of great importance for the properties of the synthetic ester with which the unit will be filled. Unfortunately, it is not possible to completely remove mineral oil, especially from the impregnated cellulose. During the transformer operation, this remaining oil may slowly diffuse to a synthetic ester. Thereby, some of the synthetic ester properties will be affected.

It is estimated that after the retro-filling process, up to 10% of the mineral oil may remain in the cellulose insulation, but also at the bottom and at the walls of the tank or on the transformer core [49–51]. Moreover, the authors of [51] estimated that for the units they analyzed, in the case of suction of mineral oil from the bottom of the tank and its partial removal from the surface of transformer elements, the oil content in the synthetic ester ranges from 1.2% to 1.7%. On the other hand, when the oil is drained only to the drain valve level, the content ranges from 8.6 to 9.7%. The ranges shown may vary depending on the fluid exchange technique used, but also depending on factors such as, e.g., the ratio of cellulosic materials to insulating liquid. To reduce the residual mineral oil in the tank, the rinsing procedure of core and windings with a heated ester should be used.

The presence of mineral oil in the synthetic ester strongly influences the fire properties such as flash and fire points. Figure 14 shows the dependency of the flash and fire points depending on the content of mineral oil in synthetic esters [42]. As can be seen, the fire properties of the liquid decreases with the increasing mineral oil content in the synthetic ester. The decrease in both the flash and fire points is greatest in the range of up to 10% of the mineral oil content in the synthetic ester. Therefore, it is so important to remove during the process of retro-filling as much mineral oil as possible.

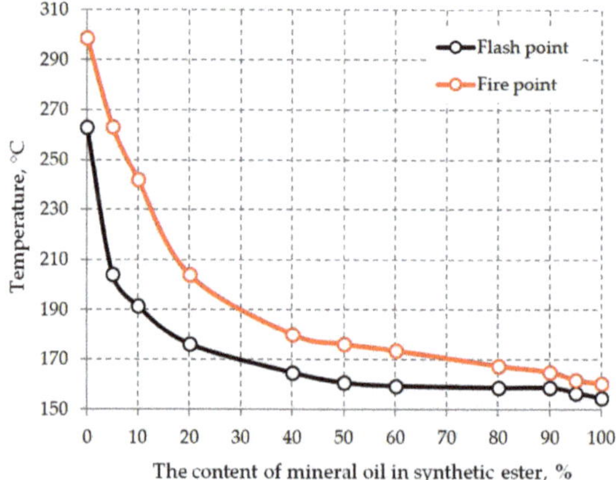

Figure 14. The flash and fire points depending on the content of mineral oil in synthetic ester, on the basis of data from [42].

In accordance with the information from [53], the application of synthetic ester Midel 7131 for retro-filling a transformer depends on the voltage and power rating of the unit. In the case of the distribution transformers up to 33 kV, no problems relating to the retro-filling process are anticipated. In turn, for power transformers above 33 kV, the additional steps to prepare the transformer for liquid exchange treatment may be necessary. The general guideline for the retro-filling of transformers above 33 kV is given in [53].

6.3. Drying

Due to their high water solubility, the ester fluids can be used to dry the transformer cellulose insulation [54,55,138]. The authors of [139] proved that retro-filling a mineral oil-filled power transformer with a natural ester will decrease the cellulose moisture content in cellulose. A better drying effect is to be expected in the case of using a synthetic ester for retro-filling, which can be explained by its greater water solubility (S = 3404 ppm at 60 °C) compared to the natural ester (S = 1672 ppm at 60 °C) [32]. In [55], it was proven that the drying of synthetic esters can be effectively carried out for cellulose insulation impregnated with mineral oil. Drying takes place because the ester-cellulose system strives to reach the moisture equilibrium state. After the dry ester is introduced into the tank, the water migrates from the moist cellulose insulation to the ester. The dynamics of water migration, and hence the drying efficiency, is influenced by several factors, including [57]:

- Drying temperature;
- Thickness of cellulose materials;
- Moisture content in the synthetic ester;
- Moisture content in cellulose;
- Weight ratio of ester to cellulose.

The high temperature significantly improves the drying process. The higher temperature insulation, the more water can be dissolved in the ester (Figure 9) and the less water can be adsorbed by cellulose (Figure 10). The drying temperature is limited, inter alia, by the temperature class of the materials, which for cellulose is equal to 105 °C, while for thermally upgraded Kraft is 120 °C [57]. Ensuring a high drying temperature may be difficult, especially in winter conditions. In such a case, a thermal insulation of the transformer tank should be applied [57].

The results of the research presented in [56] show that the amount of water removed from the solid insulation decreases significantly as the thickness of the cellulose material grows. This dependence was observed for cellulose materials up to about 3 mm thick. Above this value, the effect of the thickness on drying was not observed. This can be explained by the fact that for a short drying time (7 days and 70 °C), it is possible to dry only the outer layers of thick cellulose materials.

The influence of the other factors (water content in ester and cellulose or weight ratio of ester to cellulose) on the drying process efficiency depends on the applied drying method. Drying the insulation may result from retro-filling or from the use of synthetic esters for circulating the drying of the transformer insulation system.

In the case of retro-filling, the degree of cellulose insulation drying is limited primarily to the solubility of water in the synthetic ester. After retro-filling, the drying process continues until the migration of water from the cellulose insulation to the ester ceases, that is, until the moisture equilibrium in the cellulose-ester system (Figure 10) is reached. Using the equilibrium curves shown in Figure 10, it is possible to estimate the water content in the transformer after the retro-filling process. The result of such an estimation is presented below, assuming that:

- The moisture content in cellulose materials before retro-filling is equal to 3.0%;
- The average operation temperature of the insulation is equal to 60 °C;
- The initial water content in synthetic ester is equal to 60 ppm;
- The weight ratio of the cellulose insulation to the electro-insulating liquid equals 0.0715 for the chosen transformer with a capacity of 10 MVA, a mean of 522 kg of cellulose and 7300 kg of mineral oil.

For the conditions so assumed, the use of synthetic ester for the retro-filling process will cause the reduction of water content in cellulose by about 0.75 percentage points, up to the level of 2.25%. It must be pointed out that a single oil replacement with a synthetic ester may give a better or worse drying effect than in the presented calculation example, mainly depending on the weight ratio of cellulose to insulating liquid and the transformer operation temperature.

It is also possible to apply a hot ester circulation method to dry cellulose insulation. This method consists of three stages [57]:

- Replacement of mineral oil with a synthetic ester
- Drying the cellulose insulation by means of the ester, which circulates between the transformer and the equipment for ester drying and heating
- Filling back the transformer tank with mineral oil.

Due to the continuous drying of the synthetic ester, the circulation method allows for a much better drying effect of the transformer solid insulation than in the case of retro filling. To make drying with circulating ester profitable, it is necessary to develop a method for an industrial system for the treatment of the synthetic ester. This system will allow for reducing the mineral oil concentration in the synthetic ester. The treated ester can be reused many times in the drying process.

7. DGA-Based Diagnosis of Synthetic Ester-Filled Transformers

The dissolved gas analysis (DGA) method is one of the most widely used methods for diagnosis on mineral oil-filled transformers, especially power transformers. It makes it possible to detect the beginning of a fault and its location through the detection of certain gases, thereby preventing breakdowns that could lead to significant damage. Indeed, the occurrence of partial discharges (PD) and arcing, overheating and overloading can result in chemical decomposition of the constituents of insulating systems, resulting in the formation of different by-products, especially combustible and noncombustible gases dissolved within the insulating liquid. The danger of these gases for the normal operation of the transformer depends on their nature, concentrations, relative proportions and the progression of their concentration. Among the gases used as fault markers, (Hydrogen (H_2), Carbon monoxide (CO), Carbon dioxide (CO_2), Methane (CH_4), Acetylene (C_2H_2), Ethylene (C_2H_4) and Ethane (C_2H_6)), Nitrogen (N_2) and (Oxygen (O_2) are also two gases useful for the diagnosis. Guides and standards using these gases, their concentrations and the relationships existing between them are regularly updated. According to international guidelines for the interpretation of DGA, i.e., IEC 60599 [140] and IEEE C57.104 [141], the three main methods used for mineral oils are: (1) the ratio method; (2) the Duval triangle method; (3) the method of key gases. These are based on the gas kind, the value of the gas and the fault type.

The DGA method has been used for several decades for the diagnosis of transformers filled with mineral oils. Different methods of interpretation are used. However, with the increasing use of liquid insulating substitutes for mineral oils and more particularly for esters (natural or synthetic), the question which arises is that of the extension of the DGA method for the diagnosis of power transformers filled with these alternative oils.

Perrier et al. [70] compared the thermal effect of electrical faults on DGA in esters and mineral oils. For thermal and electrical faults, they focused their studies on the effect of low thermal fault and low energy discharges, respectively. They observed that the same kind of gases (Hydrogen and Acetylene) in comparatively similar ratios are created in esters and mineral oils for electrical faults, and that the classic Duval triangle 1 fits well with this type of fault. For thermal faults as well as for stray gassing issue, Ethane (associated with Hydrogen) is the key gas in the case of natural esters. These authors also showed that new Duval triangles 4 and 6 work well for mineral and ester oils, respectively [69,142,143]. However, with esters, some interpretation care should be taken during vacuum extraction as the results may be underestimated because of the high viscosity of the esters, which could interfere with the results [70].

The DGA of canola-based esters at the discharges of the creepage type reveal that both in the case of esters and mineral oils, similar gases are generated. However, while the quantity of dissolved gases is high in the ester compared to the mineral oil, the key dissolved gas was found to be Hydrogen. H. Mnisi and G. Nyamupangedengu [144] indicated that the Duval triangle and the IEEE-based key gas diagnostic technique with great correctness levels identified the fault while the IEC 60599-based approach failed to identify the fault. According to the authors, the IEC 60599 requires modifications when used in the transformers filled with canola oil.

Gómez et al. [145] investigated two faults, namely discharge and overheating. They considered four different commercial insulating liquids, one mineral oil and three natural esters. They compared five key dissolved hydrocarbon gases—hydrogen(H_2), methane (CH_4), ethane (C_2H_6), ethylene (C_2H_4) and acetylene (C_2H_2). They applied in the analysis classic Duval triangle 1, modified Duval triangle 3, Roger's ratio method, the Dornenburg method and the IEC 60599-based method to predict the type of fault. The findings confirmed that modifications are needed to use these method for non-mineral oils when a fault recognition is expected.

Additionally, in [146–148], the authors found that although the gases generated by faults of electrical and thermal origin are the same for mineral oil and esters, it should be pointed out that there is a considerable difference in the amount of the generated gases and the different solubility of these gases in insulating liquids which undoubtedly affects the way the results are interpreted. The example here can be carbon monoxide and dioxide, which are generated in large amounts in the case of thermal exposure to esters [146,147]. Another example is the high portion of Propane in the sum of the gases generated mainly in the synthetic esters during thermal faults type T2 (above 300 °C) [148].

According to Lashbrook et al. [149], the methods of dissolved gases analysis applied to mineral oils may be still valid for esters in most cases. However, it is necessary to be vigilant with the stray gassing of ethane in natural esters. This is not seen in mineral oils or in synthetic esters.

8. Static Electrification of Synthetic Esters

The formation of a charge double layer is a natural process for any solid–liquid interphase. When liquid comes into contact with a solid as a result of physicochemical reactions at the interphase, the originally electrically neutral system becomes polarized. The source of the charge are dissociation processes of impurities or additives in the oil. The resulting ions of both signs may react electrochemically with the surface of the solid body in contact with the insulating liquid. These corrosive processes are also a source of electric charge carriers. As a result, a charge layer is created on the surface of the solid and a charge of the opposite sign is formed in the liquid layer near the solid surface. These two layers of charge are called electric double layers (EDL). The EDL formation phenomenon was initially described by Helmholtz. In the Helmholtz model, it is assumed that the charge of the counter-ions forms a monolayer in the liquid. This model was then developed by Gouy and Chapman by assuming the distribution of ions in the liquid as continuous, resulting in a diffusion layer of finite thickness. The Stern model based on both of these models takes into account the existence of both the Helmholtz and Gouy-Chapman layers.

In transformers with a forced oil flow cooling, the EDL layer becomes broken, which results in movement and the local accumulation of space charge. These phenomena in turn are the cause of partial discharges, leading to serious failures of power transformers. This problem has already been noticed in the 1970s [150], and led to an interest in the phenomenon of static electrification of mineral oils [151–154]. Growing ecological requirements made the transformer industry turn to biodegradable oils and oils made from renewable materials. The more common use of a new type of insulating liquid in increasingly larger units is associated with the risk of their failure resulting from static electrification.

The research carried out by various authors indicates a general tendency towards stronger static electrification of esters in relation to mineral oils [67,68,155]. As is shown by the research presented by Paillat et al. [156], the electrification current tends to increase with the increasing ester oil rate of movement relative to the solid phase as well as the function of flow time and temperature. Kolcunova et al. [155] explained the increase in the value of current with an increase in the speed of phase movement by the changes in the thickness of the laminar sub-layer in relation to the Debye-length. This is a general property also confirmed for mineral oil by researchers [157,158]. However, the current values are higher than for mineral oil.

A comparison of the electrification of mineral oil and esters, presented by Beroual et al. [67], confirms the fact of generating the highest values of electrification currents in the case of ester oil. Moreover, these authors showed that the type of paper or pressboard with which both mineral and ester oil comes into contact has a similar effect on the value of the electrification current. The flow of mineral oil at the surface of the pressboard generates the higher values of the electrification current than in the case of the transformer paper surface, regardless of the flow velocity. Ester oil shows a greater ability to electrification to interact with TU (Thermally Upgraded) paper, for higher rates of phase shift.

The dependence of the value of the electrification current on the degree of roughness of the solid at which the double layer is formed has been reported by Jaroszewski et al. [159] Even a slight increase in the roughness of the dielectric surface entails an increase in the generation of the electrification current. Zdanowski et al. [160] reported similar dependencies.

In addition to the abovementioned factors affecting the level of flow electrification of transformer oils, N'Cho et al. [161] indicated another very important factor increasing electrification currents, which is the contamination by oil aging byproducts.

Although researches clearly showed a higher charge generation by ester oils compared to mineral oils, Zelu et al. [162] indicated a lower risk of ester oils in filled transformers from exposure to electrostatic hazards due to the faster decay of space charge.

9. Summary

This article reviews the state of the art of the selected issues that are important for consideration when synthetic esters are applied in high voltage applications. The literature data as well as the critical comments were presented to be a kind of guide for researchers and industry representatives working with synthetic esters on a daily basis. This was done in a comparative form with relation to the data concerning mineral oil for which the knowledge in each field is quite well recognized. The following issues were raised by the authors: AC and DC breakdown voltage of synthetic esters, LI-BDV and pre-breakdown phenomena of synthetic esters, synthetic ester-based nanofluids, combined paper-synthetic ester-based insulating systems, application of synthetic esters for retro-filling and drying of mineral oil-immersed transformers, DGA-based diagnosis of synthetic ester-filled transformers and static electrification of synthetic esters. In general, the analyzed state-of-the-art methods indicated great development opportunities for synthetic esters. However, researchers still face many challenges, so that the synthetic esters can be applied on a larger scale not only in distribution transformers, but also in the power transformers of a high level of nominal voltages.

Author Contributions: All authors contributed equally in the preparation of the article. All authors have read and agreed to the published version of the manuscript.

Funding: This research received no external funding.

Conflicts of Interest: The authors declare no conflict of interest.

References

1. Oommen, T.; Claiborne, C.; Mullen, J. Biodegradable electrical insulation fluids. In Proceedings of the Electrical Insulation Conference and Electrical Manufacturing and Coil Winding Conference, Rosemont, IL, USA, 25 September 1997; pp. 465–468. [CrossRef]
2. Gockenbach, E.; Borsi, H. Natural and Synthetic Ester Liquids as alternative to mineral oil for power transformers. In Proceedings of the 2008 Annual Report Conference on Electrical Insulation and Dielectric Phenomena, Quebec City, QC, Canada, 26–29 October 2008; pp. 521–524. [CrossRef]
3. Perrier, C.; Beroual, A. Experimental investigations on insulating liquids for power transformers: Mineral, ester, and silicone oils. *IEEE Electr. Insul. Mag.* **2009**, *25*, 6–13. [CrossRef]
4. Martin, R.; Athanassatou, H.; Duart, J.C.; Perrier, C.; Sitar, I.; Walker, J.; Claiborne, C.; Boche, T.; Cherry, D.; Darwin, A.; et al. *Experiences in Service with New Insulating Liquids*; Cigré Technical Brochure 436; International Council on Large Electric Systems (CIGRE): Paris, France, 2010.
5. Fofana, I. 50 years in the development of insulating liquids. *IEEE Electr. Insul. Mag.* **2013**, *29*, 13–25. [CrossRef]
6. Fernández, I.; Ortiz, A.; Delgado, F.; Renedo, C.; Pérez, S. Comparative evaluation of alternative fluids for power transformers. *Electr. Power Syst. Res.* **2013**, *98*, 58–69. [CrossRef]
7. Bathina, V.; Sood, Y.R.; Jarial, R.K. Ester Dielectrics: Current Perspectives and Future Challenges. *IETE Tech. Rev.* **2016**, *34*, 448–459. [CrossRef]
8. Rao, U.M.; Fofana, I.; Jaya, T.; Rodriguez-Celis, E.M.; Jalbert, J.; Picher, P. Alternative Dielectric Fluids for Transformer Insulation System: Progress, Challenges, and Future Prospects. *IEEE Access* **2019**, *7*, 184552–184571. [CrossRef]
9. Oommen, T. Vegetable oils for liquid-filled transformers. *IEEE Electr. Insul. Mag.* **2002**, *18*, 6–11. [CrossRef]
10. McShane, C.; Corkran, J.; Rapp, K.; Luksich, J. Natural Ester Dielectric Fluid Development. In Proceedings of the 2005/2006 IEEE/PES Transmission and Distribution Conference and Exhibition, Dallas, TX, USA, 21–24 May 2006; pp. 18–22. [CrossRef]
11. Liao, R.; Hao, J.; Chen, G.; Ma, Z.; Yang, L. A comparative study of physicochemical, dielectric and thermal properties of pressboard insulation impregnated with natural ester and mineral oil. *IEEE Trans. Dielectr. Electr. Insul.* **2011**, *18*, 1626–1637. [CrossRef]
12. Rapp, K.J.; Luksich, J.; Sbravati, A. Application of natural ester insulating liquids in power transformers. In Proceedings of the My Transfo 2014, Turin, Italy, 18–19 November 2014; pp. 1–7.

13. Sitorus, H.B.; Setiabudy, R.; Bismo, S.; Beroual, A. Jatropha curcas methyl ester oil obtaining as vegetable insulating oil. *IEEE Trans. Dielectr. Electr. Insul.* **2016**, *23*, 2021–2028. [CrossRef]

14. Haegele, S.; Vahidi, F.; Tenbohlen, S.; Rapp, K.J.; Sbravati, A. Lightning Impulse Withstand of Natural Ester Liquid. *Energies* **2018**, *11*, 1964. [CrossRef]

15. Huang, Z.; Chen, X.; Li, J.; Wang, F.; Zhang, R.; Mehmood, M.A.; Liang, S.; Jiang, T. Streamer characteristics of dielectric natural ester-based liquids under long gap distances. *AIP Adv.* **2018**, *8*, 105129. [CrossRef]

16. Tokunaga, J.; Nikaido, M.; Koide, H.; Hikosaka, T. Palm fatty acid ester as biodegradable dielectric fluid in transformers: A review. *IEEE Electr. Insul. Mag.* **2019**, *35*, 34–46. [CrossRef]

17. Szewczyk, R.; Vercesi, G. Innovative insulation materials for liquid-immersed transformers. *Presented at the DuPont Webinar*, 13 March 2015.

18. Rozga, P.; Stanek, M. Comparative analysis of lightning breakdown voltage of natural ester liquids of different viscosities supported by light emission measurement. *IEEE Trans. Dielectr. Electr. Insul.* **2017**, *24*, 991–999. [CrossRef]

19. Rozga, P. Comparative assessment of impregnation efficiency of insulating pressboard by selected dielectric esters and mineral oil using capillary action test. *Przegląd Elektrotech.* **2018**, *10*, 69–72. (In Polish) [CrossRef]

20. Available online: https://www.cargill.com/bioindustrial/dielectric-fluids (accessed on 12 October 2020).

21. Available online: https://www.midel.com/ (accessed on 12 October 2020).

22. Russel, M.; Lashbrook, M.; Satija, N. Synthetic esters for Power transformers AT > 100 kV. *ITMA J.* **2013**.

23. Dohnal, D.; Frotscher, F. The use of natural and synthetic esters in tap changers for power transformers. *CEPSI* **2014**, *E9*, 1–9.

24. IEC 61099. *Insulating Liquids—Specifications for Unused Synthetic Organic Esters for Electrical Purposes*; International Electrotechnical Commission (IEC): Geneva, Switzerland, 2010.

25. Stanek, M. The Use of Measuring the Intensity of Light Emitted by Electrical Discharges Developing in Biodegradable Insulating Esters to Assess Their Lightning Strength. Ph.D. Thesis, Lodz University of Technology, Lodz, Poland, 2019.

26. Liu, Q.; Wang, Z.D. Breakdown and withstand strengths of ester transformer liquids in a quasi-uniform field under impulse voltages. *IEEE Trans. Dielectr. Electr. Insul.* **2013**, *20*, 571–579. [CrossRef]

27. Pukel, G.J.; Schwarz, R.; Baumann, F.; Muhr, H.M.; Eberhardt, R.; Wieser, B.; Chu, D. Power transformers with environmentally friendly and low flammability ester liquids. *Cigre Sess.* **2013**, 1–6. [CrossRef]

28. Asano, R.; Page, S.A. Reducing Environmental Impact and Improving Safety and Performance of Power Transformers with Natural Ester Dielectric Insulating Fluids. *IEEE Trans. Ind. Appl.* **2013**, *50*, 134–141. [CrossRef]

29. IEC 60814. *IEC Insulating Liquids—Oil-Impregnated Paper and Pressboard—Determination of Water by Automatic Coulometric Karl Fischer Titration*; International Electrotechnical Commission (IEC): Geneva, Switzerland, 1997.

30. Atanasova-Höhlein, I.; Končan-Gradnik, M.; Gradnik, T.; Čuček, B.; Przybylek, P.; Siodla, K.; Liland, K.B.; Leivo, S.; Liu, Q. *Moisture Measurement and Assessment in Transformer Insulation—Evaluation of Chemical Methods and Moisture Capacitive Sensors*; Cigré Technical Brochure 741; International Council on Large Electric Systems (CIGRE): Paris, France, 2018.

31. Przybyłek, P. Water solubility in synthetic ester and mixture of ester with mineral oil in aspect of cellulose insulation drying. *Prz. Elektrotech.* **2016**, *10*, 92–95. [CrossRef]

32. Przybylek, P. Water saturation limit of insulating liquids and hygroscopicity of cellulose in aspect of moisture determination in oil-paper insulation. *IEEE Trans. Dielectr. Electr. Insul.* **2016**, *23*, 1886–1893. [CrossRef]

33. Aralkellian, V.G.; Fofana, I. Water in Oil-Filled, High-Voltage Equipment, Part I: States, Solubility, and Equilibrium in Insulating Materials. *IEEE Electr. Insul. Mag.* **2007**, *23*, 15–27. [CrossRef]

34. Jovalekic, M.; Kolb, D.; Tenbohlen, S.; Bates, L.; Szewczyk, R. A methodology for determining water saturation limits and moisture equilibrium diagrams of alternative insulation systems. In Proceedings of the 2011 IEEE International Conference on Dielectric Liquids, Trondheim, Norway, 26–30 June 2011. [CrossRef]

35. Water solubility algorithms for dielectric liquids. In *DOMINO*^{TM} *Application Bulletin*; Serial number: MKT-AB 11 Rev B; Doble: Marlborough, MA, USA, 1999.

36. Dombek, G.; Nadolny, Z.; Przybylek, P.; Lopatkiewicz, R.; Marcinkowska, A.; Druzynski, L.; Boczar, T.; Tomczewski, A. Effect of Moisture on the Thermal Conductivity of Cellulose and Aramid Paper Impregnated with Various Dielectric Liquids. *Energies* **2020**, *13*, 4433. [CrossRef]

37. IEC 61203. *Synthetic Organic Esters for Electrical Purposes—Guide for Maintenance of Transformer Esters in Equipment*; International Electrotechnical Commission (IEC): Geneva, Switzerland, 1992.

38. Martins, M.A.G. Vegetable oils, an alternative to mineral oil for power transformers- experimental study of paper aging in vegetable oil versus mineral oil. *IEEE Electr. Insul. Mag.* **2010**, *26*, 7–13. [CrossRef]

39. Martins, M.A.G.; Gomes, A.R. Comparative study of the thermal degradation of synthetic and natural esters and mineral oil: Effect of oil type in the thermal degradation of insulating kraft paper. *IEEE Electr. Insul. Mag.* **2012**, *28*, 22–28. [CrossRef]

40. IEC 60156. *Insulating Liquids—Determination of the Breakdown Voltage at Power Frequency—Test Method*; International Electrotechnical Commission (IEC): Geneva, Switzerland, 2018.

41. Scigala, K. Analysis of the Influence of the Used Dielectric Liquid on the Distribution of the Electric Field in the Insulation System-Solid Insulation-Dielectric Liquid. Master's Thesis, Lodz University of Technology, Lodz, Poland, 2018. (In Polish).

42. Dombek, G.; Gielniak, J. Fire safety and electrical properties of mixtures of synthetic ester/mineral oil and synthetic ester/natural ester. *IEEE Trans. Dielectr. Electr. Insul.* **2018**, *25*, 1846–1852. [CrossRef]

43. Khaled, U.; Beroual, A. AC dielectric strength of synthetic ester-based Fe_3O_4, Al_2O_3 and SiO_2 nanofluids—Conformity with normal and weibull distributions. *IEEE Trans. Dielectr. Electr. Insul.* **2019**, *26*, 625–633. [CrossRef]

44. Mohamad, M.S.; Zainuddin, H.; Ghani, S.A.; Chairul, I.S. AC breakdown voltage and viscosity of palm fatty acid ester. *J. Electr. Eng. Technol.* **2017**, *12*, 2333–2341. [CrossRef]

45. Mentlik, V.; Trnka, P.; Hornak, J.; Totzauer, P. Development of a Biodegradable Electro-Insulating Liquid and Its Subsequent Modification by Nanoparticles. *Energies* **2018**, *11*, 508. [CrossRef]

46. Beroual, A.; Khaled, U.; Alghamdi, A.M. DC Breakdown Voltage of Synthetic Ester Liquid-Based Nanofluids. *IEEE Access* **2020**, *8*, 125797–125805. [CrossRef]

47. Khaled, U.; Beroual, A. DC breakdown voltage of natural ester oil-based Fe_3O_4, Al_2O_3, and SiO_2 nanofluids. *Alex. Eng. J.* **2020**. [CrossRef]

48. Mahidhar, G.D.P.; Sarathi, R.; Taylor, N.; Edin, H. Study on performance of silica nanoparticle dispersed synthetic ester oil under AC and DC voltages. *IEEE Trans. Dielectr. Electr. Insul.* **2018**, *25*, 1958–1966. [CrossRef]

49. Fafana, I.; Wasserberg, V.; Borsil, H.; Gockenbach, E. Retrofilling conditions of high voltage transformers. *IEEE Electr. Insul. Mag.* **2001**, *17*, 17–30. [CrossRef]

50. Fofana, I.; Wasserberg, V.; Borsi, H.; Gockenbach, E. Preliminary investigations for the retrofilling of perchlorethylene based fluid filled transformer. *IEEE Trans. Dielectr. Electr. Insul.* **2002**, *9*, 97–103. [CrossRef]

51. Fatyga, P.; Morańda, H. Evaluation of the Mineral Oil and Synthetic Ester Percentage Composition after Replacing Oil with Ester Fluid in Power Transformer. In Proceedings of the International Conference Transformation, Toruń, Poland, 9–11 May 2017. (In Polish).

52. Retrofiling–Upgrading Your Transformer. Available online: https://www.midel.com/midel-in-use/retrofilling/ (accessed on 25 August 2020).

53. Insulect-Energy Blog. Available online: https://insulect.com/energy-blog/can-i-use-midel-fluids-to-retrofill-transformers-at-high-voltages (accessed on 25 August 2020).

54. Wasserberg, V.; Borsi, H.; Gockenbach, E. A new method for drying the paper insulation of power transformers during service. In Proceedings of the Conference Record of the 2000 IEEE International Symposium on Electrical Insulation (Cat. No.00CH37075), Anaheim, CA, USA, 5 April 2002; pp. 251–254. [CrossRef]

55. Przybylek, P. Drying transformer cellulose insulation by means of synthetic ester. *IEEE Trans. Dielectr. Electr. Insul.* **2017**, *24*, 2643–2648. [CrossRef]

56. Przybylek, P.; Moranda, H.; Moscicka-Grzesiak, H.; Szczesniak, D. Application of Synthetic Ester for Drying Distribution Transformer Insulation—The Influence of Cellulose Thickness on Drying Efficiency. *Energies* **2019**, *12*, 3874. [CrossRef]

57. Przybylek, P.; Moranda, H.; Moscicka-Grzesiak, H.; Cybulski, M. Laboratory Model Studies on the Drying Efficiency of Transformer Cellulose Insulation Using Synthetic Ester. *Energies* **2020**, *13*, 3467. [CrossRef]

58. Przybylek, P.; Moranda, H.; Moscicka-Grzesiak, H.; Cybulski, M. Analysis of factors affecting the effectiveness of drying cellulose materials with synthetic ester. *IEEE Trans. Dielectr. Electr. Insul.* **2020**, *27*, 1538–1545. [CrossRef]

59. Dang, V.-H.; Beroual, A.; Perrier, C. Comparative study of streamer phenomena in mineral, synthetic and natural ester oils under lightning impulse voltage. In Proceedings of the 2010 International Conference on High Voltage Engineering and Application, New Orleans, LA, USA, 11–14 October 2010; pp. 560–563. [CrossRef]
60. Liu, Q.; Wang, Z.D. Streamer characteristic and breakdown in synthetic and natural ester transformer liquids under standard lightning impulse voltage. *IEEE Trans. Dielectr. Electr. Insul.* **2011**, *18*, 285–294. [CrossRef]
61. Rozga, P. Streamer propagation in small gaps of synthetic ester and mineral oil under lightning impulse. *IEEE Trans. Dielectr. Electr. Insul.* **2015**, *22*, 2754–2762. [CrossRef]
62. Rozga, P.; Stanek, M.; Rapp, K. Lightning properties of selected insulating synthetic esters and mineral oil in point-to-sphere electrode system. *IEEE Trans. Dielectr. Electr. Insul.* **2018**, *25*, 1699–1705. [CrossRef]
63. Dang, V.-H.; Beroual, A.; Perrier, C. Investigations on streamers phenomena in mineral, synthetic and natural ester oils under lightning impulse voltage. *IEEE Trans. Dielectr. Electr. Insul.* **2012**, *19*, 1521–1527. [CrossRef]
64. Rozga, P. Studies on behavior of dielectric synthetic ester under the influence of concentrated heat flux. *IEEE Trans. Dielectr. Electr. Insul.* **2016**, *23*, 908–914. [CrossRef]
65. Livesey, P.; Lashbrook, M.; Martin, R. Investigation of the factors affecting the dielectric dissipation factor of synthetic and natural esters. In Proceedings of the 2019 IEEE 20th International Conference on Dielectric Liquids (ICDL), Roma, Italy, 23–27 June 2019. [CrossRef]
66. Graczkowski, A.; Gielniak, J. Influence of impregnating liquids on dielectric response of impregnated cellulose insulation. In Proceedings of the 2010 10th IEEE International Conference on Solid Dielectrics, Potsdam, Germany, 4–9 July 2010. [CrossRef]
67. Beroual, A.; Sadaoui, F.; Coulibaly, M.-L.; Perrier, C. Investigation on static electrification phenomenon of ester liquids and mineral oil. In Proceedings of the 2015 IEEE Conference on Electrical Insulation and Dielectric Phenomena (CEIDP), Ann Arbor, MI, USA, 18–21 October 2015; pp. 391–394. [CrossRef]
68. Zdanowski, M. Streaming electrification of mineral insulating oil and synthetic ester MIDEL. *IEEE Trans. Dielectr. Electr. Insul.* **2014**, *21*, 1127–1132. [CrossRef]
69. Duval, M.; Hoehlein, I.; Scatiggio, F.; Cyr, M.; Grisaru, M.; Frotscher, R.; Martins, M.; Bates, L.; Boman, P.; Hall, A.C.; et al. *DGA in Non-Mineral Oils and Load Tap Changers and Improved DGA Diagnosis Criteria*; CIGRE brochure 443; International Council on Large Electric Systems (CIGRE): Paris, France, 2010.
70. Perrier, C.; Marugan, M.; Beroual, A. DGA comparison between ester and mineral oils. *IEEE Trans. Dielectr. Electr. Insul.* **2012**, *19*, 1609–1614. [CrossRef]
71. Martin, D.; Wang, Z.D. Statistical analysis of the AC breakdown voltages of ester based transformer oils. *IEEE Trans. Dielectr. Electr. Insul.* **2008**, *15*, 1044–1050. [CrossRef]
72. Wang, X.; Wang, Z.; Noakhes, J. Motion of conductive particles and the effect on AC breakdown strengths of esters. In Proceedings of the 2011 IEEE International Conference on Dielectric Liquids, Trondheim, Norway, 26–30 June 2011. [CrossRef]
73. Dang, V.-H.; Beroual, A.; Perrier, C. Comparative study of statistical breakdown in mineral, synthetic and natural ester oils under AC voltage. *IEEE Trans. Dielectr. Electr. Insul.* **2012**, *19*, 1508–1513. [CrossRef]
74. Reffas, A.; Moulai, H.; Béroual, A. Comparison of dielectric properties of olive oil, mineral oil, and other natural and synthetic ester liquids under AC and lightning impulse stresses. *IEEE Trans. Dielectr. Electr. Insul.* **2018**, *25*, 1822–1830. [CrossRef]
75. Beroual, A.; Khaled, U.; Noah, P.S.M.; Sitorus, H. Comparative Study of Breakdown Voltage of Mineral, Synthetic and Natural Oils and Based Mineral Oil Mixtures under AC and DC Voltages. *Energies* **2017**, *10*, 511. [CrossRef]
76. Xiang, J.; Liu, Q.; Wang, Z.D. Streamer characteristic and breakdown in a mineral oil and a synthetic ester liquid under DC voltage. *IEEE Trans. Dielectr. Electr. Insul.* **2018**, *25*, 1636–1643. [CrossRef]
77. Reffas, A.; Beroual, A.; Moulai, H. Comparison of breakdown voltage of vegetable olive with mineral oil, natural and synthetic ester liquids under DC voltage. *IEEE Trans. Dielectr. Electr. Insul.* **2020**, *27*, 1691–1697. [CrossRef]
78. Hestad, Ø.L.; Ingebrigsten, S.; Lundgaard, L. Streamer initiation in cyclohexane, midel 7131 and nytro 10X. In Proceedings of the IEEE International Conference on Dielectric Liquids, Coimbra, Portugal, 26 June–1 July 2005; pp. 123–126. [CrossRef]

79. Liu, Q.; Wang, Z.; Perrot, F. Impulse breakdown voltages of ester-based transformer oils determined by using different test methods. In Proceedings of the 2009 IEEE Conference on Electrical Insulation and Dielectric Phenomena, Virginia Beach, VA, USA, 18–21 October 2009; pp. 608–612. [CrossRef]

80. Ngoc, M.N.; Lesaint, O.; Bonifaci, N.; Denat, A.; Hassanzadeh, M. A comparison of breakdown properties of natural and synthetic esters at high voltage. In Proceedings of the 2010 Annual Report Conference on Electrical Insulation and Dielectic Phenomena, West Lafayette, IN, USA, 17–20 October 2010. [CrossRef]

81. Rozga, P.; Stanek, M.; Cieslinski, D. Comparison of properties of electrical discharges developing in natural and synthetic ester at inception voltage. In Proceedings of the 2013 Annual Report Conference on Electrical Insulation and Dielectric Phenomena, Shenzhen, China, 20–23 October 2013; pp. 891–894. [CrossRef]

82. Denat, A.; Lesaint, O.; Mc Cluskey, F. Breakdown of liquids in long gaps: Influence of distance, impulse shape, liquid nature, and interpretation of measurements. *IEEE Trans. Dielectr. Electr. Insul.* **2015**, *22*, 2581–2591. [CrossRef]

83. Rozga, P. Using the light emission measurement in assessment of electrical discharge development in different liquid dielectrics under lightning impulse voltage. *Electr. Power Syst. Res.* **2016**, *140*, 321–328. [CrossRef]

84. Rozga, P. Streamer Propagation and Breakdown in a Very Small Point-Insulating Plate Gap in Mineral Oil and Ester Liquids at Positive Lightning Impulse Voltage. *Energies* **2016**, *9*, 467. [CrossRef]

85. Liu, Q.; Wang, Z.D. Streamer characteristic and breakdown in synthetic and natural ester transformer liquids with pressboard interface under lightning impulse voltage. *IEEE Trans. Dielectr. Electr. Insul.* **2011**, *18*, 1908–1917. [CrossRef]

86. Wang, K.; Wang, F.; Shen, Z.; Lou, Z.; Han, Q.; Li, J.; Trnka, P.; Rozga, P. Breakdown and streamer behavior in homogeneous synthetic trimethylolpropane triesters insulation oil. *IEEE Trans. Dielectr. Electr. Insul.* **2020**, *27*, 1501–1507. [CrossRef]

87. IEC 60897. *Methods for the Determination of the Lightning Impulse Breakdown Voltage of Insulating Liquids*; International Electrotechnical Commission (IEC): Geneva, Switzerland, 1987.

88. Tobazéon, R. Prebreakdown phenomena in dielectric liquids. *IEEE Trans. Dielectr. Electr. Insul.* **1994**, *1*, 1132–1147. [CrossRef]

89. Lesaint, O.; Massala, G. Positive streamer propagation in large oil gaps: Experimental characterization of propagation modes. *IEEE Trans. Dielectr. Electr. Insul.* **1998**, *5*, 360–370. [CrossRef]

90. Yamashita, H.; Amano, H. Prebreakdown phenomena in hydrocarbon liquids. *IEEE Trans. Electr. Insul.* **1988**, *23*, 739–750. [CrossRef]

91. Lesaint, O. Prebreakdown phenomena in liquids: Propagation 'modes' and basic physical properties. *J. Phys. D Appl. Phys.* **2016**, *49*, 144001. [CrossRef]

92. Rao, U.M.; Fofana, I.; Beroual, A.; Rozga, P.; Pompili, M.; Calcara, L.; Rapp, K.J. A review on pre-breakdown phenomena in ester fluids: Prepared by the international study group of IEEE DEIS liquid dielectrics technical committee. *IEEE Trans. Dielectr. Electr. Insul.* **2020**, *27*, 1546–1560. [CrossRef]

93. Choi, S. Enhancing thermal conductivity of fluids with nanoparticles, Developments and Applications of Non-Newtonian flows. *ASME J. Fluids Eng.* **1995**, *231*, 99–105.

94. Xuan, Y.; Li, Q. Heat transfer enhancement of nanofluids. *Int. J. Heat Fluid Flow* **2000**, *21*, 58–64. [CrossRef]

95. Eastman, J.A.; Choi, S.U.-S.; Li, S.; Yu, W.; Thompson, L.J. Anomalously increased effective thermal conductivities of ethylene glycol-based nanofluids containing copper nanoparticles. *Appl. Phys. Lett.* **2001**, *78*, 718–720. [CrossRef]

96. Choi, S.-S.; Zhang, Z.G.; Yu, W.; Lockwood, F.E.; Grulke, E.A. Anomalous thermal conductivity enhancement in nanotube suspensions. *Appl. Phys. Lett.* **2001**, *79*, 2252–2254. [CrossRef]

97. Godson, L.; Raja, B.; Lal, D.M.; Wongwises, S. Enhancement of heat transfer using nanofluids—An overview. *Renew. Sustain. Energy Rev.* **2010**, *14*, 629–641. [CrossRef]

98. Puspitasari, P.; Permanasari, A.A.; Shaharun, M.S.; Tsamroh, D.I. Heat transfer characteristics of NiO nanofluid in heat exchanger. *AIP Conf. Proc.* **2020**, *2228*, 030023. [CrossRef]

99. Rafiq, M.; Lv, Y.; Li, C. A Review on Properties, Opportunities, and Challenges of Transformer Oil-Based Nanofluids. *J. Nanomater.* **2016**, 1–23. [CrossRef]

100. Segal, V.; Hjortsberg, A.; Rabinovich, A.; Nattrass, D.; Raj, K. AC (60 Hz) and impulse breakdown strength of a colloidal fluid based on transformer oil and magnetite nanoparticles. In Proceedings of the Conference Record of the 1998 IEEE International Symposium on Electrical Insulation (Cat. No.98CH36239), Arlington, VA, USA, 7–10 June 1998; pp. 619–622. [CrossRef]

101. Zhou, Y.X.; Wang, Y.S.; Tian, J.H.; Sha, Y.C.; Jiang, X.X.; Gao, S.Y.; Sun, Q.H.; Nie, Q. Breakdown characteristics in transformer oil modified by nanoparticles. *High Volt. Eng.* **2010**, *36*, 1155–1159.
102. Saidur, R.; Leong, K.; Mohammed, H. A review on applications and challenges of nanofluids. *Renew. Sustain. Energy Rev.* **2011**, *15*, 1646–1668. [CrossRef]
103. Mergos, J.A.; Athanassopoulou, M.D.; Argyropoulos, T.G.; Dervos, C.T. Dielectric properties of nanopowder dispersions in paraffin oil. *IEEE Trans. Dielectr. Electr. Insul.* **2012**, *19*, 1502–1507. [CrossRef]
104. Du, Y.; Lv, Y.; Li, C.; Chen, M.; Zhong, Y.; Zhou, J.; Li, X.; Zhou, Y. Effect of semiconductive nanoparticles on insulating performances of transformer oil. *IEEE Trans. Dielectr. Electr. Insul.* **2012**, *19*, 770–776. [CrossRef]
105. Taha-Tijerina, J.; Narayanan, T.N.; Gao, G.; Rohde, M.; Tsentalovich, D.A.; Pasquali, M.; Ajayan, P.M. Electrically Insulating Thermal Nano-Oils Using 2D Fillers. *ACS Nano* **2012**, *6*, 1214–1220. [CrossRef] [PubMed]
106. Kharthik, R.; Raja, T.; Madavan, R. Enhancement of critical characteristics of transformer oil using nanomaterials. *Arab. J. Sci. Eng.* **2013**, *38*, 2725–2733. [CrossRef]
107. Lv, Y.; Wang, W.; Ma, K.; Zhang, S.; Zhou, Y.; Li, C.; Wang, Q. Nanoparticle Effect on Dielectric Breakdown Strength of Transformer Oil-Based Nanofluids. In Proceedings of the 2013 Annual Report Conference on Electrical Insulation and Dielectric Phenomena, Shenzhen, China, 20–23 October 2013; pp. 680–682. [CrossRef]
108. Lv, Y.Z.; Zhou, Y.; Li, C.R.; Wang, Q.; Qi, B. Recent progress in nanofluids based on transformer oil: Preparation and electrical insulation properties. *IEEE Electr. Insul. Mag.* **2014**, *30*, 23–32. [CrossRef]
109. Mdavan, R.; Balaraman, S. Investigation on effects of different types of nanoparticles on critical parameters of nano-liquid insulation systems. *J. Mol. Liq.* **2017**, *230*, 437–444. [CrossRef]
110. Huang, Z.; Li, J.; Yao, W.; Wang, F.; Wan, F.; Tan, Y.; Mehmood, M.A. Electrical and thermal properties of insulating oil-based nanofluids: A comprehensive overview. *IET Nanodielectr.* **2019**, *2*, 27–40. [CrossRef]
111. Khaled, U.; Beroual, A. AC Dielectric Strength of Mineral Oil-Based Fe_3O_4 and Al_2O_3 Nanofluids. *Energies* **2018**, *11*, 3505. [CrossRef]
112. Khaled, U.; Beroual, A. Statistical Investigation of AC Dielectric Strength of Natural Ester Oil-Based Fe_3O_4, Al_2O_3, and SiO_2 Nano-Fluids. *IEEE Access* **2019**, *7*, 60594–60601. [CrossRef]
113. Peppas, G.D.; Danikas, M.G.; Bakandritsos, A.; Charalampakos, V.P.; Pyrgioti, E.C.; Gonos, I.F. Statistical investigation of AC breakdown voltage of nanofluids compared with mineral and natural ester oil. *IET Sci. Meas. Technol.* **2016**, *10*, 644–652. [CrossRef]
114. Peppas, G.D.; Charalampakos, V.P.; Pyrgioti, E.C.; Bakandritsos, A.; Polykrati, A.D., Gonos, I.F. A study on the Breakdown Characteristics of Natural Ester Based Nanofluids with Magnetic Iron Oxide and SiO_2 Nanoparticles. In Proceedings of the 2018 IEEE International Conference on High Voltage Engineering and Application (ICHVE), Athens, Greece, 10–13 September 2018. [CrossRef]
115. Makmud, M.Z.H.; Illias, H.A.; Ching, Y.C.; Sarjadi, M.S. Influence of Conductive and Semi-Conductive Nanoparticles on the Dielectric Response of Natural Ester-Based Nanofluid Insulation. *Energies* **2018**, *11*, 333. [CrossRef]
116. Li, J.; Zhang, Z.; Zou, P.; Grzybowski, S.; Zahn, M. Preparation of a vegetable oil-based nanofluid and investigation of its breakdown and dielectric properties. *IEEE Electr. Insul. Mag.* **2012**, *28*, 43–50. [CrossRef]
117. Beroual, A.; Khaled, U. Statistical Investigation of Lightning Impulse Breakdown Voltage of Natural and Synthetic Ester Oils-Based Fe_3O_4, Al_2O_3 and SiO_2 Nanofluids. *IEEE Access* **2020**, *8*, 112615–112623. [CrossRef]
118. Khaled, U.; Beroual, A. Lightning impulse breakdown voltage of synthetic and natural ester liquids-based Fe_3O_4, Al_2O_3 and SiO_2 nanofluids. *Alex. Eng. J.* **2020**, *59*, 3709–3713. [CrossRef]
119. Chen, G.; Li, J.; Yin, H.; Huang, Z.; Wang, Q.; Liu, L.; Sun, J.; He, J. Analysis of Dielectric Properties and Breakdown Characteristics of Vegetable Insulating Oil with Modified by ZnO Nanoparticles. In Proceedings of the 2018 IEEE International Conference on High Voltage Engineering and Application (ICHVE), Athens, Greece, 10–13 September 2018. [CrossRef]
120. Du, Y.; Lv, Y.; Li, C.; Chen, M.; Zhou, J.; Li, X.; Zhou, Y.; Tu, Y. Effect of electron shallow trap on breakdown performance of transformer oil-based nanofluids. *J. Appl. Phys.* **2011**, *110*, 104104. [CrossRef]
121. Hwang, J.G.; Zahn, M.; O'Sullivan, F.M.; Pettersson, L.A.A.; Hjortstam, O.; Liu, R. Effects of nanoparticle charging on streamer development in transformer oil-based nanofluids. *J. Appl. Phys.* **2010**, *107*, 014310. [CrossRef]

122. Miao, J.; Dong, M.; Ren, M.; Wu, X.; Shen, L.; Wang, H. Effect of nanoparticle polarization on relative permittivity of transformer oil-based nanofluids. *J. Appl. Phys.* **2013**, *113*, 204103. [CrossRef]

123. Sima, W.; Shi, J.; Yang, Q.; Huang, S.; Cao, X. Effects of conductivity and permittivity of nanoparticle on transformer oil insulation performance: Experiment and theory. *IEEE Trans. Dielectr. Electr. Insul.* **2015**, *22*, 380–390. [CrossRef]

124. Ibrahim, M.E.; Abd-Elhady, A.M.; Izzularab, M.A. Effect of nanoparticles on transformer oil breakdown strength: Experiment and theory. *IET Sci. Meas. Technol.* **2016**, *10*, 839–845. [CrossRef]

125. Madawan, R.; Kumar, S.S.; Iruthyarajan, M.W. A comparative investigation on effects of nanoparticles on characteristics of natural esters-based nanofluids. *Colloids Surf. A* **2018**, *556*, 30–36. [CrossRef]

126. Beroual, A. Electronic and gaseous processes in the prebreakdown phenomena of dielectric liquids. *J. Appl. Phys.* **1993**, *73*, 4528–4533. [CrossRef]

127. Beroual, A. Pre-breakdown mechanisms in dielectric liquids and predicting models. In Proceedings of the 2016 IEEE Electrical Insulation Conference (EIC), Montreal, QC, Canada, 19–22 June 2016. [CrossRef]

128. Beroual, A.; Zahn, M.; Badent, A.; Kist, K.; Schwabe, A.J.; Yamashita, H.; Yamazawa, K.; Danikas, M.; Chadband, W.D.; Torshin, Y. Propagation and structure of streamers in liquid dielectrics. *IEEE Electr. Insul. Mag.* **1998**, *14*, 6–17. [CrossRef]

129. Oommen, T.V. Moisture equilibrium in paper-oil insulation systems. In Proceedings of the 1983 EIC 6th Electrical/Electronical Insulation Conference, Chicago, IL, USA, 3–6 October 1983; pp. 162–166.

130. Liao, R.; Lin, Y.; Guo, P.; Liu, H.; Xia, H. Thermal aging effects on the moisture equilibrium curves of mineral and mixed oil-paper insulation systems. *IEEE Trans. Dielectr. Electr. Insul.* **2015**, *22*, 842–850. [CrossRef]

131. Koch, M.; Tenbohlen, S.; Stirl, T. Diagnostic Application of Moisture Equilibrium for Power Transformers. *IEEE Trans. Power Deliv.* **2010**, *25*, 2574–2581. [CrossRef]

132. IEC 60422. *Mineral Insulating Oils in Electrical Equipment—Supervision and Maintenance Quidance*; International Electrotechnical Commission (IEC): Geneva, Switzerland, 2013.

133. Dai, J.; Wang, Z.D. A Comparison of the Impregnation of Cellulose Insulation by Ester and Mineral oil. *IEEE Trans. Dielectr. Electr. Insul.* **2008**, *15*, 374–381. [CrossRef]

134. IEC 60243-1. *Electric Strength of Insulating Materials-Test Methods-Part 1: Tests at Power Frequencies*; International Electrotechnical Commission (IEC): Geneva, Switzerland, 2013.

135. Rozga, P. Using the three-parameter Weibull distribution in assessment of threshold strength of pressboard impregnated by different liquid dielectrics. *IET Sci. Meas. Technol.* **2016**, *10*, 665–670. [CrossRef]

136. Sbravati, A.; Rapp, K.; Schmitt, P.; Krause, C. Transformer insulation structure for dielectric liquids with higher permittivity. In Proceedings of the 2017 IEEE 19th International Conference on Dielectric Liquids (ICDL), Manchester, UK, 25–29 June 2017. [CrossRef]

137. Dombek, G.; Nadolny, Z.; Marcinkowska, A. Effects of Nanoparticles Materials on Heat Transfer in Electro-Insulating Liquids. *Appl. Sci.* **2018**, *8*, 2538. [CrossRef]

138. Villarroel, R.; García, B.; García, D.F.; Burgos, J.C. Assessing the Use of Natural Esters for Transformer Field Drying. *IEEE Trans. Power Deliv.* **2014**, *29*, 1894–1900. [CrossRef]

139. Moore, S.; Rapp, K.; Baldyga, R. Transformer insulation dry out as a result of retrofilling with natural ester fluid. In Proceedings of the IEEE PES Transmission & Distribution Conference & Exposition, Orlando, FL, USA, 7–10 May 2012. [CrossRef]

140. IEC 60599. *Mineral Oil-Impregnated Electical Equipment in Service–Guide to the Interpretation of Dissolved and Free Gases Analysis*, 2nd ed.; International Electrotechnical Commission (IEC): Geneva, Switzerland, 1999.

141. IEEE C57.104. *IEEE Guide for the Interpretation of Gases Generated in Oil-Immersed Transformers*; IEEE: New York, NY, USA, 2008.

142. Duval, M. The duval triangle for load tap changers, non-mineral oils and low temperature faults in transformers. *IEEE Electr. Insul. Mag.* **2008**, *24*, 22–29. [CrossRef]

143. Duval, M.; Baldyga, R. Stray gassing of FR3 oils in transformers in service. In Proceedings of the 76th International Conference on Doble Clients, Boston, USA, 29 March 2009.

144. Mnisi, H.; Nyamupangedengu, C. Dissolved Gases Analysis of canola-based ester oil under creepage discharge. In Proceedings of the 2020 International SAUPEC/RobMech/PRASA Conference, Cape Town, South Africa, 29–31 January 2020. [CrossRef]

145. Gómez, N.A.; Wilhelm, H.M.; Santos, C.C.; Stocco, G.B. Dissolved gas analysis (DGA) of natural ester insulating fluids with different chemical compositions. *IEEE Trans. Dielectr. Electr. Insul.* **2014**, *21*, 1071–1078. [CrossRef]
146. Hanson, D.; Li, K.; Plascencia, J.; Beauchemin, C.; Claiborne, C.; Cherry, D.; Frimpong, G.; Luksich, J.; Lemm, A.; Martin, R. Understanding dissolved gas analysis of ester liquids: An updated review of gas generated in ester liquid by stray gassing, thermal decomposition and electrical discharge. In Proceedings of the 2016 IEEE Electrical Insulation Conference (EIC), Montreal, QC, Canada, 19–22 June 2016; pp. 138–144. [CrossRef]
147. Przybyłek, P.; Gielniak, J. Concentration analysis of gases formed in mineral oil, natural ester, and synthetic ester by discharges of high energy. *Ekspolatacja Niezawodn. Maint. Reliab.* **2018**, *20*, 435–442. [CrossRef]
148. Przybylek, P.; Gielniak, J. Analysis of Gas Generated in Mineral Oil, Synthetic Ester, and Natural Ester as a Consequence of Thermal Faults. *IEEE Access* **2019**, *7*, 65040–65051. [CrossRef]
149. Lashbrook, M.; Al-Amin, H.; Martin, R. Natural ester and synthetic ester fluids, applications and maintenance. In Proceedings of the 2017 10th Jordanian International Electrical and Electronics Engineering Conference (JIEEEC), Amman, Jordan, 16 May 2017.
150. Crofts, D. The electrification phenomena in power transformers. *IEEE Trans. Electr. Insul.* **1988**, *23*, 137–146. [CrossRef]
151. Yamada, N.; Kishi, A.; Nitta, T.; Tanaka, T.; Ima, Y. Model Approach to the Static Electrification Phenomena Induced by the Flow of Oil in Large Power Transformers. *IEEE Trans. Power Appar. Syst.* **1980**, *99*, 1097–1106. [CrossRef]
152. Shimizu, S.; Murata, H.; Honda, M. Electrostatics in Power Transformers. *IEEE Trans. Power Appar. Syst.* **1979**, *98*, 1244–1250. [CrossRef]
153. Radwan, R.; El-Dewieny, R.; Metwally, I.A. Investigation of static electrification phenomenon due to transformer oil flow in electric power apparatus. *IEEE Trans. Electr. Insul.* **1992**, *27*, 278–286. [CrossRef]
154. Peyraque, L.; Beroual, A.; Buret, F. Static electrification induced by oil flow in power transformers. *Intern. Symp. Electr. Insul.* **1996**, *2*, 745–749. [CrossRef]
155. Kolcunová, I.; Kurimský, J.; Cimbala, R.; Petráš, J.; Dolnik, B.; Džmura, J.; Balogh, J. Contribution to static electrification of mineral oils and natural esters. *J. Electrost.* **2017**, *88*, 60–64. [CrossRef]
156. Paillat, T.; Zelu, Y.; Morin, G.; Perrier, C. Ester oils and flow electrification hazards in power transformers. *IEEE Trans. Dielectr. Electr. Insul.* **2012**, *19*, 1537–1543. [CrossRef]
157. Vihacencu, M.; Dumitran, L.M.; Nothinger, P.V. Transformer Mineral Oil Electrification. In Proceedings of the 7th International Symposium on Advanced Topics in Electrical Engineering (ATEE), Bucharest, Romania, 12–14 May 2011.
158. Kedzia, J.; Willner, B. Electrification current in the spinning disk system. *IEEE Trans. Dielectr. Electr. Insul.* **1994**, *1*, 58–62. [CrossRef]
159. Jaroszewski, M. Electrification currents measurements of natural ester obtained by rotating disc method. *Przegląd Elektrotech.* **2016**, *1*, 90–93. [CrossRef]
160. Zdanowski, M.; Wolny, S.; Zmarzly, D.; Kedzia, J. The analysis and selection of the spinning disk system parameters for the measurement of static electrification of insulation oils. *IEEE Trans. Dielectr. Electr. Insul.* **2007**, *14*, 480–486. [CrossRef]
161. N'Cho, J.S.; Fofana, I.; Beroual, A. Parameters affecting the static electrification of aged transformer oils. In Proceedings of the 2011 Annual Report Conference on Electrical Insulation and Dielectric Phenomena, Cancun, Mexico, 16–19 October 2011; pp. 571–574. [CrossRef]
162. Zelu, Y.; Paillat, T.; Morin, G.; Perrier, C.; Saravolac, M. Study on flow electrification hazards with ester oils. In Proceedings of the 2011 IEEE International Conference on Dielectric Liquids, Trondheim, Norway, 26–30 June 2011. [CrossRef]

Publisher's Note: MDPI stays neutral with regard to jurisdictional claims in published maps and institutional affiliations.

Article

Laboratory Model Studies on the Drying Efficiency of Transformer Cellulose Insulation Using Synthetic Ester

Piotr Przybylek *, **Hubert Moranda**, **Hanna Moscicka-Grzesiak and Mateusz Cybulski**

Institute of Electric Power Engineering, Poznan University of Technology, Piotrowo 3A, 61-138 Poznan, Poland; hubert.moranda@put.poznan.pl (H.M.); hanna.moscicka-grzesiak@put.poznan.pl (H.M.-G.); mateusz.e.cybulski@doctorate.put.poznan.pl (M.C.)
* Correspondence: piotr.przybylek@put.poznan.pl; Tel.: +48-061-665-2018

Received: 25 May 2020; Accepted: 30 June 2020; Published: 4 July 2020

Abstract: This paper presents the results of laboratory tests of cellulose insulation drying with the use of synthetic ester. The effectiveness of the drying process was investigated depending on the initial moisture of cellulose samples (2%, 3%, and 4%), ester temperature (55, 70, and 85 °C), initial moisture of the ester (70, 140, and 220 ppm), drying time (48, 96, and 168 h), and the weight ratio of cellulosic materials to ester (0.067 and 0.033). A large influence of temperature and time of drying on the efficiency of the drying process was found. This is important information due to the application of the results in the transformers drying procedure. The heating and drying ester unit should provide the highest possible temperature. For the assumed experiment conditions the initial moisture of the ester had little effect on the drying efficiency. An ester with a moisture content below 140 ppm can still be considered as meeting the requirements for drying cellulose with significant moisture. The weight ratio of cellulose products to ester has no major effect on drying efficiency during cellulose drying by circulating dry ester.

Keywords: transformer; oil–paper insulation; moisture; drying; synthetic ester

---

1. Introduction

The main insulation of the vast majority of power transformers is based on paper/pressboard-mineral oil system. During the operation of the transformer, its insulation system is constantly aging. High temperature and the presence of an electric field significantly accelerate the aging process.

One of the effects of aging is the increase of moisture content in cellulose insulation. The main cause of the moisture contamination process is the chemical decomposition of cellulose, which is accompanied by water formation. Looseness of the transformer tank plays a smaller role in the insulation moistening process than chemical cellulose degradation. It was found that, with the passage of transformer years of use, the annual increase in cellulose moisture increases [1,2].

Moisture in the transformer's cellulose insulation depends on the transformer's power and operating conditions. Figure 1 shows the water content in cellulose insulation depending on the transformer years of operation. The figure shows data obtained from grid transformers (160, 250, 330, and 500 MVA) and generator transformers, investigated in three countries [3]. The colors on the graph correspond to the moisture classification according to [4]. The figure shows that insulation with water content above 2% should be treated as wet, and above 4% as excessively wet.

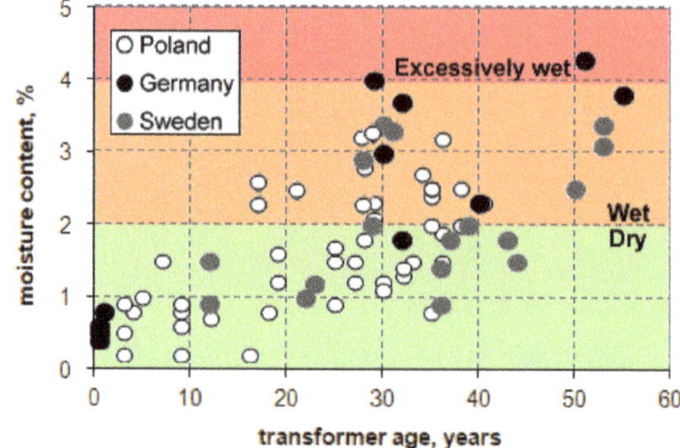

Figure 1. Dependence of water content in cellulose insulation on transformer age in the whole investigated population divided into three moisture ranges as per IEEE classification, on the basis of data from [3].

One of the dangerous effects of high moisture content in cellulose insulation is the occurrence of the bubble effect phenomenon. It involves the rapid release of water adsorbed on cellulose fibers after exceeding the critical temperature. This leads to an increase in the pressure in the transformer's tank even above its tearing strength, which can result in an explosion and fire [5,6].

2. Drying Methods for Cellulose Insulation

Drying of transformer insulation is carried out during the production and operation. During the production process, stationary devices in the factory are used. On the other hand, drying of insulation during operation is carried out at the place of the unit installation with the use of a mobile device or in a repair plant—if the general condition of the transformer requires major renovation [7].

The drying methods used so far require heating the insulation and creating a vacuum in the tank. The insulation is heated with hot oil, hot air, or electric current flow. The LFH (low frequency heating) method has the best opinion. In this method, three-phase high voltage winding is supplied and the low voltage winding is shorted. The frequency of the supply voltage is reduced to the lowest possible value at which the transformation effect still occurs. It is usually between 0.4 and 2 Hz [8]. The tank design must allow for the creation of an appropriate vacuum. The method is very effective, but unfortunately very expensive. For a network transformer, the service costs are about € 150,000.

3. Drying Cellulose Insulation Using Synthetic Ester

The use of an ester as a drying medium is possible due to the huge solubility of water in the ester. Figure 2 shows a comparison of the water saturation limit of three selected liquids used for filling transformers. Such great possibilities of dissolving water in the ester result from the fact that one molecule of the ester is able to attach, on the basis of hydrogen bonds, as many as four molecules of water.

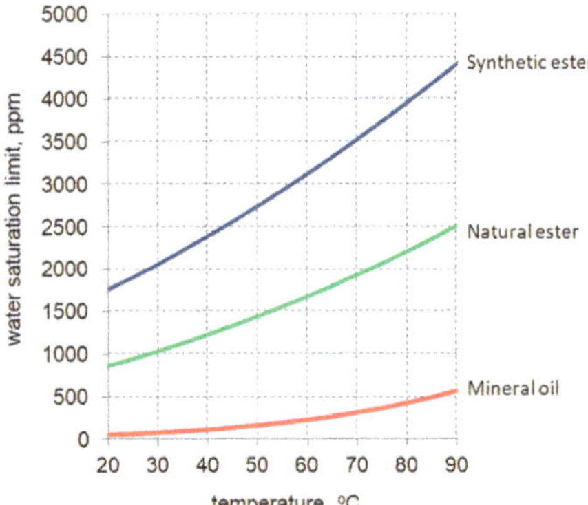

Figure 2. Comparison of the water saturation limit of four selected insulating liquids, on the basis of data from [9].

If wet cellulose is surrounded by dry ester, then there is an intensive migration of water from cellulose to the ester until the moisture equilibrium between cellulose and ester is achieved. The closer to equilibrium, the smaller the dynamics of water migration. The dynamics of water migration depend, to varying degrees, on many factors. The main ones are: initial moisture content in cellulose insulation, moisture of the ester, the mass ratio of cellulosic materials to liquid, insulation temperature, and thickness of cellulosic materials [10].

4. Research Results and Discussion

4.1. Introduction

In the earlier research, described in publication [11], the effect of cellulose sample thickness on the efficiency and dynamics of their drying with the use of synthetic ester was investigated. During the experiment 0.05, 0.5, 3.0, and 5.0 mm thickness samples were tested.

In this experiment, the drying process of 3 mm thick pressboard samples with three different moisture levels, about 2%, 3%, and 4%, was studied. The effects of drying temperature (55, 70, and 85 °C), initial moisture of the ester (70, 140, and 220 ppm) and the mass ratio of cellulose products to synthetic ester (0.067 and 0.033) were studied. These mass ratios are found in grid and distribution transformers.

4.2. Sample Preparation

Sample preparation was multi-stage. In the first stage the pressboard samples were dried at 85–95 °C for 26–29 h in a vacuum of 0.2–0.4 mbar to reduce the water content to less than 0.1%. Then the samples were conditioned in a climatic chamber to achieve the assumed water content. In the third stage, pressboard samples were degassed and impregnated with mineral oil. The air trapped in the samples was rapidly released from the cellulose materials, causing oil bubbling, which stopped after about 15 min from the moment of applying the vacuum. Next, the pressboard samples were placed in the climatic chamber for 144 h for their further impregnation and stabilization. In final stage, the water content (WCP_i) in the samples was measured by Karl Fischer titration (KFT) method [12]. All parameters of sample preparations and sample moisture contents are presented in Table 1. The samples prepared in this way were subjected to a drying process using a synthetic ester.

Table 1. Samples preparation parameters.

Assumed Water Content in Samples (%)	Drying Conditions (Vacuum Chamber)	Moisture Conditions (Climatic Chamber *)	Mineral Oil Impregnation Conditions (Vacuum Chamber)	Stabilization Conditions (Climatic Chamber *)	Water Content in Samples Determined by KFT (%)
2		$T = 80\,°C$ $RH = 10\%$ $t = 264\,h$	$T = 80\,°C$ $p = 60\,mbar$ $t = 15\,min$	$T = 80\,°C$ $RH = 10\%$ $t = 144\,h$	2.07
3	$T = 85–95\,°C$ $p = 0.2–0.4$ $mbar$ $t = 26–29\,h$	$T = 40\,°C$ $RH = 12\%$ $t = 150\,h$	$T = 40\,°C$ $p = 8\,mbar$ $t = 15\,min$	$T = 40\,°C$ $RH = 12\%$ $t = 144\,h$	3.32
4		$T = 40\,°C$ $RH = 21\%$ $t = 316\,h$	$T = 40\,°C$ $p = 10\,mbar$ $t = 15\,min$	$T = 40\,°C$ $RH = 21\%$ $t = 144\,h$	4.09

(*) temperature fluctuation $\leq 1.3\,°C$; relative humidity fluctuation $\leq 2.5\%$ RH.

Different water content in the synthetic ester was obtained by mixing in an appropriate proportion the dry ester (50 ppm) with the moistened ester (950 ppm). In this way, a synthetic ester with an initial moisture level (WCE_i) of 70, 140, and 220 ppm was prepared.

4.3. Experimental Procedure

The experimental procedure is presented in Figure 3. At the beginning the cellulose samples of initial water content WCP_i impregnated with mineral oil were placed in a glass bottles filled with synthetic ester of water content WCE_i. The bottles were placed in a thermal chamber, which kept the drying temperature T constant (Figure 4). The motion of the ester inside the bottles was forced by a magnetic stirrer.

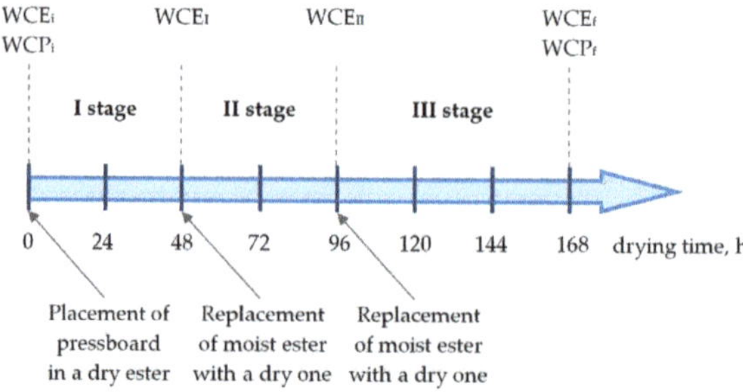

Figure 3. Drying procedure; WCE—water content in ester; WCP—water content in pressboard.

The first drying stage lasted 48 h. After this time, the water content in synthetic ester WCE_I was measured using the KFT method (Figure 5), and the moistened ester was replaced with a dry one. After a further 48 h of drying, the water content in ester WCE_{II} was measured again and the ester was replaced with dry one. The third and final drying stage ended after 168 h. After this time, the water content in ester WCE_{III} was measured last time. The final water content in the pressboard (WCP_f) was also measured using the KFT method.

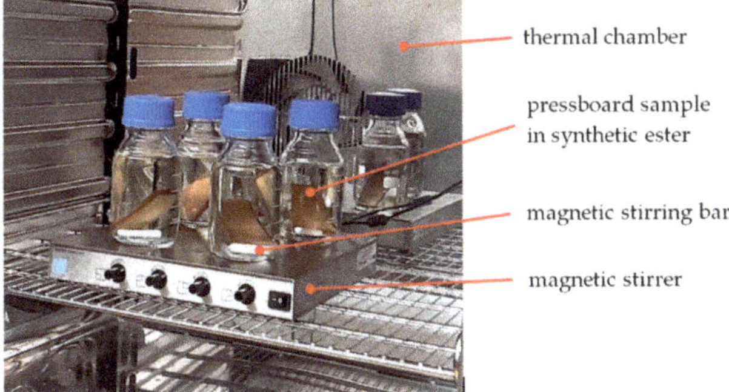

Figure 4. Measurement setup used for drying pressboard samples by means of synthetic ester.

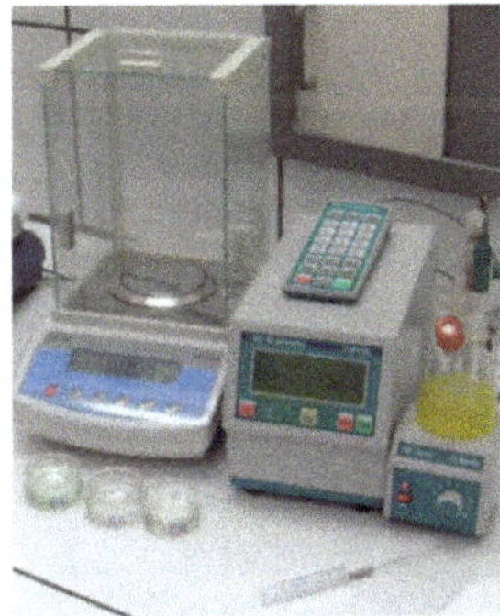

Figure 5. Instrument for measuring water content by Karl Fischer coulometric titration method.

Based on the measured water content in the synthetic ester (WCE$_I$, WCE$_{II}$) and the mass of cellulose and the mass of ester it was possible to calculate the water content in pressboard samples after drying time equal to 48 h and 96 h, respectively.

4.4. Investigation Results

Figure 6 shows the water loss in cellulose during its drying, depending on the initial moisture content of the cellulose. The curves are made for three values of drying temperature (55, 70, and 85 °C). The thickness of the samples was 3 mm. Dry synthetic ester (with a moisture content of 50 to 70 ppm) was introduced into the cellulose sample chambers three times. The first portion of the ester was kept in the chamber in the time range of 0–48 h, the second in the time range of 48–96 h, and the third in the range of 96–168 h. It should be noted that the introduction of the ester three times, according to our estimates, gives a similar effect to drying continuously. Under real conditions, continuous drying corresponds to the circulation of the ester between the transformer tank and the ester heating and drying unit. Figure 6a refers to the mass ratio of cellulose to ester 0.033, while Figure 6b to the mass ratio 0.067. A very large influence of temperature on the intensity of the drying process can be seen. The amount of water removed from cellulose in the described conditions is approximately proportional to its initial moisture. The effect of the mass ratio of cellulose to ester on the intensity of the drying process is small.

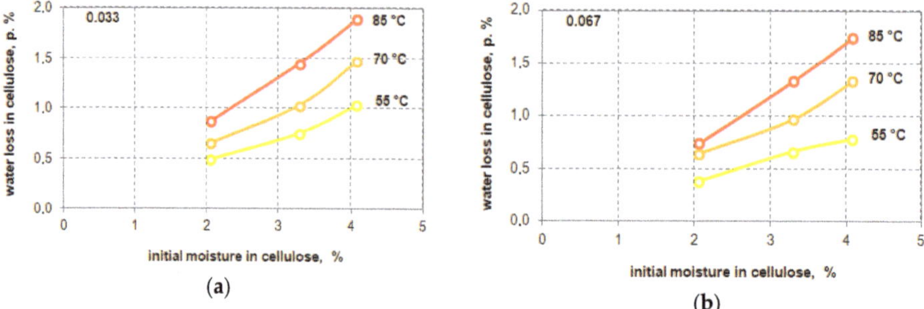

Figure 6. Water loss in samples depending on the initial moisture content of the cellulose at different temperatures; sample thickness 3 mm; initial moisture of the ester 70 ppm; total drying time 168 h—with the three times introduction of the ester; mass ratio of cellulose to ester of 0.033 (**a**) and 0.067 (**b**).

Figure 7 shows the water content of cellulose samples depending on the drying time, with three times the introduction of a dry synthetic ester with an initial moisture content of 70 ppm. The temperature was 85 °C. A decrease can be observed in the dynamics of the drying process with the time and reduction of moisture in cellulose.

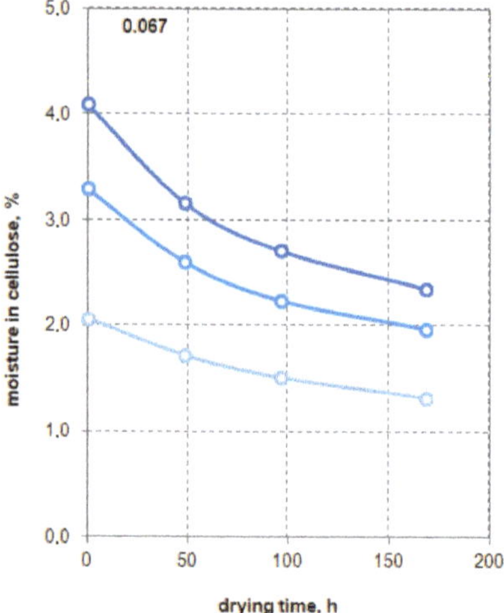

Figure 7. Water content in cellulose samples depending on drying time for three different values of initial moisture of cellulose; mass ratio of cellulose to ester 0.067, samples thickness 3 mm, temperature 85 °C; triple introduction of an ester with moisture content of 70 ppm.

When starting the research, it was assumed that the initial moisture content of the ester has a significant impact on the cellulose drying process efficiency. Figure 8 shows the water loss in cellulose samples, in percentage points, depending on the initial moisture content of the ester. The weight ratio of cellulose to ester was 0.067, the thickness of the samples was 3 mm, and the drying time was 168 h, while the temperature was 70 °C (Figure 8a) and 85 °C (Figure 8b).

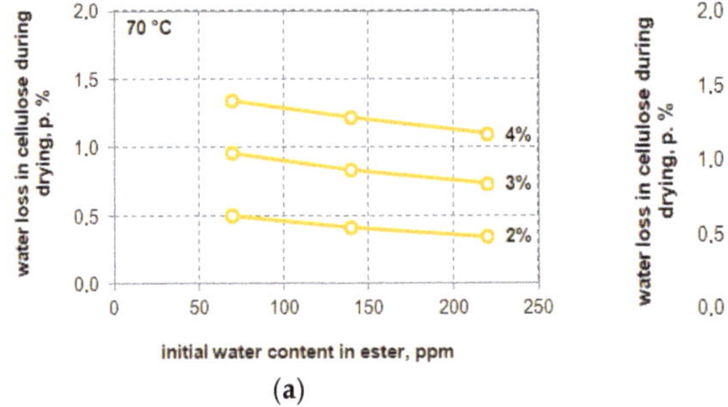
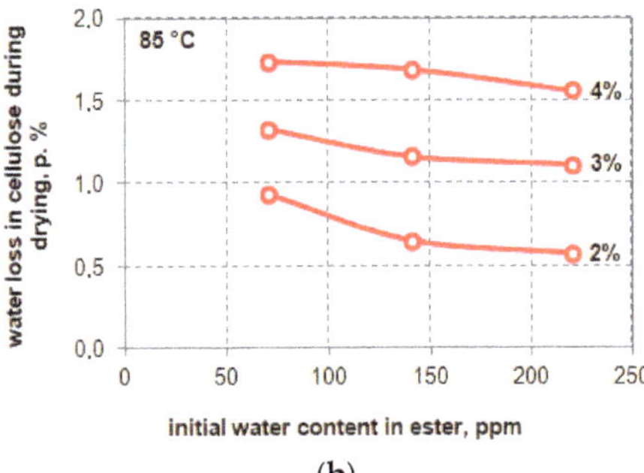

(a)

(b)

Figure 8. Water loss in cellulose in percentage points depending on the initial moisture of the three times introduced ester; initial moisture of cellulose 2%, 3%, and 4%, insulation thickness 3 mm, mass ratio of cellulose to ester 0.067, drying time 168 h; temperature 70 °C (**a**) and 85 °C (**b**).

It is easier to assess the drying efficiency by analyzing the percentage water loss, as shown in Figure 9. For example, a 3% moisture cellulose sample dried for 168 h with a 70 ppm moisture ester at 85 °C loses about 40% of water, and if a 220 ppm moisture ester is used, cellulose may lose about 34% of water. The influence of different ester moistures within the tested limits (70–220 ppm) is not significant. This can be explained by using Figure 10, prepared on the basis of CIGRE documents [2]. This figure shows the moisture of the ester in equilibrium state depending on the cellulose moisture. It can be seen that the moisture content of the ester used in the experiment is far from equilibrium state at which the cellulose drying process ends.

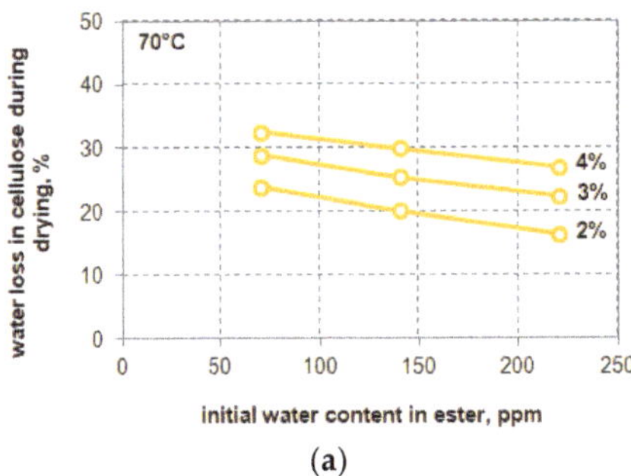
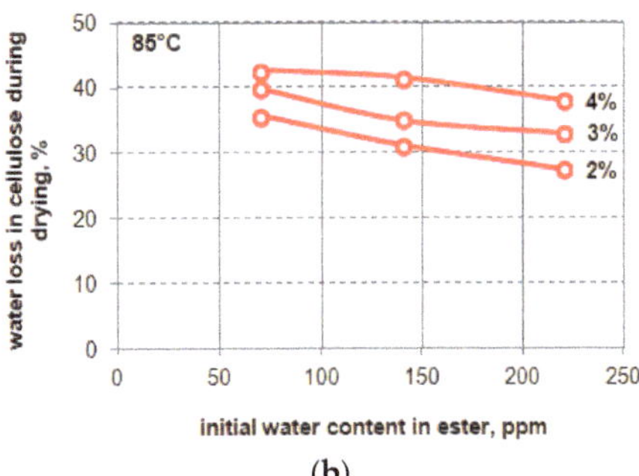

(a)

(b)

Figure 9. Percentage water loss in cellulose depending on the initial moisture of the three times introduced ester; initial moisture of cellulose 2%, 3%, and 4%, insulation thickness 3 mm, mass ratio of cellulose to ester 0.067, total drying time 168 h; temperature 70 °C (**a**) and 85 °C (**b**).

A good illustration of the large impact of temperature on the efficiency of the drying process is Figure 11. However, Figure 12 shows that the effectiveness of this process is slightly dependent on the moisture of the ester. Therefore, when planning the transformer drying procedure, the highest possible temperature should be provided. On the other hand, an ester with a moisture content of even 220 ppm will fulfill its task—under the condition that we ensure its high temperature (at least 85 °C).

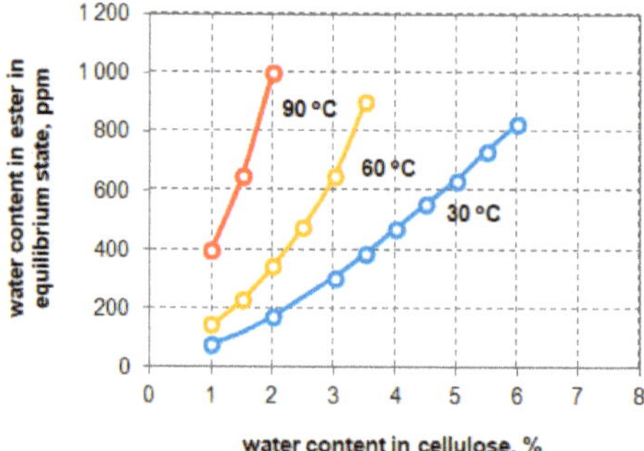

Figure 10. Moisture of the ester in the equilibrium state of the cellulose-ester system depending on the moisture content of cellulose for various temperatures; prepared on the basis of data from [2].

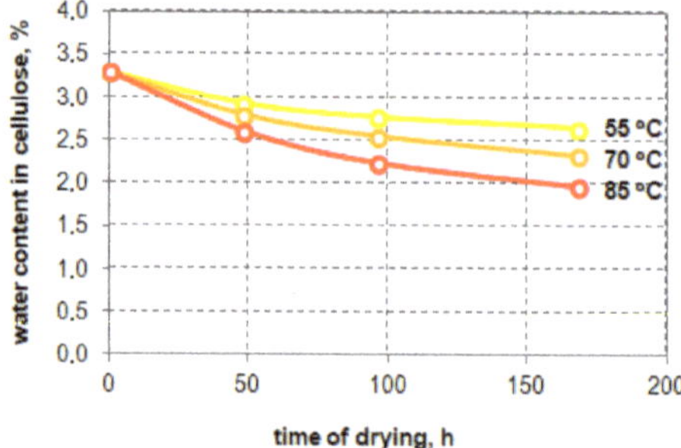

Figure 11. Cellulose moisture depending on the drying time for three temperatures (55, 70, and 85 °C); sample thickness 3 mm, ester with a moisture content of 70 ppm was introduced three times, the mass ratio of cellulose to ester 0.067.

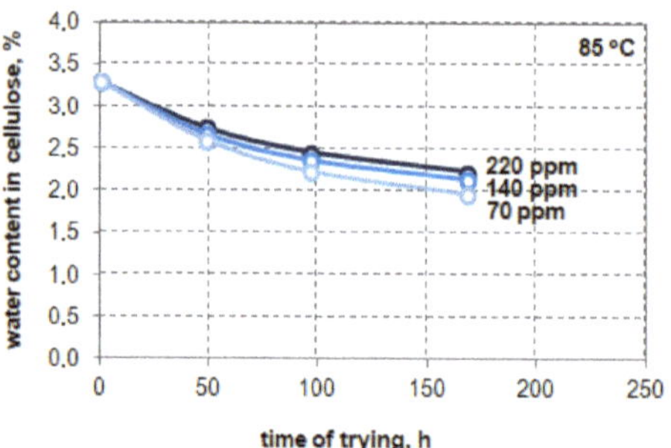

Figure 12. Cellulose moisture depending on the drying time for three initial values of moisture of the ester introduced three times (70, 140, and 220 ppm); sample thickness 3 mm, temperature 85 °C, mass ratio of cellulose to ester 0.067.

The experiments are summarized in Figure 13. The moisture content of cellulose at various drying times (48, 96, and 168 h) is shown for the worst drying parameters (220 ppm, 55 °C) and the best (70 ppm, 85 °C). For initial moisture of cellulose, about 3% and 4%, after the whole drying process (168 h), differences in cellulose moisture are at the level of about 1 percent point. This fact should be taken into account when planning the transformer drying procedure, paying particular attention to ensure high temperature of the drying process.

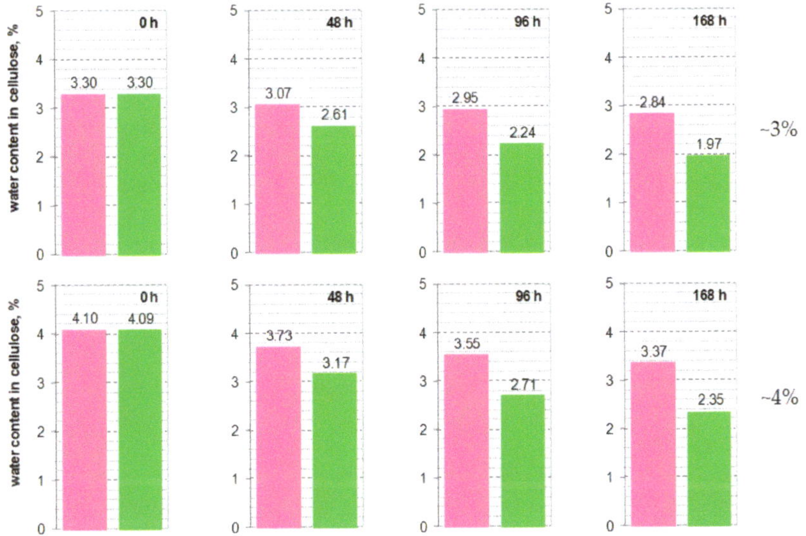

Figure 13. Cellulose moisture in the drying process depending on the drying time—comparison for the worst (pink color: 220 ppm, 55 °C) and best (green color: 70 ppm, 85 °C) drying process conditions; initial moisture of cellulose ~3% and ~4%, thickness of samples 3 mm, mass ratio of cellulose to ester 0.067.

5. Conclusions

There are two groups of methods that allow on-site transformer drying. The first group includes methods consisting of heating the insulation by means of oil, air, solvent vapors, or low-frequency current and applying a vacuum [8,13,14]. To sufficiently dry the transformer insulation a vacuum equal to 1 mbar have to be applied [8,15]. The necessity of vacuum application is the biggest disadvantage of these methods. In cases involving many distribution transformers an underpressure of just 500 mbar can be used, due to the low mechanical strength of their tanks. Moreover, the application of the vacuum may result in deimpregnation of cellulose materials, which leads to deterioration of dielectric parameters of insulation.

The second group of methods is based on the use of mineral oil as a drying medium. These methods are much safer for transformers than vacuum ones, while their biggest drawback is their low drying efficiency. The authors of [16] proved the higher drying efficiency when applying natural ester. It should be pointed out that the water solubility in synthetic ester is much higher than in mineral oil and natural ester; therefore, better drying efficiency should be expected for this liquid.

The method proposed by authors, based on the use of synthetic ester for drying cellulose insulation, is free from disadvantages of methods based on applying vacuum and is much more efficient than drying the insulation by means of mineral oil which was proved in our previous research [10].

The drying efficiency of cellulose insulation using a synthetic ester depends, to varying degrees, on cellulose moisture, insulation temperature, drying time, mass ratio of cellulosic materials to ester, and initial water content in ester which was the subject of the research described in this article. The main conclusions from this research are given below.

The initial moisture content in cellulose is very important. At the temperature used in the experiment, equal to 55, 70, and 85 °C, there was water loss (in percentage points) during the drying process, approximately proportional to the initial moisture of cellulose.

A very large influence of temperature on the efficiency of the drying process was found. For example, with a mass ratio cellulose to ester of 0.067, increasing the temperature from 55 to

85 °C results in at least a two-fold increase in cellulose water loss for all investigated initial moisture values of cellulose (2%, 3%, and 4%).

With the time of drying its dynamics decreases, which is understandable, as the cellulose-ester system approaches the moisture equilibrium state.

The conducted research shows that when drying a transformer using a synthetic ester, special attention should be paid to ensuring the highest possible process temperature, which may be difficult in winter conditions. In such a situation, thermal insulation of the transformer tank should be used.

It was found that the moisture content of the ester from 70–220 ppm—with significant moisture content of the transformer cellulose insulation, equal to 2%, 3%, and 4%—does not have a significant impact on the intensity of the drying process. However, with a relatively low moisture content of insulation, below 2% at low temperature (below 50 °C), the effect of moisture on the ester may be noticeable.

In case of low temperature and thick cellulose insulation, the drying time may exceed 168 h due to low water migration within this material. It is related with moisture diffusion coefficients which were described in [17,18].

On the basis of obtained results, the next study related to drying cellulose by means of ester circulating between the transformer tank and a drying unit is planned. The final result of these investigations will be finding a means to achieve optimal drying conditions on the basis of which a mobile system for on-site drying of transformer's insulation will be developed.

Author Contributions: Conceptualization and methodology, P.P., H.M.-G., and H.M.; Validation, P.P. and H.M.; Formal analysis, P.P., H.M.-G., and H.M.; Investigation, P.P., H.M., and M.C.; Resources, P.P. and H.M.; Data curation, P.P.; Writing—original draft, H.M. and H.M.-G.; Writing—review and editing, P.P. and H.M.; Visualization, H.M.; Supervision, P.P. and H.M.; Project administration, H.M.; Funding acquisition, H.M. and P.P. All authors have read and agreed to the published version of the manuscript.

Funding: This research was funded by the Polish National Center for Research and Development from the funds of Subactivity 4.1.2 "Regional research and development agendas" under the project POIR.04.01.02-00-0045/17-00 entitled "Mobile insulation drying system for distribution transformers using a liquid medium"; the total value of the project is PLN 7,677,957, including co-financing from the National Center for Research and Development PLN 6,084,569.

Conflicts of Interest: The authors declare no conflict of interest.

References

1. Moser, H.P.; Dahinden, V. *Transformerboard*; Scientia Electra: Vermont, VT, USA, 1979; pp. 1–120.
2. Athanassatou, H.; Duart, J.C.; Perrier, C.; Sitar, I.; Walker, J.; Claiborne, C.; Boche, T.; Cherry, D.; Darwin, A.; Gockenbach, E.; et al. *Experiences in Service with New Insulating Liquids*; Cigré Technical Brochure 436; International Council on Large Electric Systems (CIGRE): Paris, France, 2010.
3. Gielniak, J.; Graczkowski, A.; Moranda, H.; Przybylek, P.; Walczak, K.; Nadolny, Z.; Moscicka-Grzesiak, H.; Feser, K.; Gubanski, S.M. Moisture in cellulose insulation of power transformers—Statistics. *IEEE Trans. Dielectr. Electr. Insul.* **2013**, *20*, 982–987. [CrossRef]
4. IEEE Std. 62-1995. *IEEE Guide for Diagnostic Field Testing of Electric Power Apparatus—Part 1: Oil Filled Power Transformers, Regulators, and Reactors*; IEEE: Piscataway, NJ, USA, 1995. [CrossRef]
5. Oommen, T.V.; Lindgren, S.R. Bubble evolution from transformer overload. In Proceedings of the Transmission and Distribution Conference Exposition, Atlanta, GA, USA, 2 November 2001; pp. 137–142. [CrossRef]
6. Przybylek, P. Investigations of the temperature of bubble effect initiation oil-paper insulation. *Prz. Elektrotechniczny* **2010**, *86*, 166–169.
7. Walczak, K. Drying Methods of Power Transformer's Solid Insulation in the Place of its Installation. In Proceedings of the International Transformer Conference Transformator'13, Gdansk, Poland, 6–8 June 2013; pp. 1–9.
8. Koestinger, P.; Aronsen, E.; Boss, P.; Rindlisbacher, G. *Practical Experience with the Drying of Power Transformers in the Field, Applying the LFH Technology*; International Council on Large Electric Systems (CIGRE): Paris, France, 2004.

9. Przybylek, P. Water saturation limit of insulating liquids and hygroscopicity of cellulose in aspect of moisture determination in oil-paper insulation. *IEEE Trans. Dielectr. Electr. Insul.* **2016**, *23*, 1886–1893. [CrossRef]
10. Przybylek, P. Drying transformer cellulose insulation by means of synthetic ester. *IEEE Trans. Dielectr. Electr. Insul.* **2017**, *24*, 2643–2648. [CrossRef]
11. Przybylek, P.; Moranda, H.; Moscicka-Grzesiak, H.; Szczesniak, D. Application of Synthetic Ester for Drying Distribution Transformer Insulation—The Influence of Cellulose Thickness on Drying Efficiency. *Energies* **2019**, *12*, 3874. [CrossRef]
12. IEC 60814. *Insulating Liquids—Oil-Impregnated Paper and Pressboard—Determination of Water by Automatic Coulometric Karl Fischer Titration*; International Electrotechnical Commission (IEC): Geneva, Switzerland, 1997.
13. Bosigner, J. The Use of Low Frequency Heating Techniques in the Insulation Drying Process for Liquid Filled Small Power Transformers. In Proceedings of the 2001 IEEE/PES Transmission and Distribution Conference and Exposition Developing New Perspectives, Atlanta, GA, USA, 2 November 2001; pp. 688–692. [CrossRef]
14. Przybylek, P. A new concept of applying methanol to dry cellulose insulation at the stage of manufacturing a transformer. *Energies* **2018**, *11*, 1658. [CrossRef]
15. Betie, A.; Meghnefi, F.; Fofana, I.; Yeo, Z. Modeling the insulation paper drying process from thermogravimetric analyses. *Energies* **2018**, *11*, 517. [CrossRef]
16. Villarroel, R.; Garcia, B.; Garcia, D.F.; Burgos, J.C. Assessing the use of natural esters for transformer field drying. *IEEE Trans. Power Deliv.* **2004**, *29*, 1894–1900. [CrossRef]
17. Garcia, D.F.; Villarroel, R.; Garcia, B.; Burgos, J.C. Effect of the thickness on the water mobility inside transformer cellulose insulation. *IEEE Trans. Power Deliv.* **2016**, *31*, 955–962. [CrossRef]
18. Garcia, D.F.; Garcia, B.; Burgos, J.C. A review of moisture diffusion coefficients in transformer solid insulation-part 1: Coefficients for paper and pressboard. *IEEE Electr. Insul. Mag.* **2013**, *29*, 46–54. [CrossRef]

Article

Streaming Electrification of Nycodiel 1255 Synthetic Ester and Trafo EN Mineral Oil Mixtures by Using Rotating Disc Method

Maciej Zdanowski 

Faculty of Electrical Engineering, Automatic Control and Informatics, Opole University of Technology, Prószkowska 76, 45-758 Opole, Poland; m.zdanowski@po.edu.pl

Received: 19 October 2020; Accepted: 21 November 2020; Published: 24 November 2020

Abstract: Power transformers are the main element of an electric power system. The service life of these devices depends to a large extent on the technical condition of their insulation system. Replacing mineral oils with natural or synthetic ester (retrofilling process) may increase the efficiency and operational safety of transformers, and also limit their adverse environmental impact. It is technically unfeasible to completely remove mineral oil from a transformer. Its small residues form a mixture with fluid ester, with different physicochemical and electric properties. Streaming electrification is one of the phenomena which, under unfavorable conditions, may damage the insulation system of a forced oil cooled transformer. It is necessary to run prophylactic tests for the ECT (electrostatic charging tendency) of insulating liquid mixtures from the point of view of transformer retrofilling, which is being used more often than before. The article presents the results of studies on selected physicochemical, and electrical properties, and the ECT of mixtures of fresh and aged Trafo EN mineral oil with Nycodiel 1255 synthetic ester. In this regard, the density, the kinematic viscosity, the conductivity, and the relative dielectric constant were measured. The molecular diffusion coefficient was determined using Adamczewski's empirical dependency. The streaming electrification was tested in a rotating disc system. The impact of the rotation time, the diameter, and the disc's rotation speed on the amount of the electrification current generated were analyzed. In addition, the co-relation between the electrification current and the composition of the mixture was determined using fresh and aged mineral oil. On the basis of the electrification model, the volume density of the q_w charge was calculated, which is a parameter defining the ECT of insulating liquids. Based on the results, it was concluded that the synthetic ester is characterized by a higher susceptibility to electrification than the mineral oil. However, combining synthetic ester with a small amount (up to 20%) of fresh or aged mineral oil significantly reduces its ECT, which is beneficial from the point of view of retrofilling power transformers.

Keywords: mineral oil; natural ester; synthetic ester; insulation liquid mixtures; power transformers; retrofilling; streaming electrification; ECT; rotating disc system

---

1. Introduction

1.1. Research of Insulating Liquids Mixtures

Power transformers are a key element of the equipment in power plants, transmission systems, electro-power companies, and large industrial plants. Replacing or repairing these devices as a result of breakdowns involves large financial expenditures [1,2]. A transformer's long and reliable operation strictly depends on the technical condition of its insulation system [3–6]. For economic reasons, the vast majority of transformers is filled with mineral oils with well-known properties [7–9].

Due to tighter fire safety and environmental protection regulations, more often than before the power industry is interested in alternative electro-insulating fluids, which include natural, and synthetic esters [10–13]. Liquid esters are used to fill new or modernized transformers. The process of removing mineral oils from transformers to replace them with liquid esters is defined as retrofilling [14–17]. For technical reasons, it is not possible to completely remove mineral oil from the inside of a transformer. Its small amount (4–7%) usually remains in the paper insulation, the windings, the core, and in places hard to reach in a transformer [18]. Replacing mineral oil with liquid ester creates in a transformer a mixture with properties different than each base fluid. It is necessary to test the properties of insulating liquids generated as a result of retrofilling to ensure long, effective, and safe operation of power transformers. In this regard, the physicochemical and electrical properties, ageing processes, the flammability parameters and the thermal properties of insulating liquid mixtures were examined. Fofana at al. [19,20] presented the research results of two insulating liquids mixtures, proposed as alternatives to mineral oil. The first part of the investigation presented a comparison of the mixed liquids properties with those of pure liquids. The second part of the studies, evaluates the compatibility of the mixed liquids with insulating papers used in power transformers. Perrier et al. [21] showed the test results of heat transfer, breakdown voltage, aging stability and electrostatic charging tendency of different mixtures based on the mineral oil, silicon oil and synthetic ester. It was shown that the best mixture enabling optimization of the power transformer insulation is a mixture containing 20% synthetic ester and 80% mineral oil. Rao et al. [22] presented research results of the 80% mineral oil and 20% synthetic ester mixture in the field of thermal aging processes. The main purpose of the work was to lower the cost of transformer insulation liquid having a good thermal performance and improved oxidation stability. The authors observed that the oxidation rate of the blend of mineral oil with synthetic ester is lower than mineral oil. Yu at al. [23] presented the research results of the electrical and physicochemical parameters of dielectric liquid mixtures based on Envirotemp FR3 natural ester and Karamay No. 25 mineral oil. The tests showed that, with an increase of the natural ester concentration in the mixture, the pour point, acidy, dynamic viscosity, and AC breakdown voltage increased. Beroual et al. [24] showed the test results of AC and DC breakdown voltage of different mixtures based on Midel 7131 synthetic ester, Envirotemp FR3 natural ester and mineral oil. The authors confirmed that the increase in the concentration of both types of esters in a mixture with mineral oil always increases the electrical strength. It has been suggested that transformer retrofilling can be considered with mixtures composed of ester oil (80%) and mineral oil (20%). Hamadi et al. [25] presented a study on the electrical and thermal stability of mixtures of Borak 22 mineral oil and Midel 7131 synthetic ester. The authors showed that the use of synthetic ester in a mixture with mineral oil effectively slows down aging processes of insulating liquids. Dombek and Gielniak [26] investigated the effect of the mixture composition of Nynas Draco mineral oil and Midel 7131 synthetic ester with Envirotemp FR3 natural ester on the flash point, fire point, net calorific value, breakdown voltage, relative permittivity, dissipation factor and conductivity. The authors showed that the content of the mixture significantly determines the change of the tested parameters. Nadolny et al. [27,28] presented the research results of thermal properties of Nytro Taurus mineral oil with Envirotemp FR3 natural ester and Midel 7131 synthetic ester mixtures. The authors showed the measurement results of thermal conductivity, viscosity, specific heat, density, and thermal expansion of the created mixtures in a wide temperature range. The authors showed that the increase in the concentration of esters in the mixture with mineral oil deteriorates the transformer cooling.

1.2. Streaming Electrification Measurement Methods

Another group of issues are ECT tests of the insulating liquids. For this purpose, a number of measurement methods were prepared to evaluate the risk of streaming electrification in the insulation of transformers. Gao et al. [29] have developed a measurement system with a closed insulating liquid circulation, to simulate the oil flow in a transformer. The electrification current was generated in an insulation system model. The current value was changing depending on the flow rate and the temperature of the mineral oil, and also depending on the values of the DC voltage applied. Zdanowski [30] has prepared a flow system that makes it possible to test the streaming electrification depending on the type of the metering pipe material, the flow rate, and the temperature of the insulating fluid. Zmarzły and Frącz [31,32] created measuring systems with a plate and an oscillating cylinder. In these systems, the electrification current can be measured in relation to the oscillation frequency of the measuring electrode. Ren et al. [33] used a modified mini-static tester to test electrification. In this system, the electrification current was generated as a result of a flowing sample of mineral oil with a volume of 50 mL at a speed of 1.2 m/s through a cellulose filter. Leblanc et al. [34] used a measurement system with a capillary pipe, developed at the PPRIME institute (Poitiers, France). In this case, two electrometers were used, one of which would measure the leakage of electrical charges from the capillary pipe, and the other from a metal vessel placed in a Faraday cage. Zmarzły and Kędzia [35] prepared a system with a rotating electrometer, which makes it possible to record 1/f noise generated by the streaming electrification phenomenon. The electrification current signal obtained could be analyzed in the time domain, frequency domain, and in the time-frequency domain. Zdanowski et al. [36] developed a measurement system with a rotating disc to evaluate the ECT in insulating liquids. The electrification current was measured with an electrometer connected to a tank, in which the disc was placed submerged in the insulating liquid. Electrical charges were generated as a result of the disc being put into rotary motion. In the system, it is possible to measure the electrification current in relation to the rotation time, the diameter, and the rotation speed of the disc. For the disc system, an electrification model was developed, which makes it possible to determine the q_w charge volume density at the interface between the disc and the liquid. This parameter is a material indicator used to evaluate and compare the ECT of insulating liquids. The main advantages of the disc system is its simple design, the ease of balancing the disc dynamically, the small volume of the liquid being tested, and the possibility of measuring the current continuously.

1.3. Purpose and Scope of Research

This paper is another stage of the author's research works related to the streaming electrification of insulating liquid mixtures. Zdanowski and Maleska [37,38] showed a high correlation between the electrification current and the composition of Trafo EN mineral oil mixtures with Envirotemp FR3 natural ester, Midel 1204 natural ester and Midel 7131 synthetic ester. The electrification current characteristics significantly depended on the type of mineral oil used in the tests (fresh or aged). The minimum and maximum of the electrification current were observed. This time, the subject was to evaluate the ECT of mixtures of Nycodiel 1255 synthetic ester with fresh and aged Trafo EN mineral oil. Their main purpose of the paper was to specify the most beneficial composition of the mixture in terms of retrofilling transformers. The tests were performed in a rotating disc system, designed and built by the author. In the first stage of the study, the impact of the mixture's composition on its density, kinematic viscosity, molecular diffusion coefficient, conductivity, and relative dielectric constant was determined. Then, the impact of the rotation time, the rotation speed, and the disc's diameter on the electrification current generation in the insulating liquid was examined. In the next stage, the impact of the composition of the mixture of synthetic ester with fresh and aged mineral oil on the form of the electrification current was analyzed. Based on the measurements of the physicochemical and electrical properties, and of the electrification current, the volume density of the q_w charge was determined using the electrification model, which is a parameter defining the ECT of insulating liquids. The key conclusion from the studies is that a small amount (up to 20%) of fresh and aged Trafo EN effectively reduces the electrification of Nycodiel 1255 synthetic ester. The results show that replacing mineral oil with synthetic ester reduces the risk of streaming electrification in the insulation system of transformers. Running diagnostic tests for the streaming electrification phenomenon may help reduce the effects of possible failures, and thus shorten the time and reduce the costs of repairing power transformers.

2. Materials and Methods

To prepare the insulating liquid mixtures, Trafo EN (MO) mineral oil from Orlen Oil (Krakov, Poland), and Nycodiel 1255 (SE) synthetic ester from Cargil (Paris, France) were used. In order to simulate the ageing processes that take place in the course of a power transformer's operation, the mineral oil was subjected to accelerated thermal ageing according to standard IEC 1125 (Method C: 164 h, 120 C, copper catalyst—1144 cm^2/kg of oil, air—0.15 L/h). The mixtures based on the synthetic ester, and on the fresh and aged mineral oil were prepared at an ambient temperature, and then cured in tightly sealed bottles for a month. The liquids was mixed with the aid of a magnetic stirrer at a speed of 1100 rpm for about 60 min, according to the procedure proposed by Beroual et al. [24]. The mixture's volume was 1000 mL, and its composition was changing by 10%. The fluid's density was determined by means of a universal glass densimeter according to standard ISO 3675. The kinematic viscosity was measured with a Brookfield DV-II + Pro viscometer on the basis of standard ISO 2555. The electrical parameters were determined according to standard IEC 60,247 with the use of a tri-electrode capacitor. The conductivity was determined by measuring the resistivity with an MR0-4c meter. The relative dielectric constant was determined by measuring the electrical capacity with a Hioki 3522-50 LCR HiTester meter. The molecular diffusion coefficient was determined based on the empirical dependency (1) specified by Adamczewski [39]:

$$D_m = \frac{3.93 \times 10^{-14} \cdot T}{v_k \rho} \tag{1}$$

where T is temperature, v_k kinematic viscosity, and ρ density.

Figure 1 presents the rotating disc system for measuring the electrification current of insulating liquids. The system consisted of an air-tight measurement vessel with a diameter of 100 mm, filled with 500 mL of the liquid, placed on a Teflon insulator. In the vessel, metal discs were installed with diameters of 50, 60, 70, and 80 mm. The discs' rotation speed within the range from 100 to 500 rpm was regulated by means of a type 57BYGH step motor, combined with a system with a controller and a display. Each disc was connected with the motor by means of a metal mandrel, and an isolating clutch, serving as a mechanical and electric separator. The whole system was placed in a metal casing, serving the role of a Faraday cage. The rotary motion of the disc generated electrostatic charges, and their leakage to the ground was measured with a Keithley 6517A electrometer. The half-rack-sized model 6517A has a special low current input amplifier with an input bias current of <3 fA with just 0.75 fA p-p (peak-to-peak) noise and <20 µV burden voltage on the lowest range. The electrometer allows the measurement of current in the range 1 fA to 20 mA. The measurement process was controlled by means of a dedicated software application [40] installed on a portable computer. The measurement process consisted in washing and de-greasing the disc carefully, and then in seasoning it for two hours in the liquid being tested, to stabilize the hydrodynamic and electric conditions in the measuring system. During the tests, the air temperature in the lab was at a level of approx. 20 °C. Each point on the current characteristics chart is the average of the 500 values obtained from five measurement series, each of which lasted for 100 s. The error bars were determined based on the average electrification current, the standard deviation, and the significance level $\alpha = 0.05$. The streaming electrification phenomenon in the measuring system starts to grow as a result of an increase in the disc's rotation speed. Due to the liquid's very complex motion in the measurement vessel, during the disc's activation, transient states are observed, characterized by considerable fluctuations in the form of the electrification current. The transitional processes result from a fast increase in the disc's speed in respect to the adjacent liquid. This results in a lot of slippage, a consequence of which is a rapid growth in the electrification current. This condition disappears within seconds after the speed of the disc and of the adjacent liquid layer become equal. Zmarzły [35] has analyzed the electrification current signal in the transient state. Because of the limited metering capabilities of the Keithley 6517A electrometer applied, the article presents the measurement results, which were recorded after the transient states disappeared.

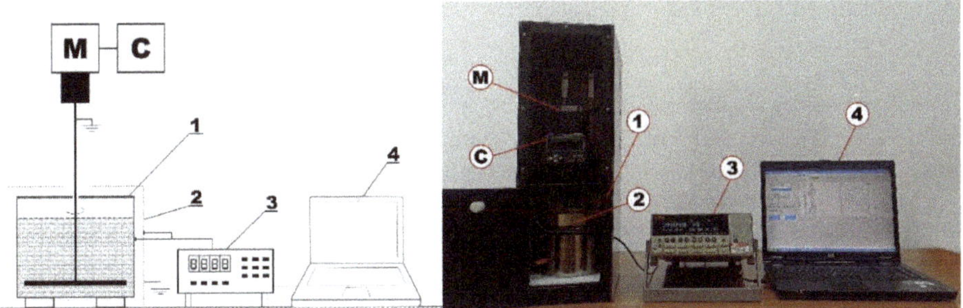

Figure 1. Rotating disc system for the investigation of electrification current of insulation liquids: 1—measuring vessel with liquid and disc, 2—Faraday cage, 3—Keithley 6517A electrometer, 4—portable computer, M—stepper motor and C—controller with display.

Mathematical Model of the Streaming Electrification Phenomenon

The impact of the hydrodynamic conditions and of the physicochemical properties of the insulation liquids on the electrification process in the rotating disc system is described with the use of the model developed by Kędzia [41]. The liquid's motion in the system being discussed is very complex. The liquid adjacent to the disc moves with the disc's angular speed. The centrifugal forces create the liquid's component speeds in the radial direction. At the measurement vessel's wall, the liquid moves vertically. Then, the liquid moves over the disc's upper and lower surface towards its axis. The current in the measurement circuit is the sum of the diffusion current from the q_w charge spread across the vessel's and the disc's surface, the diffusion current from the q_0 charge introduced by the disc's rotational movement into the volume of the liquid, and the conductivity current, flowing in the liquid as a result of the electric field created by the charge in the liquid. In the model, it was assumed that the liquid's speed profile is fully developed. It was also assumed that the volumetric density of the q_w charge in the double electrical layer depends only on the properties of the liquid and of the solid, and does not depend on the hydrodynamic conditions. For this reason, the q_w parameter can be used to determine and compare the ECT of insulating liquids. The electrification current in the disc system is described with the use of the Equation (2):

$$I = \frac{\sigma q_w V}{\varepsilon_0 \varepsilon_r A} + \frac{D_m q_w}{\delta A} S_d - \frac{D_m q_w}{\delta} S_d \tag{2}$$

Constant A is given by the Equation (3):

$$A = \frac{q_w}{q_0} = 1 + \frac{\delta V}{\lambda^2 S} \tag{3}$$

The thickness of the laminar sublayer is described by the Equation (4):

$$\delta = \frac{B v_k}{S_n^{\frac{1}{c}} \left(\frac{\tau_w}{\rho}\right)^{0.5}} \tag{4}$$

The Debye length can be determined from Equation (5):

$$\lambda = \sqrt{\frac{D_m \varepsilon_0 \varepsilon_r}{\sigma}} \tag{5}$$

The shear stresses are given by Equation (6):

$$\tau_w = 0.0396 Re^{-0.25} \rho v^2 \tag{6}$$

The Reynolds number is given by Equation (7):

$$Re = \frac{\omega R^2}{\nu_k} \tag{7}$$

where I denotes the electrification current, q_w—volume charge density on the phase border, q_0—volume charge density in the liquid volume, σ—conductivity, ρ—density, ν_k—kinematic viscosity, ε_0—vacuum electric permittivity, ε_r—relative dielectric constant, D_m—molecular diffusion coefficient, δ—laminar sublayer thickness, τ_w—shearing stress, λ—Debye length, ω—angular velocity, v—average velocity, Re—Reynolds number, V—liquid volume, R—disc radius, S_d—disc surface, S—disk and container surface, S_n—Schmidt number ($S_n = \nu_k/D_m$), and A, B, C—constant ($A = q_w/q_0$, $B = 11.7$; $C = 3$).

3. Results

The density and viscosity are among the most important indicators describing the physicochemical properties of liquid dielectrics. The density of an insulating liquid is used in design calculations to determine a transformer's weight, and the viscosity determines the cooling effectiveness of its active parts. Liquids with lower viscosity are more effective at removing heat from the inside of a transformer into the environment. The conductivity and viscosity depend to a significant extent on the temperature, and thus they affect other properties of the insulating liquids, which include the carrier mobility, molecular diffusion coefficient, relative dielectric constant, and charge relaxation time. The relative dielectric constant influences the stress distribution in a transformer's insulation system, and thus it determines its electric durability [26–28]. When analyzing the data contained in Tables 1 and 2, it can be observed that the ageing processes affect to a different extent the physicochemical and electrical properties of Trafo EN mineral oil. The changes in the density (ρ) and in the relative dielectric constant (ε_r) do not exceed 1%. The kinematic viscosity (ν_k) is characterized by a 9% increase, and the molecular diffusion coefficient (D_m) shows a 10% drop. The largest differences are visible in the conductivity (σ), the value of which increases by nearly two orders of magnitude. And when comparing Nycodiel 1255 synthetic ester with Trafo EN mineral oil, we can see that the ester is characterized by a higher density, kinematic viscosity, conductivity, relative dielectric constant, and a lower molecular diffusion coefficient.

Table 1. Properties of Nycodiel 1255 synthetic ester and fresh Trafo EN mineral oil mixtures (20 °C).

Mixture Content	ρ (kg/m^3)	ν_k (m^2/s)	D_m (m^2/s)	σ (S/m)	ε_r (–)
SE 100%	970	6.21×10^{-5}	1.31×10^{-11}	7.85×10^{-12}	3.19
90% SE + 10% MO	963	5.63×10^{-5}	1.45×10^{-11}	6.41×10^{-12}	3.09
80% SE + 20% MO	952	4.95×10^{-5}	1.67×10^{-11}	4.90×10^{-12}	2.97
70% SE + 30% MO	941	4.55×10^{-5}	1.84×10^{-11}	4.02×10^{-12}	2.88
60% SE + 40% MO	932	3.99×10^{-5}	2.11×10^{-11}	3.02×10^{-12}	2.76
50% SE + 50% MO	923	3.63×10^{-5}	2.35×10^{-11}	2.58×10^{-12}	2.66
40% SE + 60% MO	911	3.19×10^{-5}	2.70×10^{-11}	2.01×10^{-12}	2.57
30% SE + 70% MO	902	2.93×10^{-5}	2.97×10^{-11}	1.55×10^{-12}	2.45
20% SE + 80% MO	891	2.69×10^{-5}	3.28×10^{-11}	1.32×10^{-12}	2.36
10% SE + 90% MO	881	2.37×10^{-5}	3.77×10^{-11}	9.88×10^{-13}	2.26
MO 100%	871	2.21×10^{-5}	4.11×10^{-11}	7.97×10^{-13}	2.19

Table 2. Properties of Nycodiel 1255 synthetic ester and aged Trafo EN mineral oil mixtures (20 °C).

Mixture Content	ρ (kg/m^3)	v_k (m^2/s)	D_m (m^2/s)	σ (S/m)	ε_r (–)
SE 100%	970	6.21×10^{-5}	1.31×10^{-11}	7.85×10^{-12}	3.19
90% SE + 10% MO	962	5.65×10^{-5}	1.45×10^{-11}	8.29×10^{-12}	3.11
80% SE + 20% MO	951	5.29×10^{-5}	1.56×10^{-11}	8.61×10^{-12}	2.99
70% SE + 30% MO	940	4.71×10^{-5}	1.78×10^{-11}	9.21×10^{-12}	2.92
60% SE + 40% MO	931	4.35×10^{-5}	1.94×10^{-11}	9.72×10^{-12}	2.81
50% SE + 50% MO	922	3.99×10^{-5}	2.14×10^{-11}	1.02×10^{-11}	2.72
40% SE + 60% MO	910	3.62×10^{-5}	2.40×10^{-11}	1.07×10^{-11}	2.63
30% SE + 70% MO	901	3.21×10^{-5}	2.72×10^{-11}	1.14×10^{-11}	2.52
20% SE + 80% MO	889	2.98×10^{-5}	2.97×10^{-11}	1.19×10^{-11}	2.41
10% SE + 90% MO	880	2.75×10^{-5}	3.25×10^{-11}	1.27×10^{-11}	2.35
MO 100%	870	2.41×10^{-5}	3.77×10^{-11}	1.33×10^{-11}	2.29

For example, Figures 2a–c and 3a,b present the dependencies that describe the impact of the percentage share of fresh Trafo EN mineral oil and Nycodiel 1255 synthetic ester in the mixtures on the change in the physicochemical and electric parameters. An increase in the content of the mineral oil in the mixture with the synthetic ester caused a linear drop in the density (Figure 2a), and in the relative dielectric constant (Figure 3b), and a non-linear drop in the kinematic viscosity (Figure 2b), and in the conductivity (Figure 3a), and also a non-linear increase in the molecular diffusion coefficient (Figure 2c). Similar parameters are demonstrated by both mixtures of Envirotemp FR3 natural ester with Nytro Taurus [27] and Trafo EN [37] mineral oils, and also mixtures of Midel 7131 synthetic ester with Nytro Draco [26] and Nytro Taurus [28] mineral oils.

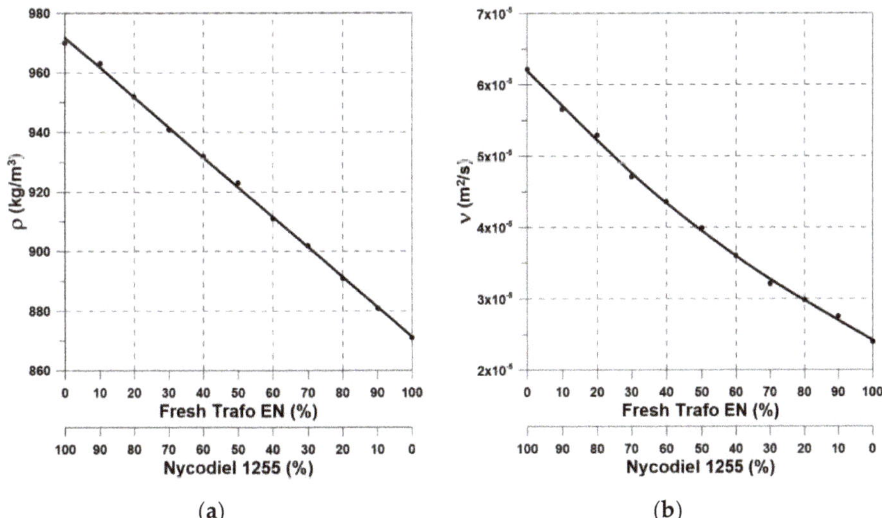

Figure 2. *Cont.*

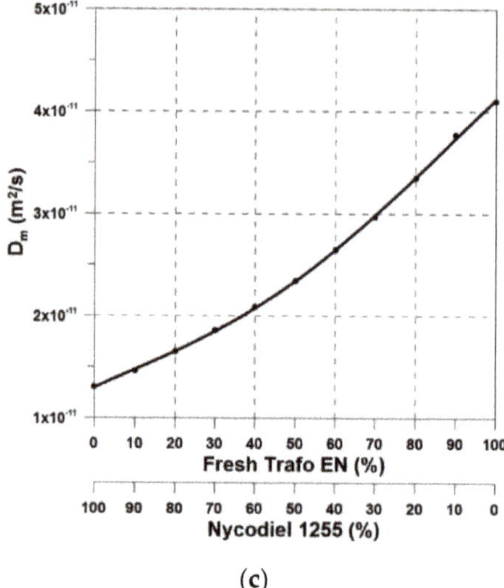

(c)

Figure 2. (**a**) Density, (**b**) kinematic viscosity and (**c**) molecular diffusion coefficient vs. mixing of Nycodiel 1255 synthetic ester with fresh Trafo EN mineral oil.

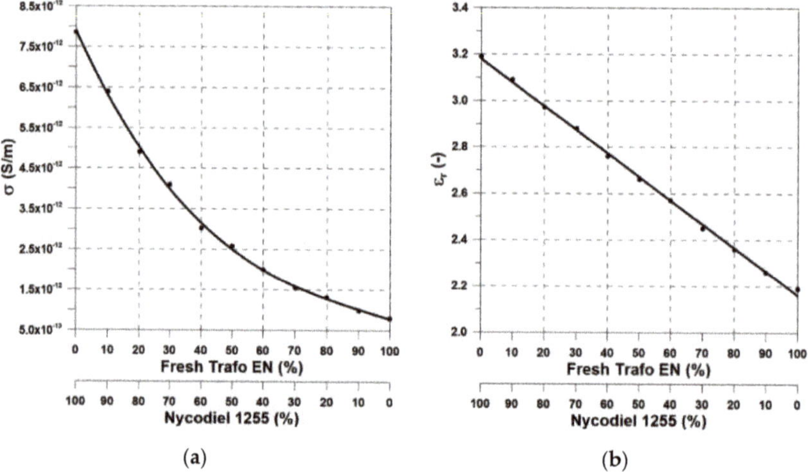

(a) (b)

Figure 3. (**a**) Conductivity and (**b**) relative dielectric constant vs. mixing of Nycodiel 1255 synthetic ester with fresh Trafo EN mineral oil.

To sum up, it should be concluded that comparing the properties of synthetic esters with those of mineral oils is important from the point of view of retrofilling power transformers. It is difficult to clearly indicate, which of the insulating liquids discussed is better or worse for use in transformers. The synthetic esters and mineral oils show a number of desired properties, unfortunately they are not free from disadvantages. Undoubtedly, the use of the synthetic esters is supported by environmental and fire safety aspects, and thus the interest in these liquids increases.

Figure 4a presents the electrification current characteristics of the synthetic ester in relation to the rotation time of the discs with different diameters (50, 60, 70, 80 mm) at a speed of 500 rpm. A change in the disc's rotation speed within the range from 50 to 500 rpm results in a non-linear increase in the electrification current of Nycodiel 1255 insulating liquid (Figure 4b). It was observed that the diameter and the rotation speed of the discs are factors that significantly influence the hydrodynamic conditions in the measurement system, and thus they affect the generation of the streaming electrification phenomenon.

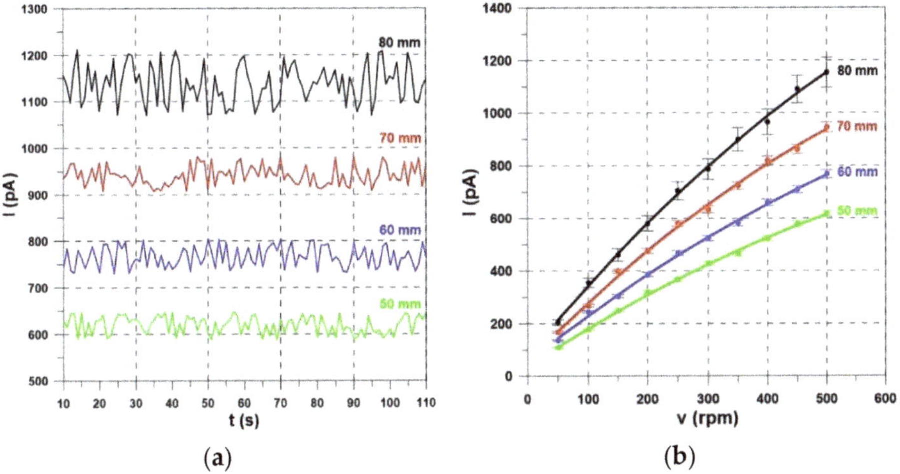

Figure 4. Electrification current of Nycodiel 1255 synthetic ester vs. (**a**) rotating time (v = 500 rpm) and (**b**) rotational velocity for discs of different diameters (d = 50, 60, 70, 80 mm).

Figure 5a presents the dependencies between the electrification current in fresh and aged Trafo EN mineral oil, and Nycodiel 1255 synthetic ester depending on the rotation speed of the disc with the diameter of 80 mm. For each of the liquids, a non-linear increase in the electrification current is observed. The fresh mineral oil electrifies the least. The ageing processes in the mineral oil increase its electrification tendency. The synthetic ester shows an electrification nearly three times higher than that of the fresh Trafo EN oil. At the highest rotation speed (v = 500 rpm) of the disc (d = 80 mm), the electrification current was 417 pA (fresh Trafo EN oil), 854 pA (aged Trafo EN oil), and 1154 pA (Nycodiel 1255), respectively. Figure 5b presents the impact of the disc's rotation speed on the volume density of the q_w charge. The average q_w valueis 0.036 C/m^3 (fresh Trafo EN oil), 0.065 C/m^3 (aged Trafo EN oil), 0.192 C/m^3 (Nycodiel 1255), respectively. When analyzing the charts, it may be stated that this parameter depends to a significant extent on the insulating liquid type. On the other hand, it does not depend on the hydrodynamic conditions prevailing in the measuring system, which is confirmed by its usefulness in determining the ECT of liquid dielectrics. As the study shows, the chemical structure of a liquid and the ageing processes can have a significant effect on the streaming electrification phenomenon in power transformers. In addition, as a result of retrofilling a transformer, a mixture is generated inside with electrostatic features that can be different than those of each base fluid.

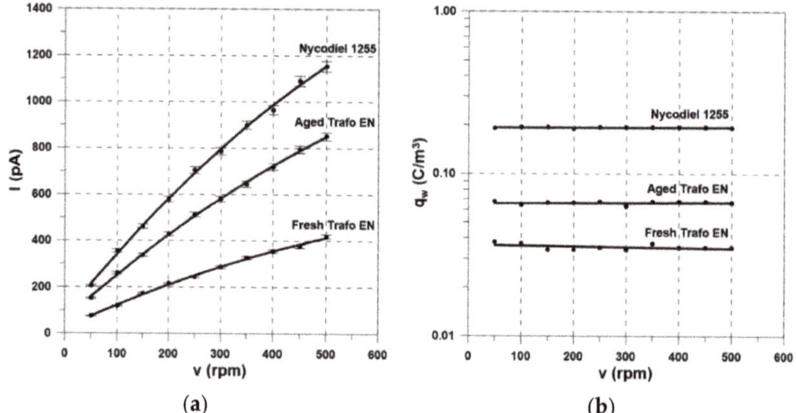

Figure 5. (**a**) Electrification current and (**b**) volume charge density q_w of insulating liquids vs. rotational velocity of the disc ($d = 80$ mm).

Figure 6a presents the impact of the composition of the mixtures of Nycodiel 1255 synthetic ester with Trafo EN mineral oil on the electrification current form. In the measurements, a disc with a diameter of 80 mm was used, rotating at a speed of 500 rpm. Significant differences in the form of the current characteristics were demonstrated, depending on the type of the mineral oil admixture (fresh or aged). In the first case, the electrification current decreases and reaches its minimum value (1011 pA), when the content of fresh oil in the mixture is 20%. Increasing the volume of oil in the mixtures results in an intensive growth in the electrification current value. The current characteristic reaches its maximum value (2613 pA) in the mixture consisting of 80% of mineral oil and 20% of synthetic ester. In the second case, the admixture of aged mineral oil of up to approx. 20% leads to lowering the electrification current value. Increasing the percentage share of mineral oil does not lead to any significant changes in the electrification current form anymore. Figure 6b presents the dependencies between the q_w charge volume density and the mixture's composition. The characteristics of the q_w charge have a different form than the current characteristics, since the electrification model, apart from the electrification current, takes additionally account of the physical-chemical and electrical properties of the liquids being studied.

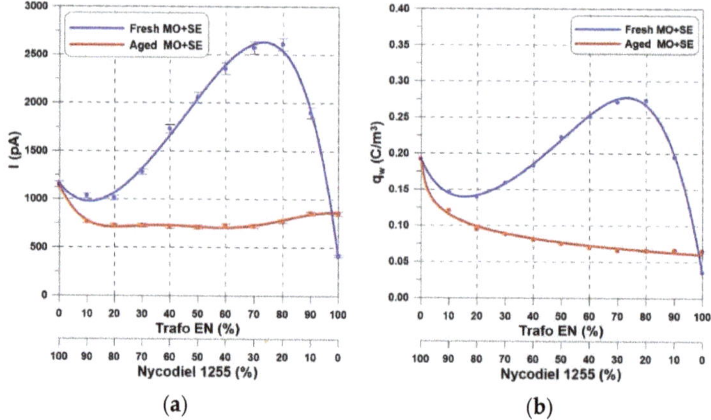

Figure 6. (**a**) Electrification current and (**b**) volume charge density q_w vs. mixing of Nycodiel 1255 synthetic ester with fresh and aged Trafo EN mineral oil ($v = 500$ rpm, $d = 80$ mm).

This leads to a significant conclusion that, in order to specify the ECT of insulating liquids, it is necessary to know both the electrification current and the properties of the liquids. In order to visualize the electrification current value, and the q_w charge value for selected liquids, their results are presented in the form of bar charts (Figure 7a,b).

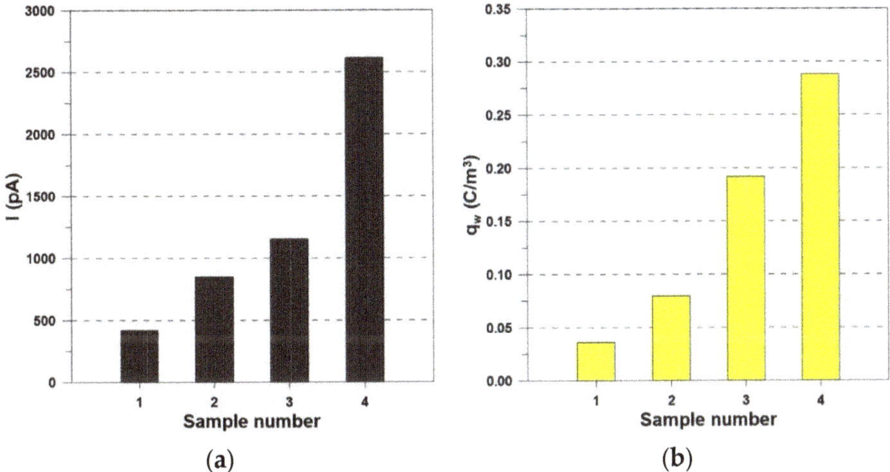

Figure 7. (**a**) Electrification current and (**b**) volume charge density q_w of: 1—fresh Trafo EN, 2—aged Trafo EN, 3—Nycodiel 1255, 4—20% (SE)—80% (fresh MO) mixture.

The tests performed demonstrated that the type of the mineral oil used (fresh or aged) and the percentage share of the particular components in the mixtures substantially contribute to the generation of the streaming electrification phenomenon. Similar dependencies between the parameters being studied were obtained for the mixtures of Trafo EN mineral oil with Envirotemp FR3 natural ester [37], and with Midel 1204 natural ester and Midel 7131 synthetic ester [38]. Rajab et al. [42] presented the results of their studies on the electrification in mixtures of PFAE (palm fatty acid ester) with fresh and used mineral oil. The authors also observed the characteristic maximum of the electrification current, when the content of PFAE was 80%. It is difficult to clearly explain this phenomenon, since for the density, the kinematic viscosity, the conductivity, or the relative dielectric constant, no maximum or minimum values are observed. The reasons for this type of phenomena can result from different chemical structures of mineral oil and synthetic ester. These liquids differ in, for example, the capacity to absorb water from the environment, which may influence the structure of the electrical double layer, where the charge responsible for electrification in insulating liquids is created.

4. Conclusions

Due to their complex structure and the role they serve in an electric power system, power transformers should be properly diagnosed, which may be reflected in their effective, reliable, and long operation. For this reason, more often than before, when modernizing transformers, mineral oil is being replaced with synthetic or natural esters. The primary purpose of the paper was to specify the ECT of mixtures of fresh and aged Trafo EN mineral oil with Nycodiel 1255 synthetic ester in the context of retrofiling power transformers. The ECT of the insulating liquids were determined on the basis of the volume density of the q_w charge created in the double electrical layer at the solid - liquid interface. The q_w parameter was determined using the electrification model for the disc system, based on the measurements of the density, the kinematic viscosity, the molecular diffusion factor, the conductivity, the relative dielectric constant, and the electrification current in the mixtures. The parameters that affect the size of the electrification current generated are the rotation speed, the disc diameter, and the mixtures' composition. The study demonstrated that Nycodiel 1255 synthetic ester electrifies more than fresh and aged mineral oil. It was demonstrated that the ECT of the mixtures depends to a significant extent on their composition. The forms of the electrification current in the mixtures with fresh and aged mineral oil are essentially different. In the first case, the minimal electrification current is observed at 20% of oil in the mixture, and the maximum when the oil content reaches 80%. Aged mineral oil (20%) in the mixtures reduces the electrification current value. A further increase in the aged oil content in a mixture with synthetic ester causes no further significant changes in the electrification current. In both the first and second cases, a small amount of Trafo EN mineral oil (up to 20%) effectively reduces the ECT of Nycodiel 1255 synthetic ester. The most important conclusion of the study conducted is the observation that replacing mineral oil with synthetic ester does not increase the risk of streaming electrification in the insulation system of power transformers. This phenomenon is beneficial for transformer retrofilling.

Funding: This research received no external funding.

Conflicts of Interest: The authors declare no conflict of interest.

References

1. Boss, P.; Lorin, P.; Viscardi, A.; Harley, J.W.; Isecke, J. Economical aspects and experiences of power transformer on-line monitoring. In *CIGRE Session 2000*; CIGRE: Paris, France, 2000.
2. The Council on Large Electric Systems (CIGRE). *CIGRE Technical Brochure 248*; Economics of Transformer Management; CIGRE: Paris, France, 2004.
3. Fofana, I.; Borsi, H.; Gockenbach, E.; Farzaneh, M. Aging of transformer insulating materials under selective conditions. *Electr. Energy Syst.* **2007**, *17*, 450–470. [CrossRef]
4. Martin, D.; Cui, Y.; Ekanayake, C.; Ma, H.; Saha, T. An updated model to determine the life remaining of transformer insulation. *IEEE Trans. Power Deliv.* **2015**, *30*, 395–402. [CrossRef]
5. Ruan, J.; Jin, S.; Du, Z.; Xie, Y.; Zhu, L.; Tian, Y.; Gong, R.; Li, G.; Xiong, M. Condition assessment of paper insulation in oil-immersed power transformers based on the iterative inversion of resistivity. *Energies* **2017**, *10*, 509. [CrossRef]
6. Wolny, S. Analysis of high-frequency dispersion characteristics of capacitance and loss factor of aramid paper impregnated with various dielectric liquids. *Energies* **2019**, *12*, 1063. [CrossRef]
7. Fofana, I. 50 years in the development of insulating liquids. *IEEE Electr. Insul. Mag.* **2013**, *29*, 13–25. [CrossRef]
8. N'cho, J.S.; Fofana, I.; Hadjadj, Y.; Beroual, A. Review of physicochemical-based diagnostic techniques for assessing insulation condition in aged transformers. *Energies* **2016**, *9*, 367. [CrossRef]
9. Wang, X.; Tang, C.; Huang, B.; Hao, J.; Chen, G. Review of research progress on the electrical properties and modification of mineral insulating oils used in power transformers. *Energies* **2018**, *11*, 487. [CrossRef]
10. Fernández, I.; Ortiz, A.; Delgado, F.; Renedo, C.; Pérez, S. Comparative evaluation of alternative fluids for power transformers. *Electr. Power Syst. Res.* **2013**, *98*, 58–69. [CrossRef]
11. Mohan Rao, U.; Fofana, I.; Jaya, T.; Rodriguez-Celis, E.M.; Jalbert, J.; Picher, P. Alternative dielectric fluids for transformer insulation system: Progress, challenges, and future prospects. *IEEE Access* **2019**, *7*, 184552–184571. [CrossRef]
12. Mehta, D.M.; Kundu, P.; Chowdhury, A.; Lakhiani, V.K.; Jhala, A.S. A review on critical evaluation of natural ester vis-a-vis mineral oil insulating liquid for use in transformers: Part 1. *IEEE Trans. Dielectr. Electr. Insul.* **2016**, *23*, 873–880. [CrossRef]
13. Mehta, D.M.; Kundu, P.; Chowdhury, A.; Lakhiani, V.K.; Jhala, A.S. A review on critical evaluation of natural ester vis-a-vis mineral oil insulating liquid for use in transformers: Part II. *IEEE Trans. Dielectr. Electr. Insul.* **2016**, *23*, 1705–1712. [CrossRef]
14. Shinde, R. Condition monitoring of a retro-filled power transformer by natural ester Envirotemp FR3 fluid. In Proceedings of the 3rd International Conference on Condition Assessment Techniques in Electrical Systems (CATCON), Rupnagar, India, 16–18 November 2017; pp. 265–269.
15. Rajaram, S.; Naveen, J. Maintenance of natural ester (Envirotemp FR3 dielectric fluid) filled transformer. *J. Int. Assoc. Electr. Gener. Trans. Distr.* **2018**, *31*, 22–25.

16. Zhao, Y.; Qian, Y.; Wei, B.; Wang, R.; Rapp, K.J.; Xu, Y. In-service ageing comparison study of natural ester and mineral oil filled distribution transformers. In Proceedings of the IEEE International Conference on Dielectric Liquids (ICDL), Roma, Italy, 23–27 June 2019.

17. Breazeal, R.C.; Sbravati, A.; Robalino, D.M. Evaluation of natural ester retrofilled transformers after one year of continuous overload. In Proceedings of the Electrical Insulation Conference (EIC), Calgary, AB, Canada, 16–19 June 2019; pp. 115–119.

18. McShane, C.P.; Luksich, J.; Rapp, K.J. Retrofilling aging transformers with natural ester based dielectric coolant for safety and life extension. In Proceedings of the Cement Industry Technical Conference, Conference Record IEEE-IAS/PCA, Dallas, TX, USA, 4–9 May 2003; pp. 141–147.

19. Fofana, I.; Wasserberg, V.; Borsi, H.; Gockenbach, E. Challenge of mixed insulating liquids for use in high-voltage transformers Part 1: Investigation of mixed liquids. *IEEE Elect. Insul. Mag.* **2002**, *18*, 18–31. [CrossRef]

20. Fofana, I.; Wasserberg, V.; Borsi, H.; Gockenbach, E. Challenge of mixed insulating liquids for use in high-voltage transformers Part 2: Investigations of mixed liquid impregnated paper insulation. *IEEE Electr. Insul. Mag.* **2002**, *18*, 5–16. [CrossRef]

21. Perrier, C.; Beroual, A.; Bessede, J.L. Improvement of power transformers by using mixtures of mineral oil with synthetic esters. *IEEE Trans. Dielectr. Electr. Insul.* **2006**, *13*, 556–564. [CrossRef]

22. Rao, U.M.; Sood, Y.R.; Jarial, R.K. Oxidation stability enhancement of a blend of mineral and synthetic ester oils. *IEEE Electr. Insul. Mag.* **2016**, *32*, 43–47. [CrossRef]

23. Yu, H.; Chen, R.; Hu, X.; Xu, X.; Xu, Y. Dielectric and physicochemical properties of mineral and vegetable oils mixtures. In Proceedings of the 19th IEEE International Conference on Dielectric Liquids (ICDL), Manchester, UK, 25–29 June 2017.

24. Beroual, A.; Khaled, U.; Noah, P.S.M.; Sitorus, H. Comparative study of breakdown voltage of mineral, synthetic and natural oils and based mineral oil mixtures under AC and DC voltages. *Energies* **2017**, *10*, 511. [CrossRef]

25. Hamadi, A.; Fofana, I.; Djillali, M. Stability of mineral oil and oil-ester mixtures under thermal aging and electrical discharges. *IET Gener. Transm. Distrib.* **2017**, *11*, 2384–2392. [CrossRef]

26. Dombek, G.; Gielniak, J. Fire safety and electrical properties of mixtures of synthetic ester/mineral oil and synthetic ester/natural ester. *IEEE Trans. Dielectr. Electr. Insul.* **2018**, *25*, 1846–1852. [CrossRef]

27. Nadolny, Z.; Dombek, G. Thermal properties of mixtures of mineral oil and natural ester in terms of their application in the transformer. In Proceedings of the International Conference on Energy, Environment and Material Systems (EEMS), Polanica-Zdrój, Poland, 13–15 September 2017; pp. 1–6.

28. Nadolny, Z.; Dombek, G.; Przybylek, P. Thermal properties of a mixture of mineral oil and synthetic ester in terms of its applications in the transformer. In Proceedings of the IEEE International Conference on Electrical Insulation and Dielectric Phenomena (CEIDP), Toronto, ON, Canada, 16–19 October 2016; pp. 857–860.

29. Gao, Y.; Li, F.; Lu, W.; Zhao, W. Effect of temperature on flow electrification under DC electrical field. *J. Eng.* **2019**, *2019*, 2893–2896. [CrossRef]

30. Zdanowski, M. Streaming electrification phenomenon of electrical insulating oils for power transformers. *Energies* **2020**, *13*, 3225. [CrossRef]

31. Zmarzły, D.; Frącz, P. Streaming electrification in the swinging plate system. *IEEE Trans. Dielectr. Electr. Insul.* **2017**, *24*, 3217–3225. [CrossRef]

32. Zmarzły, D.; Frącz, P. Dynamics of impulse response of streaming electrification current in swinging cylinder system insulation. *IEEE Trans. Dielectr. Electr. Insul.* **2018**, *25*, 713–720. [CrossRef]

33. Ren, S.; Liu, Q.; Zhong, L.; Yu, Q.; Xu, Y.; Cao, X.; Hanai, M.; Yamada, S.; Mori, S. Electrostatic charging tendency and correlation analysis of mineral insulation oils under thermal aging. *IEEE Trans. Dielectr. Electr. Insul.* **2011**, *18*, 499–505.

34. Leblanc, P.; Paillat, T.; Cabaleiro, J.M.; Touchard, G. Flow electrification investigation under the effect of the flow parameters. *Int. J. Plasma Environ. Sci. Technol.* **2018**, *11*, 156–160.

35. Zmarzły, D.; Kędzia, J. A noise analyzer for monitoring static electrification current. *J. Electrost.* **2005**, *63*, 409–422. [CrossRef]

36. Zdanowski, M.; Wolny, S.; Zmarzły, D.; Kędzia, J. The analysis and selection of the spinning disk system parameters for the measurement of static electrification of insulation oils. *IEEE Trans. Dielectr. Electr. Insul.* **2007**, *14*, 480–486. [CrossRef]

Energies **2020**, *13*, 6159

37. Zdanowski, M. Electrostatic charging tendency analysis concerning retrofilling power transformers with Envirotemp FR3 natural ester. *Energies* **2020**, *13*, 4420. [CrossRef]
38. Zdanowski, M.; Maleska, M. Streaming electrification of insulating liquid mixtures. *Arch. Electr. Eng.* **2019**, *68*, 387–397.
39. Adamczewski, J. *Ionization and Conductivity of Liquid Dielectric*; Polish Scientific Publishers (PWN): Warsaw, Poland, 1965. (In Polish)
40. Zdanowski, M.; Ozon, T. Measuring system for a streaming electrification tests of insulating liquids. *Poznan Univ. Technol. Acad. J. Electr. Eng.* **2016**, *86*, 393–403. (In Polish)
41. Kędzia, J.; Willner, B. Electrification current in the spinning disc system. *IEEE Trans. Dielectr. Electr. Insul.* **1994**, *1*, 58–62. [CrossRef]
42. Rajab, A.; Gumilang, H.; Tsuchie, M.; Kozako, M.; Hikita, M.; Suzuki, T. Study on static electrification of the PFAE-mineral oil mixture. *IOP Conf. Ser. Mater. Sci. Eng.* **2019**, *602*, 012016. [CrossRef]

Publisher's Note: MDPI stays neutral with regard to jurisdictional claims in published maps and institutional affiliations.

Article

Electrostatic Charging Tendency Analysis Concerning Retrofilling Power Transformers with Envirotemp FR3 Natural Ester

Maciej Zdanowski 

Faculty of Electrical Engineering, Automatic Control and Informatics, Opole University of Technology, Prószkowska 76, 45-758 Opole, Poland; m.zdanowski@po.edu.pl

Received: 30 July 2020; Accepted: 24 August 2020; Published: 27 August 2020

Abstract: Natural and synthetic esters are liquids characterized by insulating properties, high flash point, and biodegradability. For this reason, they are more and more often used as an alternative to conventional mineral oils. Esters are used to fill new or operating transformers previously filled with mineral oil (retrofilling). It is technically unfeasible to completely remove mineral oil from a transformer. Its small residues create with esters a mixture with features significantly different from those of the base liquids. This article presents electrostatic charging tendency (ECT) tests for mixtures of fresh and aged Trafo EN mineral oil with Envirotemp FR3 natural ester from the retrofilling point of view. Under unfavorable conditions, the flow electrification phenomenon can damage the solid insulation in transformers with forced oil circulation. The ECT of the insulating liquids has been specified using the volume density of the q_w charge. This parameter has been determined using the Abedian–Sonin model on the basis of the electrification current measured in the flow system, as well as selected physicochemical properties of the liquids. It was shown that ECT is strongly dependent on the type of insulating liquid and pipe material, as well as the composition of the mixtures. The most important finding from the research is that a small amount (up to 10%) of fresh and aged mineral oil is effective in reducing the ECT of Envirotemp FR3 natural ester.

Keywords: insulating liquids; mineral oil; natural ester; synthetic ester; dielectric liquid mixtures; retrofilling of power transformers; streaming electrification; ECT; insulation aging; insulation diagnostics

---

1. Introduction

The basic request of liquid dielectrics applied in power transformers is to ensure good electrical insulation and to remove heat effectively. The insulation liquids also improve the strength of cellulose paper, make it easier to extinguish electrical arcs, and protect the solid insulation against moisture and air. Most transformers being manufactured in the world are filled with mineral oils for economic reasons. The key disadvantages of mineral oils are their limited resistance to oxidation, high toxicity and explosiveness, and poor biodegradability [1]. In recent years, due to fire protection regulations and environmental protection considerations, alternative insulating liquids are becoming more popular. Among those, the most important ones include natural and synthetic esters. The physicochemical, electrical, thermal properties, and the environmental impact of mineral oils and esters used in transformers are relatively well known and described in the relevant literature [2–6]. For many years, electrostatic charging tendency (ECT) tests have been performed for mixtures of pure hydrocarbons, mineral oils, and liquid esters [7–14]. The process of removing mineral oil from a transformer and then refilling it with another insulating liquid is called retrofilling [15–18]. It is technically impossible, however, to completely remove mineral oil from a transformer. Its small amount (4–7%) usually remains within the paper insulation, and, to a small extent, in hardly accessible places of the transformer tank [19]. As a consequence, in the transformer, a mixture of two insulating liquids with unknown properties is formed; they may also vary during the transformer's operation. In many scientific facilities around the world, intensive tests are being performed regarding the features of insulating liquid mixtures in terms of retrofilling. Beroual et al. [20] presented a comparative study of AC and DC breakdown voltage of naphthenic mineral oil, Midel 7131 synthetic ester, Envirotemp FR3 natural ester, and different mixtures based on these liquids. It showed that the ester oils always have a significantly higher breakdown voltage than mineral oil. The addition of natural or synthetic ester considerably increases the breakdown voltage of mineral oil. These authors demonstrated that transformer refilling can be considered with mixtures composed of mineral oil (20%) and ester oil (80%). Yu et al. [21] presented the research results of the physicochemical and dielectric properties of insulating mixtures based on Karamay No. 25 mineral oil and Envirotemp FR3 natural ester. The research showed that with an increase of the natural ester content in the mixture, the dynamic viscosity, acidy, pour point, and AC breakdown voltage increased. The authors observed that the fire point of the mixtures was similar to mineral oil, while the flash point increased by 11.4%. Hamadi et al. [22] presented a comparative study of the electrical and thermal stability behavior of Borak 22 and Midel 7131 synthetic ester mixtures. The authors showed that mixing synthetic ester with mineral oil efficiently reduced the aging rate. Dombek and Gielniak [23] presented the research results of the flash point, fire point, net calorific value, breakdown voltage, relative permittivity, dissipation factor, and conductivity of mixtures of the Nynas Draco mineral oil and Midel 7131 synthetic ester with Envirotemp FR3 natural ester. It was shown that the content of the mixture significantly determined the change of the tested parameters. Zdanowski and Maleska [24] observed a high correlation between the electrification current and the composition of mixtures of Trafo EN mineral oil and Midel 1204 natural ester and Midel 7131 synthetic ester. The authors found big differences in the form of the current characteristics depending on whether the oil used was fresh or aged. Rajab et al. [25] presented research on ECT results of PFAE (palm fatty acid ester) and mineral oil mixtures at various percentage ratios of PFAE. The authors showed that electrostatic charging tendency increases with the percentage ratio of PFAE to 80% but then decreases for the liquid containing only PFAE. The purpose of this paper was to specify the ECT of mixtures which may be formed in a transformer as a result of replacing mineral oil with Envirotemp FR3 natural ester. The most important goal of the work was the most favorable range of admixing Envirotemp FR3 natural ester to Trafo EN mineral oil specified for retrofilling power transformers. In the first stage of the study, the impact of the hydrodynamic conditions and the physicochemical properties of the liquids on selected parameters of the Abedian–Sonin electrification model were analyzed. In the next stage, the electrification current of the liquids in a flow system was measured with pipes made of different materials. Then, the volume density of the q_w charge, designating the ECT of the insulating liquids, was determined.

2. Materials and Methods

The base liquids used for the research were Trafo EN mineral oil (MO) produced by Orlen Oil (Kraków, Poland) and Envirotemp FR3 natural ester (NE) produced by Cargil (Minneapolis, MN, USA). The mineral oil was subject to accelerated thermal aging in accordance with IEC 1125 standard (Method C: 164 h, 120 °C, cooper—1144 cm^2/kg of oil, air—0.15 l/h). The mixtures of oil and ester were prepared at ambient temperature and then seasoned for a month in tightly sealed bottles with a capacity of 1000 mL. The volumetric composition of the mixtures varied every 10%. Density (ρ) was marked with a universal glass areometer (ISO 3675). Kinematic viscosity (ν_k) was measured with a Brookfield DV-II+Pro viscometer (ISO 2555). Conductivity (σ) was determined based on the resistivity measurement with a three-terminal capacitor and MR0-4c meter (IEC 60247). Relative dielectric constant (ε_r) was determined based on the electric capacity measurement with three-terminal capacitor and Hioki 3522-50 LCR HiTester (IEC 60247). Molecular diffusion coefficient (D_m) was determined according to Equation (1) given by Adamczewski [26]. The main characteristics of the insulating liquids used are given in Tables 1 and 2.

$$D_m = \frac{3.93 \cdot 10^{-14} \cdot T}{v_k \rho}$$

(1)

where T—liquid temperature, v_k—liquid kinematic viscosity and ρ—liquid density.

Table 1. Properties of Envirotemp FR3 natural ester and fresh Trafo EN mineral oil mixtures (20 °C).

Mixture Content	ρ (kg/m^3)	v_k (m^2/s)	σ (S/m)	ε_r (–)	D_m (m^2/s)
NE 100%	920	7.80×10^{-5}	5.11×10^{-12}	3.21	1.10×10^{-11}
90% NE + 10% MO	915	6.87×10^{-5}	4.21×10^{-12}	3.11	1.25×10^{-11}
80% NE + 20% MO	910	6.05×10^{-5}	3.50×10^{-12}	3.01	1.43×10^{-11}
70% NE + 30% MO	905	5.33×10^{-5}	2.91×10^{-12}	2.91	1.63×10^{-11}
60% NE + 40% MO	900	4.70×10^{-5}	2.42×10^{-12}	2.81	1.86×10^{-11}
50% NE + 50% MO	896	4.14×10^{-5}	2.01×10^{-12}	2.72	2.12×10^{-11}
40% NE + 60% MO	891	3.64×10^{-5}	1.67×10^{-12}	2.62	2.42×10^{-11}
30% NE + 70% MO	886	3.21×10^{-5}	1.39×10^{-12}	2.52	2.76×10^{-11}
20% NE + 80% MO	881	2.83×10^{-5}	1.15×10^{-12}	2.43	3.15×10^{-11}
10% NE + 90% MO	876	2.49×10^{-5}	9.59×10^{-13}	2.33	3.60×10^{-11}
MO 100%	871	2.19×10^{-5}	7.97×10^{-13}	2.23	4.12×10^{-11}

Table 2. Properties of Envirotemp FR3 natural ester and aged Trafo EN mineral oil mixtures (20 °C).

Mixture Content	ρ (kg/m^3)	v_k (m^2/s)	σ (S/m)	ε_r (–)	D_m (m^2/s)
NE 100%	920	7.80×10^{-5}	5.11×10^{-12}	3.21	1.10×10^{-11}
90% NE + 10% MO	915	6.93×10^{-5}	5.61×10^{-12}	3.12	1.24×10^{-11}
80% NE + 20% MO	910	6.16×10^{-5}	6.17×10^{-12}	3.02	1.40×10^{-11}
70% NE + 30% MO	905	5.48×10^{-5}	6.79×10^{-12}	2.90	1.58×10^{-11}
60% NE + 40% MO	900	4.87×10^{-5}	7.48×10^{-12}	2.83	1.79×10^{-11}
50% NE + 50% MO	895	4.33×10^{-5}	8.23×10^{-12}	2.71	2.03×10^{-11}
40% NE + 60% MO	890	3.84×10^{-5}	9.06×10^{-12}	2.64	2.30×10^{-11}
30% NE + 70% MO	885	3.42×10^{-5}	9.97×10^{-12}	2.49	2.60×10^{-11}
20% NE + 80% MO	880	3.04×10^{-5}	1.10×10^{-11}	2.42	2.94×10^{-11}
10% NE + 90% MO	875	2.70×10^{-5}	1.21×10^{-11}	2.31	3.33×10^{-11}
MO 100%	870	2.40×10^{-5}	1.33×10^{-11}	2.27	3.76×10^{-11}

Figure 1 is the diagram of the flow system for measuring the electrification current of insulating liquids. The liquid was electrified as a result of flowing from the top container through the pipe down to the insulated bottom container placed in a Faraday cage. The electrification current was measured with a Keithley 6517A electrometer. The flow speed (0.34–1.75 m/s) was adjustable by changing the gas (nitrogen) pressure in the top tank. The time of flow (120 s) was determined with a solenoid valve. The temperature (20 °C) was stabilized using a heater with a thermostat. After the measurement had been completed, the liquid was transported from the bottom container to the top container by means of a pump. The lower container (max. 5 l) was made of acid-resistant steel. The point on the current characteristic is the average of 300 values obtained from five measurement series, carried out during 120 s. Error bars were determined using the electrification current average, standard deviation, and $\alpha = 0.05$ significance level. The measuring pipes with a length of 400 mm and a diameter of 4 mm were made of aluminum, Tertrans N cellulose paper produced by Tervakoski Oy (Tervakoski, Finland), and Nomex paper produced by Dupont (Wilmington, DE, USA). The measurement process was controlled by means of a dedicated software [27] installed on a portable computer.

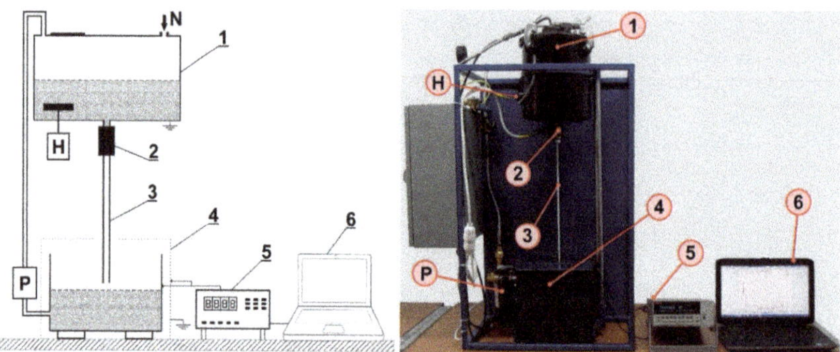

Figure 1. Flow system for the investigation of electrification current of insulation liquids: 1—upper container with liquid, 2—solenoid valve, 3—measuring pipe, 4—Faraday cage with lower container, 5—Keithley 6517A electrometer, 6—portable computer, N—nitrogen, H—heater with thermostat, and P—pump.

Insulating liquid electrification in a flow system is a very complex process. The phenomena that take place at the time are described using the electrification model prepared by Abedian and Sonin [28]. The measure of the ECT of liquid dielectrics is the volume density of the q_w charge. The q_w parameter is determined using the dependencies (2) and (3):

$$\frac{I_\infty}{q_w \pi R^2 v} = Re \frac{\tau_w \lambda^2}{\rho v^2 R^2}\left[1 - \frac{\frac{\delta}{\lambda}}{\sin h\left(\frac{\delta}{\lambda}\right)}\right] + \frac{\frac{\delta}{\lambda}}{\sin h\left(\frac{\delta}{\lambda}\right)}\left[\frac{2\lambda^2}{1 + R\frac{\delta}{2\lambda^2}}\right] \tag{2}$$

$$I = I_\infty\left[1 - e^{-t}\right] \tag{3}$$

The following are the equations that describe the Reynolds number (4), the shearing stress (5), the laminar sublayer thickness (6), and the Debye length (7):

$$Re = \frac{2Rv}{v_k} \tag{4}$$

$$\tau_w = \frac{8\rho v}{Re} \tag{5}$$

$$\delta = \frac{A v_k}{S^{\frac{1}{C}}\left(\frac{\tau_w}{\rho}\right)^{0.5}} \tag{6}$$

$$\lambda = \sqrt{\frac{D_m \varepsilon_0 \varepsilon_r}{\sigma}} \tag{7}$$

where I_∞—electrification current for infinite pipe length, q_w—volume charge density on the phase border, R—pipe radius, v—average liquid velocity, Re—Reynolds number, τ_w—shearing stress, λ—Debye length, v_k—liquid kinematic viscosity, ρ—liquid density, δ—laminar sublayer thickness, I—electrification current for any pipe length, L—characteristic length of the pipe, l—length of the pipe, D_m—molecular diffusion coefficient, ε_0—vacuum electric permittivity, ε_r—relative dielectric constant of liquid, A, C—constant ($A = 11.7$; $C = 3$), and S—Schmidt number ($S = v_k/D_m$).

3. Results

Based on the data from Tables 1 and 2, it was found that the aging processes did not cause significant changes in the density and relative dielectric constant of the Trafo EN mineral oil (below 1%). It was observed that the kinematic viscosity increased by about 9% and the conductivity by nearly two orders of magnitude (from 7.97 10^{-13} to 1.33 10^{-11}). The molecular diffusion coefficient decreased by about 10%. The change in the composition of the mineral oil and natural ester mixture caused a linear decrease in density, relative dielectric constant, and a non-linear decrease in kinematic viscosity and molecular diffusion coefficient. The conductivity when using fresh mineral oil in the mixture decreased non-linearly. When using aged oil, the conductivity increased non-linearly. From the analysis of physicochemical properties, it can be concluded that the viscosity and conductivity may have the greatest influence on the ECT of the insulating liquids.

The Reynolds number (*Re*), the shearing stress (τ_w), and the laminar sublayer thickness (δ) are parameters that describe synthetically the impact of the hydrodynamic conditions and the physicochemical properties of the liquids on the occurring electrification processes. A change in the flow rate of fresh Trafo EN oil and Envirotem FR3 natural ester between 0.34–1.75 m/s causes a linear growth in the Reynolds number (Figure 2a), in the shearing stress (Figure 2b), and a non-linear drop in the thickness of the laminar sublayer (Figure 2c). The Debye length (λ) characterizes the distribution of charges in the laminar sublayer. The λ parameter does not depend on the hydrodynamic conditions and only on the relative electrical permittivity, conductivity, and the molecular diffusion coefficient of the liquid (Figure 2d). On the basis of the Reynolds number, the type of flow (laminar or turbulent) is determined. The parameter *Re* for both liquids does not exceed the value of 2300, which indicates laminar flow. The shearing stress determines the thickness of the laminar sublayer, through which the q_w charge is diffused from the electrical double layer area into the volume of the liquid. An increase in the value of parameter τ_w reduces the thickness of the laminar sublayer and, thus, intensifies the process of the electrification current generation. The differing values of the parameters *Re*, τ_w, and δ result from the difference in the viscosity and density of the mineral oil and the natural ester.

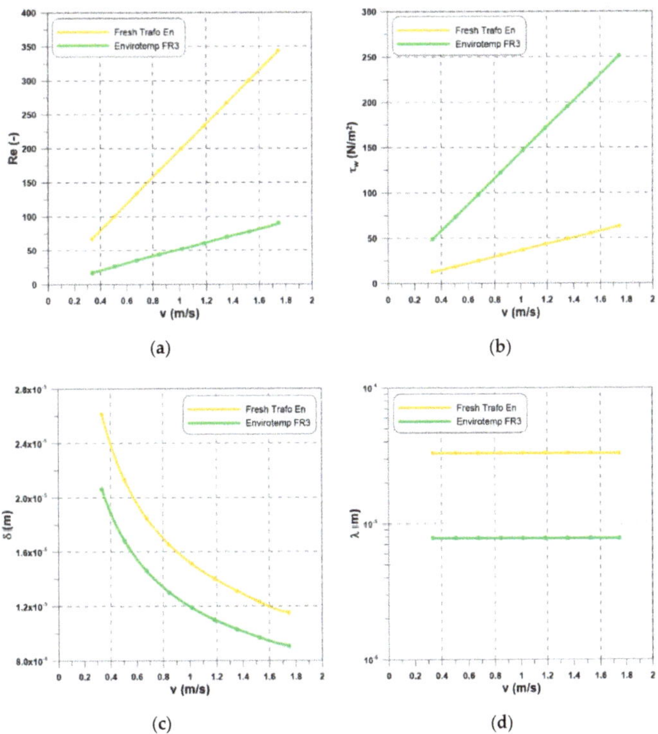

Figure 2. Selected parameters of the Abedian–Sonin model vs. flow velocity of insulating liquids: (**a**) Reynolds number; (**b**) shearing stress; (**c**) laminar sublayer thickness; (**d**) Debye length.

A percentage change in the content of oil and ester in the mixtures results in a non-linear increase in the Reynolds number (Figure 3a), the laminar sublayer thickness (Figure 3c), the Debye length (Figure 3d), and a non-linear drop in the shearing stress (Figure 3b). It results from the model analyses that the hydrodynamic conditions and the physicochemical properties of the liquids substantially affect the parameters of the Abedian–Sonin model in the flow system and, as a consequence, contribute to the generation of the q_w charge, which is the source of the flowing electrification current. Figure 4a shows the electrification current vs. flow time of fresh Trafo EN mineral oil through the aluminum pipe. The tests showed that the electrification current stabilized after about 20 s from the start of the measurement procedure. Figure 4b presents sample dependencies between the electrification current in fresh and aged Trafo EN mineral oil and Envirotemp FR3 natural ester and the speed of flowing (0.34–1.75 m/s) through an aluminum pipe. The registered characteristics are linear. The study demonstrated that natural ester electrified more than mineral oil. Figure 4c presents the impact of the flow rate of the liquids being studied on the change in the volume density of the q_w charge. The experimental tests confirm the assumptions of the Abedian–Sonin model that the q_w parameter does not depend on the hydrodynamic conditions. For this reason, it can be used as a material indicator for determining and comparing the ECT of insulating liquids.

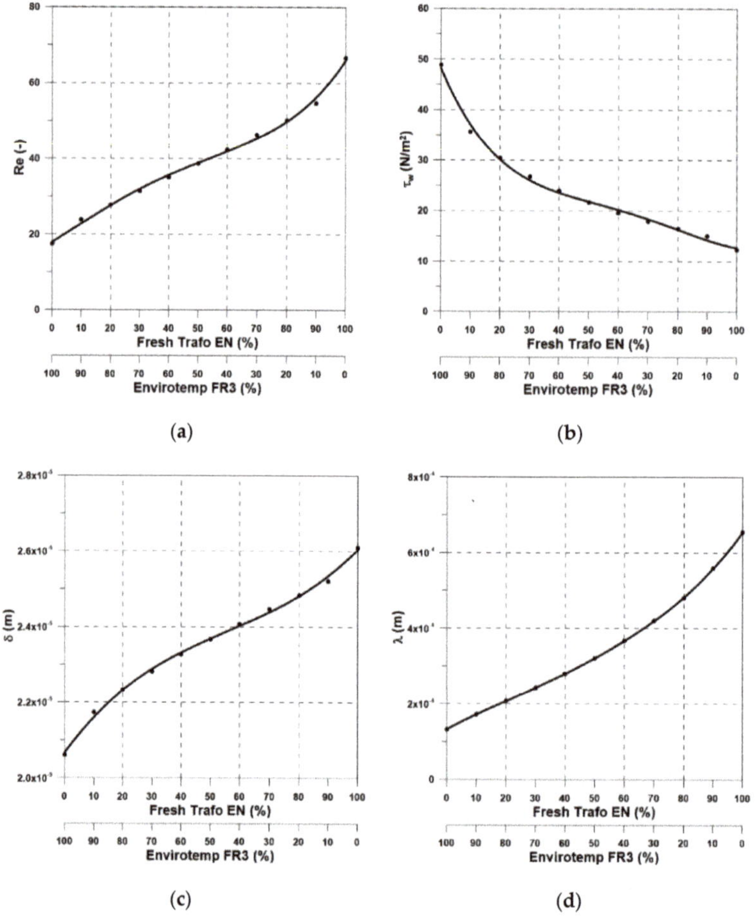

Figure 3. Selected parameters of the Abedian–Sonin model vs. mixture content: (**a**) Reynolds number; (**b**) shearing stress; (**c**) laminar sublayer thickness; (**d**) Debye length.

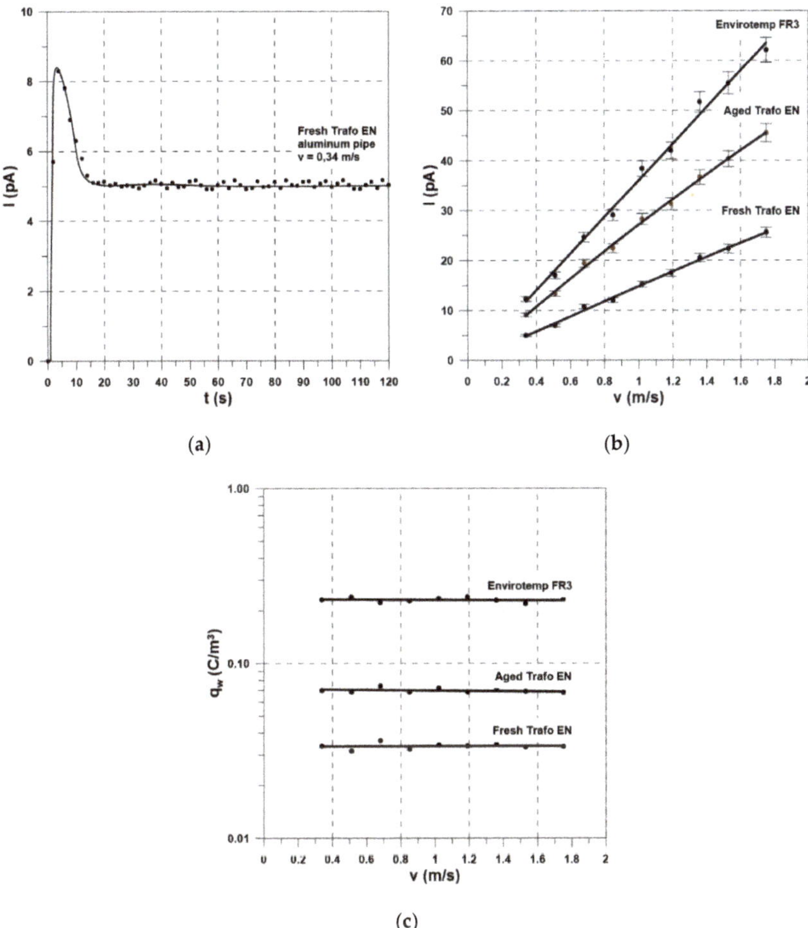

Figure 4. (**a**) Electrification current vs. flow time of Trafo EN mineral oil; (**b**) Electrification current and (**c**) volume charge density q_w vs. flow velocity of insulating liquids through an aluminum pipe.

Figure 5a,b presents the impact of the percentage content of different components in the mixtures on the generation of electrification current. In the measurements, a cellulose, an aramid, and an aluminum pipe were used. The flow rate was 0.34 m/s, and the temperature was 20 °C. The conducted research has shown that the type of pipe has a large impact on the electrification current. This is due to the type and surface roughness of the material used [9]. In both types of mixtures, a high correlation between the electrification current and the composition of the mixture and the type of the measuring pipe material was observed. In addition, a high correlation between the shape of the current characteristics and the type of mineral oil applied is present (fresh or aged oil). In the former case (Figure 5a), an increase in the concentration of fresh mineral oil in the mixtures decreases the electrification current, and its significant increase takes place. The current characteristics reach the maximum in the case of a mixture that is composed of 80% fresh mineral oil and 20% natural ester. Any further increase in the share of oil in the mixture results in a rapid drop in the electrification current. In the second case (Figure 5b), it was concluded that a small amount of aged mineral oil (up to 10%) significantly reduced the electrification of natural ester. Any further increase in the content of aged oil in the mixtures does not lead to significant changes in the generation of electrification current.

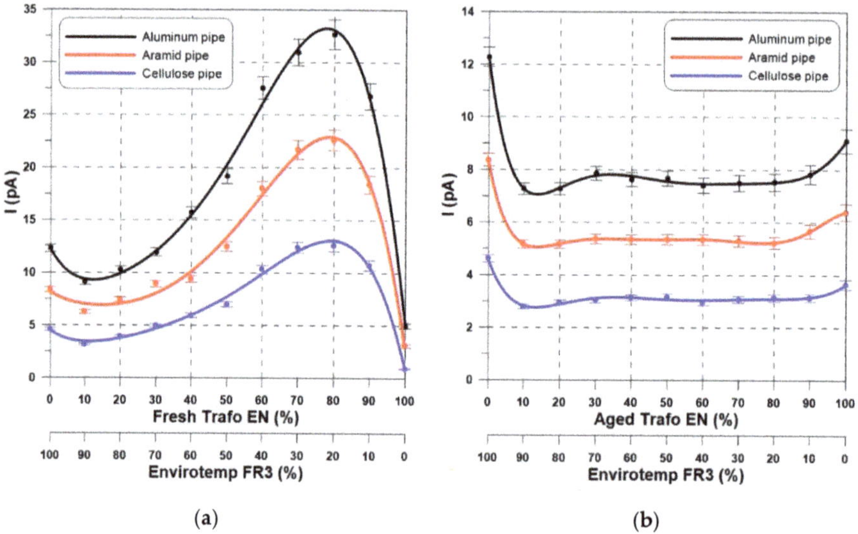

(a)   (b)

Figure 5. Electrification current vs. mixing of Envirotem FR3 natural ester with (**a**) fresh and (**b**) aged Trafo EN mineral oil.

Similarly, Figure 6a,b presents the impact of the mixtures' composition on the volume density of the q_w charge. The differences in the waveform of the characteristics of the electrification current and the q_w charge for both types of mixtures are a result of including the physicochemical properties of the liquids in the electrification model. The study proved that, in order to determine and compare the ECT of insulating liquids, it is necessary to know both their electrification current and their physicochemical properties. In order to visualize better the differences between the electrification current values measured and the q_w charge values calculated from the model, bar charts have been prepared (Figure 7a,b).

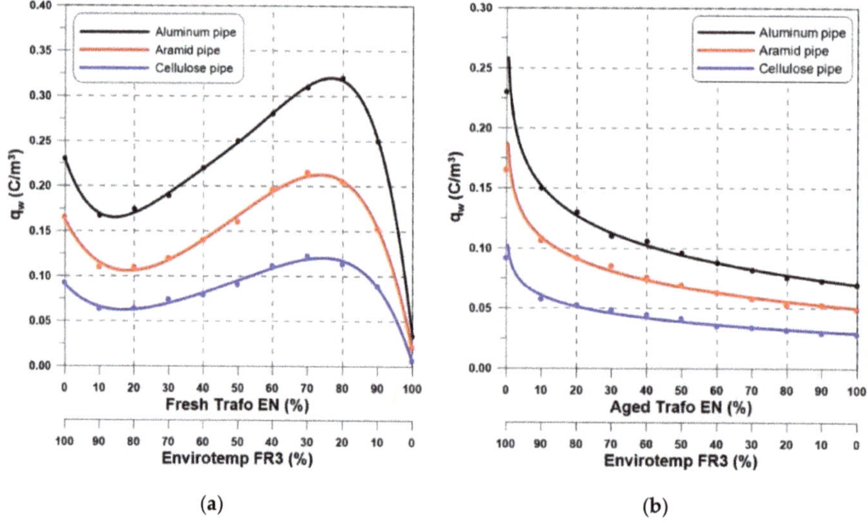

(a)   (b)

Figure 6. Volume charge density q_w vs. mixing of Envirotem FR3 natural ester with (**a**) fresh and (**b**) aged Trafo EN mineral oil.

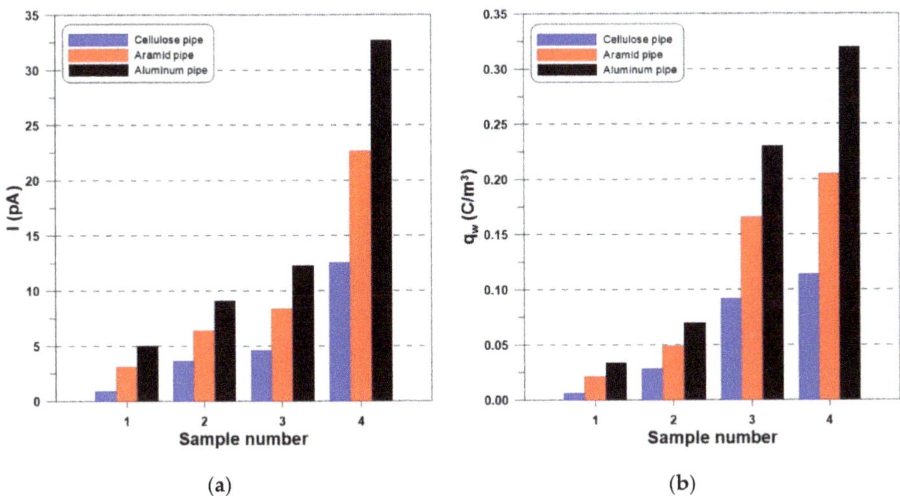

Figure 7. (**a**) Electrification current and (**b**) volume charge density q_w vs. pipe material: 1—fresh Trafo EN, 2—aged Trafo EN, 3—Envirotemp FR3, 4—20% (NE)—80% (fresh MO) mixture.

4. Conclusions

The purpose of this paper was to determine the ECT of mixtures of traditional mineral oil with natural esters in terms of retrofilling power transformers. For the experiment, it was proposed to use fresh and aged Trafo EN mineral oil and Envirotemp FR3 natural ester. Initially, selected parameters of the Abedian–Sonin electrification model were analyzed depending on the flow rate of the fluid and on the mixtures' composition. Then, the electrification current of the liquids was measured in a flow system. The ECT of the liquids was determined on the basis of the volume density of q_w charge results. This study demonstrated that natural ester electrified more intensely than both fresh and aged mineral oil. In addition, it was concluded that the ECT of the liquids was the highest when flowing through an aluminum pipe and the lowest in a cellulose pipe. The ECT of the mixtures depends substantially on the percentage content of different components and the type of mineral oil applied (fresh or aged). When using fresh oil in the mixtures, the characteristic minimum (at 10% of oil) and maximum (at 80% of oil) value of the q_w charge is observed. When aged oil is applied, a non-linear drop in the q_w charge value takes place regardless of the percentage share of both liquids in the mixtures. Comparing the change in physicochemical parameters (Tables 1 and 2) and the electrification current, no significant correlation can be seen. The change in the composition of the mixture causes minima and maxima in the characteristics of the electrification current, which cannot be seen in the case of, e.g., viscosity and conductivity. Therefore, it cannot be clearly stated which property of the insulating liquid has the greatest influence on the ECT. The most important conclusion from the study conducted is the observation that a small amount of fresh or aged mineral oil (up to 10%) significantly reduces the ECT of Envirotemp FR3 natural ester, which is to the advantage of retrofilling, making it possible to increase the efficiency and operational safety of power transformers.

Funding: This research received no external funding.

Conflicts of Interest: The author declares no conflict of interest.

References

1. Wang, X.; Tang, C.; Huang, B.; Hao, J.; Chen, G. Review of research progress on the electrical properties and modification of mineral insulating oils used in power transformers. *Energies* **2018**, *11*, 487. [CrossRef]
2. Fernández, I.; Ortiz, A.; Delgado, F.; Renedo, C.; Pérez, S. Comparative evaluation of alternative fluids for power transformers. *Electr. Power Syst. Res.* **2013**, *98*, 58–69. [CrossRef]
3. Mohan Rao, U.; Fofana, I.; Jaya, T.; Rodriguez-Celis, E.M.; Jalbert, J.; Picher, P. Alternative dielectric fluids for transformer insulation system: Progress, challenges, and future prospects. *IEEE Access* **2019**, *7*, 184552–184571. [CrossRef]
4. N'cho, J.S.; Fofana, I.; Hadjadj, Y.; Beroual, A. Review of physicochemical-based diagnostic techniques for assessing insulation condition in aged transformers. *Energies* **2016**, *9*, 367. [CrossRef]
5. Mehta, D.M.; Kundu, P.; Chowdhury, A.; Lakhiani, V.K.; Jhala, A.S. A review on critical evaluation of natural ester vis-a-vis mineral oil insulating liquid for use in transformers: Part 1. *IEEE Trans. Dielectr. Electr. Insul.* **2016**, *23*, 873–880. [CrossRef]
6. Mehta, D.M.; Kundu, P.; Chowdhury, A.; Lakhiani, V.K.; Jhala, A.S. A review on critical evaluation of natural ester vis-a-vis mineral oil insulating liquid for use in transformers: Part II. *IEEE Trans. Dielectr. Electr. Insul.* **2016**, *23*, 1705–1712. [CrossRef]
7. Zdanowski, M.; Wolny, S.; Zmarzły, D.; Boczar, T. ECT of ethanol and hexane mixtures in the spinning disc system. *J. Electrost.* **2007**, *65*, 239–243. [CrossRef]
8. Zdanowski, M.; Kędzia, J. Research on the electrostatic properties of liquid dielectric mixtures. *J. Electrost.* **2007**, *65*, 506–510. [CrossRef]
9. Zdanowski, M.; Wolny, S.; Zmarzły, D.; Kędzia, J. The Analysis and Selection of the Spinning Disk System Parameters for The Measurement of Static Electrification of Insulation Oils. *IEEE Trans. Dielectr. Electr. Insul.* **2007**, *14*, 480–486. [CrossRef]
10. Zdanowski, M. Influence of composition of dielectric liquid mixtures on electrostatic charge tendency and physicochemical parameters. *IEEE Trans. Dielectr. Electr. Insul.* **2008**, *15*, 527–532.
11. Zdanowski, M. Streaming electrification of mineral insulating oil and synthetic Ester MIDEL 7131®. *IEEE Trans. Dielectr. Electr. Insul.* **2014**, *21*, 1127–1132. [CrossRef]
12. Zdanowski, M. Streaming electrification phenomenon of electrical insulating oils for power transformers. *Energies* **2020**, *13*, 3225. [CrossRef]
13. Talhi, M.; Fofana, I.; Flazi, S. Comparative study of the electrostatic charging tendency between synthetic ester and mineral oil. *IEEE Trans. Dielectr. Electr. Insul.* **2013**, *20*, 1598–1606. [CrossRef]
14. Leblanc, P.; Paillat, T.; Cabaleiro, J.M.; Touchard, G. Flow electrification investigation under the effect of the flow parameters. *Int. J. Plasma Environ. Sci. Technol.* **2018**, *11*, 156–160.
15. Shinde, R. Condition monitoring of a retro-filled power transformer by natural ester Envirotemp FR3 fluid. In Proceedings of the 3rd International Conference on Condition Assessment Techniques in Electrical Systems (CATCON), Rupnagar, India, 16–18 November 2017; pp. 265–269.
16. Rajaram, S.; Naveen, J. Maintenance of natural ester (Envirotemp FR3 dielectric fluid) filled transformer. *J. Int. Assoc. Electr. Gener. Trans. Distr.* **2018**, *31*, 22–25.
17. Zhao, Y.; Qian, Y.; Wei, B.; Wang, R.; Rapp, K.J.; Xu, Y. In-service ageing comparison study of natural ester and mineral oil filled distribution transformers. In Proceedings of the IEEE International Conference on Dielectric Liquids (ICDL), Roma, Italy, 23–27 June 2019.
18. Breazeal, R.C.; Sbravati, A.; Robalino, D.M. Evaluation of natural ester retrofilled transformers after one year of continuous overload. In Proceedings of the Electrical Insulation Conference (EIC), Calgary, AB, Canada, 16–19 June 2019; pp. 115–119.
19. McShane, C.P.; Luksich, J.; Rapp, K.J. Retrofilling aging transformers with natural ester based dielectric coolant for safety and life extension. In Proceedings of the Cement Industry Technical Conference, Conference Record IEEE-IAS/PCA, Dallas, TX, USA, 4–9 May 2003; pp. 141–147.
20. Beroual, A.; Khaled, U.; Noah, P.S.M.; Sitorus, H. Comparative study of breakdown voltage of mineral, synthetic and natural oils and based mineral oil mixtures under AC and DC voltages. *Energies* **2017**, *10*, 511. [CrossRef]

21. Yu, H.; Chen, R.; Hu, X.; Xu, X.; Xu, Y. Dielectric and physicochemical properties of mineral and vegetable oils mixtures. In Proceedings of the 19th IEEE International Conference on Dielectric Liquids (ICDL), Manchester, UK, 25–29 June 2017.

22. Hamadi, A.; Fofana, I.; Djillali, M. Stability of mineral oil and oil-ester mixtures under thermal aging and lectrical discharges. *IET Gener. Transm. Distrib.* **2017**, *11*, 2384–2392. [CrossRef]

23. Dombek, G.; Gielniak, J. Fire safety and electrical properties of mixtures of synthetic ester/mineral oil and synthetic ester/natural ester. *IEEE Trans. Dielectr. Electr. Insul.* **2018**, *25*, 1846–1852. [CrossRef]

24. Zdanowski, M.; Maleska, M. Streaming electrification of insulating liquid mixtures. *Arch. Electr. Eng.* **2019**, *68*, 387–397.

25. Rajab, A.; Gumilang, H.; Tsuchie, M.; Kozako, M.; Hikita, M.; Suzuki, T. Study on static electrification of the PFAE-mineral oil mixture. *IOP Conf. Ser. Mater. Sci. Eng.* **2019**, *602*. [CrossRef]

26. Adamczewski, J. *Ionization and Conductivity of Liquid Dielectric*; PWN: Warszawa, Poland, 1965. (In Polish)

27. Zdanowski, M.; Ozon, T. Measuring system for a streaming electrification tests of insulating liquids. Poznan University of Technology Academic Journals. *Electr. Eng.* **2016**, *86*, 393–403. (In Polish)

28. Abedian, B.; Sonin, A.A. Theory for Electric Charging in Turbulent Pipe Flow. *J. Fluid Mech.* **1981**, *120*, 199–217. [CrossRef]

energies

Article

Statistical Analysis of AC Dielectric Strength of Natural Ester-Based ZnO Nanofluids

Hidir Duzkaya [1,2,*] and Abderrahmane Beroual [2]

[1] Department of Electrical-Electronic Engineering, Gazi University, Ankara 06560, Turkey
[2] Ecole Centrale de Lyon, University of Lyon, Ampere CNRS UMR 5005, 36 Avenue Guy Collongue, 69134 Ecully, France; Abderrahmane.Beroual@ec-lyon.fr
* Correspondence: hduzkaya@gazi.edu.tr

Abstract: Due to environmental concerns and increased energy demand, natural esters are among the alternatives to mineral oils in transformers. This study examines the electrical behavior of natural ester-based ZnO nanofluids at different concentrations in the range of 0.05–0.4 g/L. AC breakdown voltages are measured in a horizontally positioned sphere–sphere electrode system according to IEC 60156 specifications. The measurement data are analyzed using Weibull and normal distribution functions. Breakdown voltages with 1%, 10% and 50% probability are also estimated, these probabilities being of great interest for the design of power electrical components. Experimental results show that AC breakdown voltage increases with the concentration of ZnO nanoparticles, except for the concentration of 0.05 and 0.4 g/L of ZnO. Moreover, breakdown voltages at 1% and 10% probability increase by 22.7% and 13.2% when adding 0.1 g/L ZnO to natural ester, respectively.

Keywords: naturel ester oil; nanofluids; zinc oxide; AC breakdown voltage; Weibull distribution; normal distribution

Citation: Duzkaya, H.; Beroual, A. Statistical Analysis of AC Dielectric Strength of Natural Ester-Based ZnO Nanofluids. *Energies* **2021**, *14*, 99. https://dx.doi.org/10.3390/en14010099

Received: 30 November 2020
Accepted: 24 December 2020
Published: 27 December 2020

Publisher's Note: MDPI stays neutral with regard to jurisdictional claims in published maps and institutional affiliations.

1. Introduction

Insulating liquids are widely used in insulating systems for high voltage (HV) electrical components such as transformers, cables, power capacitors, reactors, circuit breakers, bushings and tap changers [1]. Insulation and heat transfer are among the main requirements that these liquids have to ensure in HV power transformers, these latter being indispensable components for the transmission and distribution electrical energy systems. In HV transformers, 75% of total faults are caused by insulation problems. These transformer failures reduce the expected life of transformers by almost half [2]. Note that the insulation system consists of insulating oil (transformer oil) and solid insulating (paper and pressboard) [3]. The main functions of transformer oils are electrical insulation, protection of solid insulation against air and moisture, improvement of solid insulation performance by penetrating cellulose, protection against corrosion and cooling [3].

The most commonly used insulating liquids in transformers are mineral oils. These latter have been marketed and used since the end of the 19th century for their relatively low cost, dielectric and cooling properties, and compatibility with cellulose-based solid insulation materials and availability [4]. Despite these common advantages, mineral oils have significant disadvantages such as flammability, low biodegradability, low moisture tolerance and corrosive sulphur compounds [5]. Low flash and fire temperatures raise heat protection problems and therefore require measures such as fire safety, firewalls and deluge systems [6]. Mineral oils consisting of different hydrocarbon compounds are a by-product of the oil industry. Oil resources will eventually be depleted; some estimates predict that oil shortages may emerge in the middle of the twenty-first century [4]. This turns out to be an important threat to the insulating liquids industry, where several billion liters are used [1]. Another environmental problem of mineral oils is their low biodegradability, below 30% for 28 days [7]. In the event of a leak or spill after an accident, it pollutes the soil and groundwater and turns into an important threat to humans and the ecosystem.

In order to avoid the problems related to these disadvantages, many studies of alternative insulating liquids were launched over more than forty years [8]. Among the requirements that these alternative liquids are expected to meet are a high dielectric strength, a good heat transfer, improved fire safety, good sustainability, environmentally friendly and extended service life [2,9]. The protection of the environment, which has become a demand/requirement nowadays, is a deciding criterion for alternative transformer oils [8]. Environmentally friendly transformer oil is defined by high biodegradability and low toxicity [2].

Alternatives to mineral oils include many types of insulating liquids in the categories of high molecular weight hydrocarbons, synthetic and natural esters [1,10]. Natural esters obtained from plants such as rapeseed, soybean, sunflower, olive, palm and jatropha, consist mainly of triglycerides, which contain unsaturated fatty acids [1,4,11]. The advantages of natural esters compared to mineral oils are high flash and fire temperatures, almost completely biodegradability, non-toxicity, high dielectric strength and high moisture tolerance [7,9]. Natural esters with fire temperatures above 300 °C can be used in applications where there is a risk of fire without taking special safety precautions [1]. Power system equipment using these less flammable oils can be positioned at lower separation/safety distances [6]. Natural esters that dissolve in nature within 28 days with a rate of over 95% also successfully meet environmental requirements [4]. These oils have higher dielectric strength than mineral oils. Due to this feature, binary mixtures with mineral oils increase the breakdown voltage in power system equipment such as transformers [7]. These properties have made it so that natural esters have been used in transformer and capacitor applications since the early 1990s [11]. The use of natural esters is increasing especially in coastal areas where transformer oil can be contaminated with water and in applications where fire risk can cause great economic damage [1,11].

The major disadvantages of natural esters are increased dielectric dissipation factor (loss factor, tan δ) at high temperatures, high pour temperature, high viscosity and poor oxidation stability [12,13]. The pour temperature is the lowest temperature at which liquid materials can maintain their fluidity properties. High pour temperature turns into a disadvantage in terms of the use of natural esters in cold climatic conditions. These disadvantages can lead to exceeding standard limits in terms of thermal, loss and electrical aspects. For example, Fofana et al. [14] have found that when the ratio of natural ester in the binary mixture with mineral oils is more than 50%, the density and viscosity parameters of the mixture exceed the standard limits. Hermetically sealed applications that prevent contact of natural esters with moisture and oxygen are widely recommended to overcome these problems [13]. In addition, reducing the ratio of unsaturated fatty acids with esterification processes of these oils is another alternative approach [11].

To circumvent and resolve the drawbacks that natural esters present, and in general to improve the thermal and electrical characteristics of insulating liquids, nanoparticles (NP) were introduced into these liquids in the mid-1990s [15]. Originally, nanoparticle-added fluids or nanofluids (NFs) aimed mainly at improving thermal characteristics such as diffusivity, conductivity, convective coefficient and heat transfer [16]. In addition to the fact that they improve thermal properties, some nanoparticles also make it possible to increase the dielectric strength of fluids. Therefore, NFs constitute ideal alternative insulators for oil-filled high voltage applications [17]. The advantages of NP-added transformer oils include better AC, DC and impulse breakdown performances, better partial discharge characteristics, less sensitivity to moisture, prolonged insulation and transformer life, increased thermal conductivity and better cooling of transformers [2].

Different types of NPs are used in the preparation of these nanofluids. Nanoparticles can be classified into three main categories: Conductive, semi-conductive and non-conductive. These classifications, which are defined in terms of electrical behavior, make it easier to define and discuss the breakdown mechanisms [16]. The most commonly used NPs in transformer oils are titanium dioxide (TiO_2), iron oxide (Fe_3O_4), aluminium oxide (Al_2O_3), silicon dioxide (SiO_2) and zinc oxide (ZnO), respectively [18]. Apart from these, NPs such as copper oxide (CuO), fullerene (C_{60}) and aluminium nitride (AlN) are also studied [1,19]. The AC and positive impulse breakdown voltages of these NP-added transformer oils could be improved by a factor of up to 50% [2,19,20]. This rate of increase in breakdown voltages differs depending on the type, size, shape and concentration of the NPs and the type of transformer oil [17]. The addition of Fe_3O_4 to mineral oil can more than double the AC breakdown voltage [16].

Hanai et al. [21] observed that the AC breakdown voltages of ZnO-based mineral oils increased by up to 8.3% compared to pure mineral oil. Bakrutheen et al. [22] found that this increase is 40.6% at 0.075% concentration for a different mineral oil. Chen et al. [20] observed that AC, positive and negative polarity impulse breakdown voltages increased up to 30.2%, 18.9% and 35.8%, respectively, in FR3 oil with 0.4 g/L ZnO.

The breakdown voltage characteristic and mechanism of ZnO, a semiconductor nanoparticle such as TiO_2, in insulating liquids has not been widely studied. Considering that almost 30% of the studies on natural ester-based nanofluids in the literature use TiO_2 [18], the examination of semiconductor ZnO nanoparticle doped nanofluids offers a potential innovation.

This study examines the AC breakdown voltage characteristics of natural ester-based ZnO nanofluids at concentrations of 0.05 to 0.4 g/L; the measurements are conducted according to the IEC 60,156 standard. The experimental results are analyzed with Weibull and normal distribution functions, and probabilities of 1%, 10% and 50% breakdown stresses are deduced.

2. Experiment

2.1. Preparation of Nanofluids

The natural ester MIDEL eN 1204 transformer oil used in this study is based on rapeseed. The physicochemical properties of this oil are shown in Table 1. The ZnO nanoparticles used in the preparation of the nanofluid are supplied from PlasmaChem Gmbh. The average diameter of these spherical particles is 25 ± 3.5 nm and the density is 5.606 g/cm^3 at 20 °C with 99.5% purity.

Table 1. Physicochemical properties of MIDEL eN 1204.

Property	MIDEL eN 1204
Density at 20 °C (g/cm^3)	0.92
Kinematic viscosity at 40 °C (mm^2/s)	8
Pour temperature (°C)	−31
Flash Point (°C)	>315
Fire Point (°C)	>350
Total acid number (mg KOH/g)	0.04
Water content (ppm)	50
Dissipation factor at 90 °C	<0.1

Two different methods are used in the preparation of NFs. In the one-step method, nanoparticles are synthesized and dispersed simultaneously in the base fluid. This method does not include the drying, storage and transportation of nanofluids. Therefore, agglomeration is minimized and fluid stability is improved [10]. Due to the high cost of large-scale production with this method, the two-step method is preferred for nanofluids based on transformer oils [2]. In this method, the fluid is firstly mixed with nanoparticles using a magnetic stirrer as depicted in Figure 1. After this step, surfactant is added to this solution and nanofluid is produced by ultrasonication [23]. In order to avoid the agglomeration problem, measurements are taken after the preparation of the nanofluid. Nanofluids prepared by the two-step method exhibit a homogeneous characteristic for several months without any agglomeration problems [10]. This method can be used on an industrial scale for almost all nanofluids. In this study, oleic acid is used as surfactant.

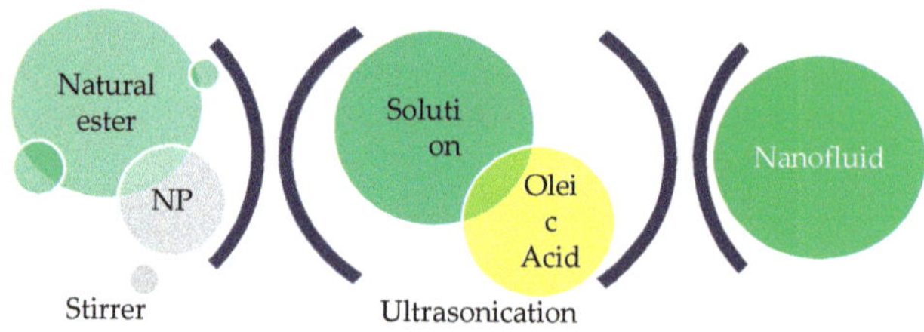

Figure 1. Diagram of two-step method for preparation of nanofluids (NFs).

Due to the high surface energies and attractive/repulsive forces of nanoparticles, the NF can become unstable. Surfactant reduces the surface tension of the fluid and increases the immersion of NPs [2]. This mechanism is also defined as steric stabilization. Oleic acid is the most widely used in NFs prepared with transformer oils. Apart from oleic acid, long-chain hydrocarbons such as hexadecyl trimethyl ammonium bromide (CTAB), sorbitan esters and sodium dodecyl sulphate (SDS) are also used, but rarely [10,23].

In this study, natural ester is purified by using a micro membrane filter and vacuum pump. ZnO NPs of five different concentrations ranging from 0.05 to 0.4 g/L are added to this fluid. Then, each sample is mixed with a magnetic stirrer for 30 min. Oleic acid is added to this solution and an ultrasonication process is applied for 2 h with an ultrasonic homogenizer. This process ensures that the NPs are homogeneously dispersed in the fluid and remain stable without aggregations/clusters. Sonics Vibra-cell sonicator used in ultrasonication has 500 W power rating, 20 kHz frequency and 60% amplitude. This ultrasonication process is applied in periods of 20 min with a 10 min waiting time between each to prevent the nanofluid from overheating. In order to eliminate possible humidity and micro air bubbles that develop during the preparation of the nanofluid, the nanofluid is kept in the oven and then under vacuum at a pressure of 1.0 Pa for 24 h.

2.2. AC Breakdown Measurement

AC breakdown voltages of pure Midel eN-1204 and these natural ester-based ZnO nanofluids are measured using the BAUR DTA 100C, according to the IEC60156 [24] using a 400 mL test cell, 12.5 mm diameter electrodes horizontally positioned at a 2.50 ± 0.05 mm electrode gap. The rest time of each sample is 30 min in order to eliminate gas bubbles in the test cell. The voltage is increased with a rise rate of 2 kV/s until breakdown occurs. The time delay between each breakdown is 2 min and the number of measured breakdown voltages in each set is 6. In order to have sufficient data for statistical analysis, five series of six measurements each, i.e., a total of 30 measurements, are carried out on each type of nanofluid sample [5,8]. After the measurement, the test cell and electrodes are cleaned with ethanol. After this stage, it is washed with hot water at a temperature of 60–80 °C and dried in an oven at 60 °C for one hour. This procedure is compatible with approaches in similar measurement studies reported in the literature [4,5].

The AC breakdown voltage characteristics are analyzed with Weibull and normal distribution functions, and the withstand voltage levels at 1%, 10% and 50% are determined.

3. Results and Discussions

The measurement results taken to check the conformity or not of their distribution with the Weibull law and the normal law are presented in Figure 2. The mean and standard deviation of these measurements are calculated using Equations (1) and (2), respectively:

$$\overline{U} = \frac{1}{n}\sum_{i=1}^{n} U_i \tag{1}$$

$$\sigma = \sqrt{\frac{1}{n}\sum_{i=1}^{n}\left(U_i - \overline{U}\right)^2} \qquad (2)$$

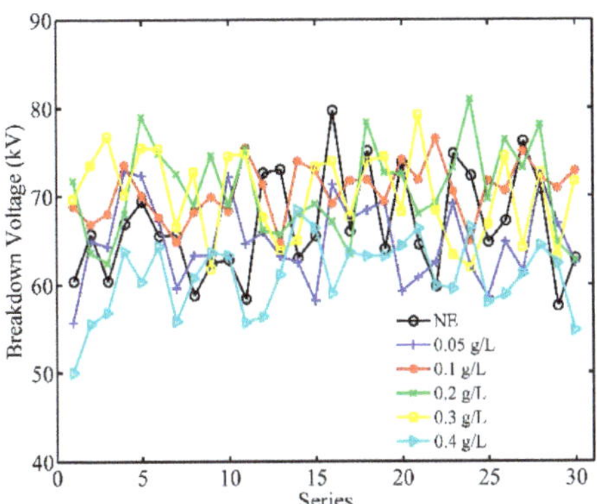

Figure 2. Distribution of AC breakdown voltages of natural ester (NE) and NFs.

It is noticed that the AC breakdown voltages of natural ester (NE) are reduced by 2.7% and 8.7% for nanofluids with a concentration of 0.05 and 0.4 g/L ZnO, respectively. This characteristic changes for 0.1, 0.2 and 0.3 g/L ZnO and increases by 5.8%, 5.8% and 5.1%, respectively.

Khaled and Beroual [8] examined the same natural ester-based Fe_3O_4, Al_2O_3 and SiO_2 NFs and observed that the best improvement in breakdown voltages did not exceed 7%. The breakdown voltage was reduced by about 15% in the 0.05 g/L-added SiO_2 nanofluid [8].

The mean and standard deviation range of these breakdown voltage measurements at different concentrations are given in Figure 3. The ratio of standard deviation to mean breakdown voltage measurements is 5.9% to the maximum for pure naturel ester and 3.1% to the minimum for 0.1 g/L ZnO concentration. The growth of this ratio is linearly related to the difference between the measurements in each concentration set.

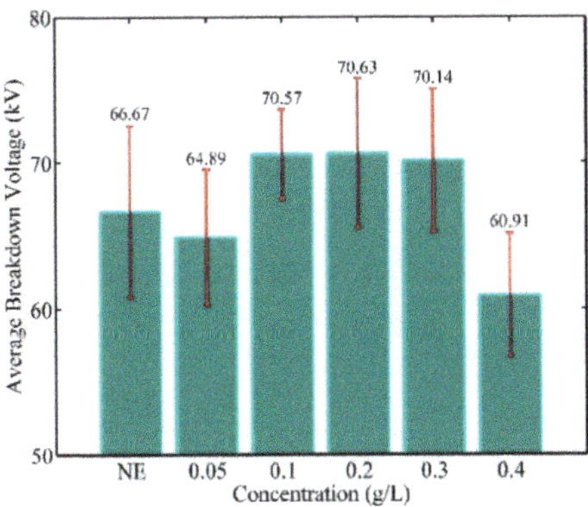

Figure 3. Average breakdown voltages of NE and ZnO nanofluids for different concentrations.

Histogram charts of NE and nanofluids are given in Figure 4. These charts include breakdown frequency at different voltage levels, mean value of breakdown voltage and standard deviation. In terms of average breakdown voltage, 0.1, 0.2 and 0.3 g/L ZnO NFs perform better than NE.

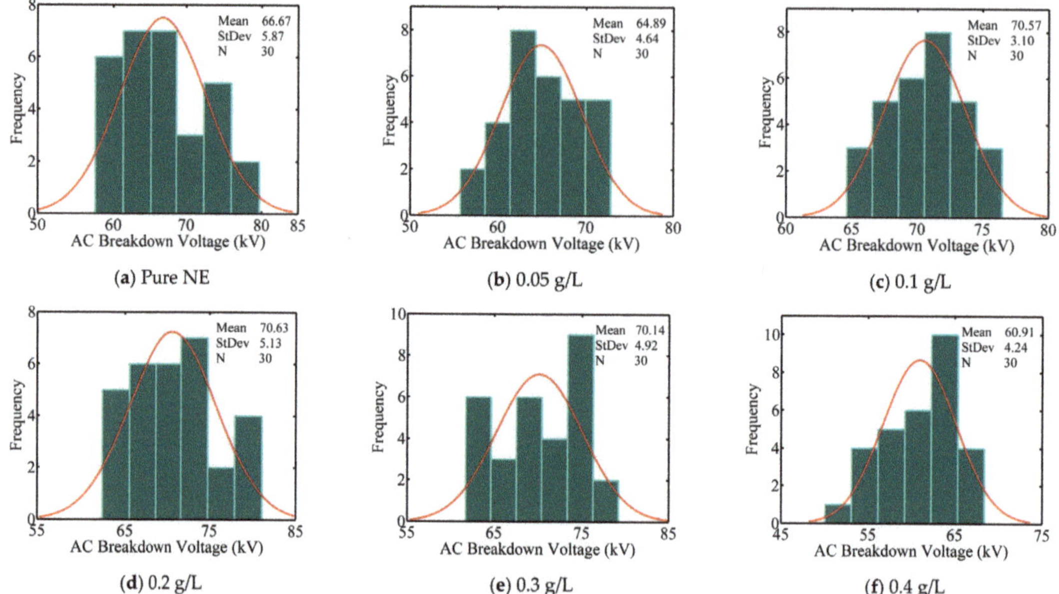

Figure 4. Histograms of NE and ZnO nanofluids for different concentrations.

In order to statistically analyze the probability of breakdown voltage by adhering to these measurements, the Weibull and normal distribution should be tested using a hypothesis. In the test of this hypothesis, which questions the distribution of measurements in the 5% significance level ($\alpha = 0.05$), Anderson–Darling and Shapiro–Wilk tests are used for Weibull and normal distributions, respectively [7,8].

The Anderson–Darling normality test can examine measurement data without grouping and is very sensitive to distributions in the tail region rather than the median [25]. The Shapiro–Wilk test is a regression-correlation based test using a sequential sample. This test, in which the normality of the samples is tested, is consistent in all alternative datasets up to 50 samples [26]. The p-value is the probability of making an error in testing the hypothesis that the measurement data conforms with the statistical law [7].

A hypothesis is accepted if the p-value obtained in these tests is greater than the significance level. Under the condition that the hypothesis is accepted, the distribution of the measurements is defined as a statistical distribution and different probability levels can be estimated [27].

W, which is also defined as the test statistic, is evaluated differently for both tests. The hypothesis is rejected due to a too large W value in the Anderson–Darling test. In order for the hypothesis to be accepted, W should be below 1.5786 at the 0.05 significance level [25]. In the Shapiro–Wilk test, in order for the hypothesis to be accepted at the 95% confidence interval, W should be in the range of 0.9303–1.0000 depending on the number of samples used [26]. The test statistics provide the necessary conditions for both tests and the normal distribution hypothesis is accepted.

The Weibull distribution of the measurements given in Figure 2 is examined using the Anderson–Darling test. According to this test, the distributions of measurements for natural ester and natural ester-based ZnO nanofluid samples are within the acceptable significance level, see Table 2. The acceptance of hypothesis tests of conformity for all samples allows the analysis of the breakdown voltage characteristics using the Weibull distribution function.

Table 2. Hypothesis test of conformity to Weibull distribution of NE and NFs.

	W	**p-Value**	**Conformity of Weibull Distribution**
NE	0.5449	0.1611	Accepted
0.05 g/L	0.3345	0.5149	Accepted
0.1 g/L	0.1782	0.9109	Accepted
0.2 g/L	0.2860	0.6007	Accepted
0.3 g/L	0.7278	0.0516	Accepted
0.4 g/L	0.3842	0.3957	Accepted

The probability curves due to the Weibull distribution are shown in Figure 5. In the 50% probability region, 0.1 to 0.3 g/L ZnO nanofluids have similar AC breakdown voltage characteristic. In this region, the AC breakdown voltage of 0.05 and 0.4 g ZnO nanofluids worsens. In the 10% probability region, the best breakdown voltage is in the 0.1 g/L ZnO nanofluid sample.

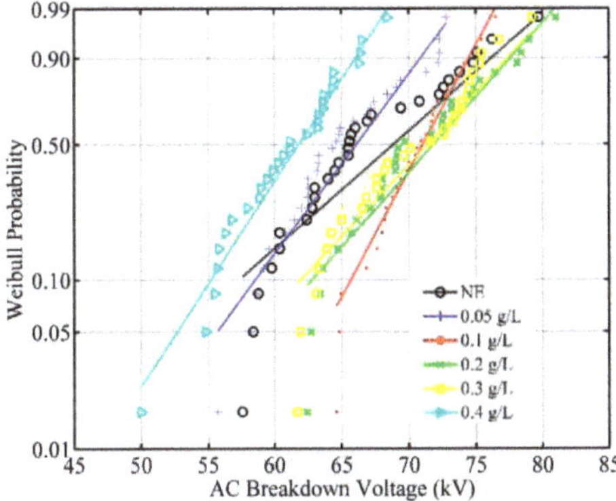

Figure 5. Weibull probability of NE and ZnO nanofluids for different concentrations.

The same sets of measurements are used to examine the compliance with the normal distribution function. The statistical distributions of these measurements are examined with the Shapiro–Wilk test and the hypothesis that the normal distribution is distributed in the significance level ($\alpha = 0.05$) is accepted in all samples, see Table 3.

Table 3. Hypothesis test of conformity to normal distribution of NE and NFs.

	W	**p-Value**	**Conformity of Normal Distribution**
NE	0.9516	0.1866	Accepted
0.05 g/L	0.9626	0.3608	Accepted
0.1 g/L	0.9756	0.7005	Accepted
0.2 g/L	0.9654	0.4231	Accepted
0.3 g/L	0.9385	0.0830	Accepted
0.4 g/L	0.9670	0.4601	Accepted

Using these normal distribution functions in which experimental measurements are analyzed, 1%, 10% and 50% probability breakdown voltages can be estimated, see Table 4. The statistical nature of the breakdown voltages complicates the design of power system equipment. In order to overcome these difficulties, the withstand voltages calculated as a statistical parameter are defined with different possibilities. The withstand voltage of the insulation is not the average value of the breakdown voltage, but as a statistical variable, a low probability of breakdown voltage, such as 1% or 10% [28]. These critical risk levels for design safety are widely studied in the literature [7,28,29]. The 1% probability of breakdown voltage is a safety factor in the design of electrical equipment and is defined as the voltage limit for operation in the safety margin [7,28].

Table 4. AC breakdown withstand voltages at different probabilities for NE and NFs.

BDV Probability	NE BDV (kV)	Type of NFs	BDV (kV)	Change (%)
		0.05 g/L	54.16	5.9
		0.1 g/L	62.78	22.7
%1	51.17	0.2 g/L	57.92	13.2
		0.3 g/L	56.83	11.1
		0.4 g/L	51.02	−0.3
		0.05 g/L	59.12	1.0
		0.1 g/L	66.27	13.2
%10	58.52	0.2 g/L	63.72	8.9
		0.3 g/L	62.94	7.5
		0.4 g/L	55.43	−5.3
		0.05 g/L	65.20	−3.5
		0.1 g/L	70.55	4.4
%50	67.55	0.2 g/L	70.85	4.9
		0.3 g/L	70.45	4.3
		0.4 g/L	60.85	−9.9

These breakdown probabilities are also considered as withstand stress levels. NFs other than that with 0.4 g/L ZnO outperform natural ester in terms of 1% withstand voltage probability. Specifically, a 22.7% increase in 0.1 g/L sample significantly improves the withstand voltage performance for this critical parameter. Similarly, this value increases for 0.2 and 0.3 g/L ZnO samples; the rate of increase is 13.2% and 11.1%, respectively. In the withstand voltage where there is a 10% probability of breakdown, the 0.1 g/L sample shows the best performance as in the previous probability level. The increase rate in this withstand voltage for 0.1 g/L nanofluid is 13%.

The electrical insulation characteristics of nanofluids with ZnO additives are improved compared to the natural ester in terms of withstand breakdown voltages at 1% and 10% probabilities.

Mechanisms of breakdown characteristics of nanofluids are not clearly defined and remain the subject of controversy. The electrical conductivity of nanoparticles seems to be an important parameter in the explanation of this mechanism. Conductive nanoparticles capture the rapidly moving electron in the fluid and turn into slow negatively charged nanoparticles. Streamer propagation slows down and therefore the breakdown voltage increases with this mechanism [16]. Conductive and nonconductive nanoparticles trap electrons by charge induction polarization, respectively. These scavenger nanoparticles reduce free electrons moving in the fluid [5].

ZnO is a semiconductor nanoparticle that traps high mobility electrons. It slows down the electrons responsible for streamer development with trapping and de-trapping processes [2]. In nanofluids using semiconductor nanoparticles, the surface trap density and charge dissipation velocity are 2.5 and 4 times higher compared to pure transformer oil, respectively [30]. With the effect of these mechanisms, AC, DC and impulse breakdown voltages can increase by 20% compared to pure oil [30]. The capture of electrons by scavenger nanoparticles increases the initial threshold voltage of the streamer, and therefore more energetic breakdown mechanisms emerge [16,27]. Due to the semiconductor property of the ZnO, the mechanism defined as bridging or tunnelling develops when the nanoparticle density exceeds a certain concentration [19,21]. In this case, layers adjacent to nanoparticles separate insulating oil and these layers act as a conductor in a very high electric field [20]. Approximately 10% reduction of breakdown voltage at a concentration of 0.4 g/L can be explained by this mechanism.

Breakdown voltages in nanofluids using ZnO nanoparticles can increase by 8.3% to 40.6% in different concentrations and fluids [21,22]. The positive and negative impulse breakdown voltage of natural ester-based ZnO nanofluids increased by 19.8% and 35.8%, respectively, for the 0.4 g/L ZnO concentration [20]. Despite the findings in these studies, the AC breakdown characteristics and withstand voltages of natural ester-based ZnO nanofluids are presented in detail with this study. The increase in AC breakdown and withstand voltages can be up to 5.8% and 22.7% for 0.1 g/L ZnO concentration, respectively.

The AC breakdown voltage averages of mineral oils commonly used in power transformers measured using the same method are 38.5 kV [16], 39.0 kV [28] and 51.6 kV [7]. The AC breakdown voltage measurement averages of synthetic esters under the same conditions are 47.0 kV [28] and 60.03 kV [5]. The electrical insulation characteristic of the natural ester, whose average AC breakdown voltage is measured as 66.67 kV, has a better performance than mineral oils and synthetic esters. This insulating performance can be improved with the addition of ZnO nanoparticles and provides a more reliable insulating medium alternative for the transformer.

The flash and fire points of natural esters are higher than mineral oils and synthetic esters. Due to these properties, the performance of power transformers using natural esters continues without deterioration even when exposed to high temperatures [6].

4. Conclusions

The main findings obtained in this study can be summarized as follows:

- In natural ester-based ZnO nanofluids, breakdown voltages decrease for 0.05 and 0.4 g/L concentrations and increase for 0.1 to 0.3 g/L concentrations. The best improvement in breakdown voltage is of about 5%; it is obtained with a concentration of 0.1 g/L ZnO.
- Breakdown voltage data of all samples comply with normal distribution. Using these distribution functions, risks of 1%, 10% and 50% probabilities of breakdown voltages are calculated. The best improvement in 1% probability withstand voltage is in 0.1 g/L ZnO nanofluid. At this concentration, the probability of breakdown voltage increases by 22.7%. The improvement at concentrations of 0.2 and 0.3 g/L for the same probability is 13.2% and 11.1%, respectively. For 10% probability, the best performance is also in the 0.1 g/L concentration. At this probability, the development of the breakdown voltage is 13%.
- This increase in breakdown voltages of ZnO-added natural esters provides the opportunity to design power system equipment, especially transformers, in smaller dimensions and to meet the increasing demand.
- It is thought that the critical value of the amount of nanoparticles is exceeded at a concentration of 0.4 g/L. Increasing nanoparticle concentration beyond this value reveals the implication of a tunnelling/bridging mechanism that leads to a reduction in breakdown voltages.

Author Contributions: Conceptualization, H.D. and A.B.; Data curation, H.D. and A.B.; Formal analysis, H.D. and A.B.; Investigation, H.D. and A.B., Methodology, H.D. and A.B.; Project administration, A.B.; Resources, H.D. and A.B.; Software, H.D.; Supervision, A.B.; Validation, H.D. and A.B.; Visualization, H.D.; Writing-original draft, H.D.; Writing-review & editing, A.B. All authors have read and agreed to the published version of the manuscript.

Funding: This research received no external funding.

Institutional Review Board Statement: Not applicable.

Informed Consent Statement: Not applicable.

Data Availability Statement: Data is contained within the article.

Acknowledgments: This work is supported by The Scientific and Technological Research Council of Turkey (TUBITAK) 2219 grant program.

Conflicts of Interest: The authors declare no conflict of interest.

References

1. Fofana, I. 50 years in the development of insulating liquid. *IEEE Electr. Insul. Mag.* **2013**, *29*, 13–25. [CrossRef]
2. Rafiq, M.; Lv, Y.; Li, C. A review on properties, opportunities, and challenges of transformer oil-based nanofluids. *J. Nanomater.* **2016**, *8371560*, 1–23. [CrossRef]
3. Wang, M.; Vandermaar, A.J.; Srivastava, K.D. Review of condition assessment of power transformers in service. *IEEE Electr. Insul. Mag.* **2002**, *18*, 12–25. [CrossRef]
4. Reffas, A.; Idir, O.; Ziani, A.; Moulai, H.; Nacer, A.; Khelfane, I.; Ouagueni, M.; Beroual, A. Influence of thermal ageing and electrical discharges on uninhibited olive oil properties. *IET Sci. Meas. Technol.* **2016**, *10*, 711–718. [CrossRef]
5. Khaled, U.; Beroual, A. AC dielectric strength of synthetic ester-based Fe_3O_4, Al_2O_3 and SiO_2 nanofluids—Conformity with normal and Weibull distributions. *IEEE Trans. Dielectr. Electr. Insul.* **2019**, *26*, 625–633. [CrossRef]
6. Stockton, D.P.; Bland, J.R.; McClanahan, T.; Wilson, J.; Harris, D.L.; McShane, P. Natural ester transformer fluids: Safety, reliability & environmental performance. In Proceedings of the IEEE Petroleum and Chemical Industry Technical Conference, Calgary, AB, Canada, 17–19 September 2007. [CrossRef]
7. Beroual, A.; Khaled, U.; Noah, P.S.M.; Sitorus, H. Comparative study of breakdown voltage of mineral, synthetic and natural oils and based mineral oil mixtures under AC and DC voltages. *Energies* **2017**, *10*, 511. [CrossRef]
8. Khaled, U.; Beroual, A. Statistical investigation of AC dielectric strength of natural ester oil-based Fe_3O_4, Al_2O_3, and SiO_2 nano-fluids. *IEEE Access* **2019**, *7*, 60594–60601. [CrossRef]
9. Mahanta, D.K.; Laskar, S. Electrical insulating liquid: A review. *J. Adv. Dielectr.* **2017**, *7*, 1730001–1/9. [CrossRef]
10. Ahmad, F.; Khan, A.A.; Khan, Q.; Hussain, M.R. State-of-art in nano-based dielectric oil: A review. *IEEE Access* **2019**, *7*, 13396–13410. [CrossRef]
11. Sitorus, H.B.H.; Setiabudy, R.; Bismo, S.; Beroual, A. Jatropha curcas methyl ester oil obtaining as vegetable insulating oil. *IEEE Trans. Dielectr. Electr. Insul.* **2016**, *23*, 2021–2028. [CrossRef]
12. Jaroszewski, M.; Beroual, A.; Golebiowski, D. Effect of temperature on dielectric loss factor of biodegradable transformer oil. In Proceedings of the IEEE International Conference on High Voltage Engineering and Application (ICHVE), Athens, Greece, 11–13 September 2018. [CrossRef]
13. Tenbohlen, S.; Koch, M.; Vukovic, D.; Weinlader, A.; Baum, J.; Harthun, J.; Schafer, M.; Barker, S.; Fotscher, R.; Dohnal, D.; et al. Application of vegetable oil-based insulating fluids to hermetically sealed power transformers. *Cigre Session* **2008**, *42*, 24–29.
14. Fofana, I.; Wasserberg, V.; Borsi, H.; Gockenbach, E. Challenge of mixed insulating liquids for use in high-voltage transformers. 1. Investigation of mixed liquids. *IEEE Electr. Insul. Mag.* **2002**, *18*, 18–31. [CrossRef]
15. Choi, S.U.S.; Eastman, J.A. Enhancing thermal conductivity of fluids with nanoparticles. In Proceedings of the ASME International Mechanical Engineering Congress and Exposition, San Francisco, CA, USA, 12–17 November 1995; pp. 66–74.
16. Khaled, U.; Beroual, A. AC dielectric strength of mineral oil-based Fe_3O_4 and Al_2O_3 nanofluids. *Energies* **2018**, *11*, 3505. [CrossRef]
17. Khaled, U.; Beroual, A. Comparative study on the AC breakdown voltage of transformer mineral oil with transformer oil-based Al_2O_3 nanofluids. In Proceedings of the IEEE International Conference on High Voltage Engineering and Application (ICHVE), Athens, Greece, 10–13 September 2018. [CrossRef]
18. Azizie, N.A.; Hussin, N. Preparation of vegetable oil-based nanofluid and studies on its insulating property: A review. *J. Phys. Conf. Ser.* **2020**, *1432*, 012025. [CrossRef]
19. Chen, J.; Sun, P.; Sima, W.; Shao, Q.; Ye, L.; Li, C. A promising nano-insulating-oil for industrial application: Electrical properties and modification mechanism. *Nanomaterials* **2019**, *9*, 788. [CrossRef]
20. Chen, G.; Li, J.; Yin, H.; Huang, Z.; Wang, Q.; Liu, L.; Sun, J.; He, J. Analysis of dielectric properties and breakdown characteristics of vegetable insulating oil with modified by ZnO nanoparticles. In Proceedings of the IEEE International Conference on High Voltage Engineering and Application (ICHVE), Athens, Greece, 10–13 September 2018. [CrossRef]
21. Hanai, M.; Hosomi, S.; Kojima, H.; Hayakawa, N.; Okubo, H. Dependence of TiO_2 and ZnO nanoparticle concentration on electrical insulation characteristics of insulating oil. In Proceedings of the Annual Report Conference on Electrical Insulation and Dielectric Phenomena, Shenzhen, China, 20–23 October 2013; pp. 780–783. [CrossRef]
22. Bakrutheen, M.; Karthik, R.; Madavan, R. Investigation of critical parameters of insulating mineral oil using semiconductive nanoparticles. In Proceedings of the International Conference on Circuits, Power and Computing Technologies (ICCPCT), Nagercoil, Indie, 20–21 March 2013; pp. 294–299. [CrossRef]
23. Kong, L.; Sun, J.; Bao, Y. Preparation, characterization and tribological mechanism of nanofluids. *Rsc Adv.* **2017**, *7*, 12599. [CrossRef]
24. IEC 60156. *Insulating Liquids—Determination of the Breakdown Voltage at Power Frequency—Test Method*; IEC 60156 Ed. 2; International Electrotechnical Commission (IEC): Geneva, Switzerland, 1995.
25. Anderson, T.W.; Darling, D.A. A test of goodness of fit. *J. Am. Stat. Assoc.* **1954**, *49*, 765–769. [CrossRef]
26. Shapiro, S.S.; Wilk, M.B. An analysis of variance test for normality (complete samples). *Biometrika* **1965**, *19*, 591–611. [CrossRef]
27. Sitorus, H.B.H.; Beroual, A.; Setiabudy, R.; Bismo, S. Statistical analysis of AC and DC breakdown voltage of JMEO (jatropha methyl ester oil), mineral oil and their mixtures. In Proceedings of the IEEE 19th International Conference on Dielectric Liquids (ICDL), Manchester, UK, 25–29 June 2017. [CrossRef]

28. Martin, D.; Wang, Z.D. Statistical analysis of the AC breakdown voltages of ester based transformer oils. *IEEE Trans. Dielectr. Electr. Insul.* **2008**, *15*, 1044–1050. [CrossRef]
29. Jian-quan, Z.; Yue-fan, D.; Mu-tian, C.; Cheng-rong, L.; Xiao-xin, L.; Yu-zhen, L. AC and lightning breakdown strength of transformer oil modified by semiconducting nanoparticles. In Proceedings of the 2011 Annual Report Conference on Electrical Insulation and Dielectric Phenomena, Cancun, Mexico, 16–19 October 2011; pp. 652–654. [CrossRef]
30. Du, Y.; Lv, Y.; Li, C.; Chen, M.; Zhong, Y.; Zhou, J.; Li, X.; Zhou, Y. Effect of semiconductive nanoparticles on insulating performances of transformer oil. *IEEE Trans. Dielectr. Electr. Insul.* **2012**, *19*, 770–776. [CrossRef]

energies

MDPI

Article

Investigation of Survival/Hazard Rate of Natural Ester Treated with Al$_2$O$_3$ Nanoparticle for Power Transformer Liquid Dielectric

Raymon Antony Raj *, Ravi Samikannu, Abid Yahya and Modisa Mosalaosi

Department of Electrical, Computer and Telecommunications Engineering, Faculty of Engineering and Technology, Botswana International University of Science and Technology, Private Bag 16, Palapye Plot 10071, Botswana; ravis@biust.ac.bw (R.S.); yahyaa@biust.ac.bw (A.Y.); mosalaosim@biust.ac.bw (M.M.)
* Correspondence: raymonhve@gmail.com

Abstract: Increasing usage of petroleum-based insulating oils in electrical apparatus has led to increase in pollution and, at the same time, the oils adversely affect the life of electrical apparatus. This increases the demand of Mineral Oil (MO), which is on the verge of extinction and leads to conducting tests on natural esters. This work discusses dielectric endurance of Marula Oil (MRO), a natural ester modified using Conductive Nano Particle (CNP) to replace petroleum-based dielectric oils for power transformer applications. The Al$_2$O$_3$ is a CNP that has a melting point of 2072 °C and a low charge relaxation time that allows time to quench free electrons during electrical discharge. Al$_2$O$_3$ is blended with the MRO and Mineral Oil (MO) in different concentrations. The measured dielectric properties are transformed into mathematical equations using the Lagrange interpolation polynomial functions and compared with the predicted values either using Gaussian or Fourier distribution functions. Addition of Al$_2$O$_3$ indicates that 0.75 g/L in MRO has an 80% survival rate and 20% hazard rate compared to MO which has 50% survival rate and 50% hazard rate. Considering the measured or interpolated values and the predicted values, they are used to identify the MRO and MO's optimum concentration produces better results. The test result confirms the enhancement of the breakdown voltage up to 64%, kinematic viscosity is lowered by up to 40% at 110 °C, and flash/fire points of MRO after Al$_2$O$_3$ treatment enhanced to 14% and 23%. Hence the endurance of Al$_2$O$_3$ in MRO proves to be effective against electrical, physical and thermal stress.

Keywords: power transformer; mineral oil; natural ester; interpolation; mathematical modeling

Citation: Raj, R.A.; Samikannu, R.; Yahya, A.; Mosalaosi, M. Investigation of Survival/Hazard Rate of Natural Ester Treated with Al$_2$O$_3$ Nanoparticle for Power Transformer Liquid Dielectric. *Energies* **2021**, *14*, 1510. https://doi.org/10.3390/en14051510

Academic Editor: Pawel Rozga

Received: 10 February 2021
Accepted: 1 March 2021
Published: 9 March 2021

Publisher's Note: MDPI stays neutral with regard to jurisdictional claims in published maps and institutional affiliations.

1. Introduction

The power transformer (PT) is vulnerable and is the most expensive power system network equipment. It provides functions such as stepping up and stepping down the voltage between electrical power generating stations. The PT works at 100% load for 24 h throughout the day whereas the distribution transformer works at 50% or 70% of the full load [1]. Increase in the power demand across the world is persistent and it is expected to be double in the forthcoming years as per the predicted data of International Energy Associations [2]. This demand has caused the rise in PT numbers in the power system network. The report released by allied market research indicates that countries in the Asia-Pacific (APAC) are the biggest market for the future transformer industries [3]. Here, a projected growth of USD 3 million is expected by 2025, which almost strikes a peak of 6.9% Compound Annual Growth Rate (CAGR) from the year 2020 [4]. This growth is brought about by the huge demand for power, rapid urbanization, and PT's replacement. The report published by the Allied Market Research estimated that the consumption of MO in 2014 was nearly 1437.8 million litres [5]. It further predicted that MO's sale was to reach 3.4 billion United States dollar by 2020 which represents a 6.3% increase from 2015. Upon extrapolating the data with the same growth rate up to 2030, the consumption of MO will be 2815.56 million litres. MO will dominate the insulating oil market up to the year 2025. Even though there are several alternatives to MO in the market, bio-based insulating fluids have their shortcomings on the dielectric properties [6]. Insulating oil is an integral part of the power utility with applications in power and distribution transformers, power cables, bushings, and circuit breakers. Natural ester proves to be a suitable candidate with higher moisture tolerance than MO [7]. The use of mineral oil in the transformer as a coolant and insulator is well documented [8,9]. MO's availability is limited since it is extracted from a non-renewable source. Its lack of biodegradability has also led to a debate regarding its usage in recent times [10]. Mineral oil is a fossil fuel (crude oil) made up of hydrocarbons and compounds which are refined through a distillation process by boiling and refining for suitable use in transformers [11]. Numerous oils, like paraffin, naphtha, aromatic, to name a few, are used in various ratios and components [12–14]. Silicones are inert synthetic liquids, known as polydimethylsiloxane, are thermally stable with a composition exhibiting electrical properties like mineral oils with a similar structure as the methyl organic group [15,16]. High molecular coolants, either natural or synthetic are classified by National Electric Code (NEC) as less inflammable with a fire-point below 300 °C [5]. Synthetics are generally polymerized olefins popularly known as polyalphaolefins or PAOs. Both silicones and high molecular coolants are comparable in performance with MO. However, they differ in their temperature endurance, viscosity, and pour point [16,17].

Esters are alternatives to the mineral oils and are generally either synthesized chemically from organic precursors or natural oils [18–20]. Researchers further added alternative insulating fluids from natural esters, food-grade synthetic (Butylated Hydroxy Toluene (BHT), Butylated Hydroxy Anisole (BHA), Propyl Gallate (P.G.)), natural antioxidants (α, β-Tocopherols (αT, βT)) and synergists (Citric Acid (C.A.), Ascorbic Acid (A.A.), Rosemary extracts (R.A.)) which contribute significantly [21]. Among the several synthetic esters, Pentaerythritol, a polyol, is found to have suitable dielectric properties, good thermal stability, low-temperature tolerance and more biodegradable as compared to mineral oils [22,23]. Natural esters compose of chemically stable fatty acids with high viscosity [24]. The commercialization of vegetable oil as coolant resulted from significant research to find a suitable, fully biodegradable insulating medium. The use of vegetable oil as biodegradable, moisture absorbent, coolant and insulator in transformers started as a research project as early as the 1990s. Vegetable oils are known to be highly biodegradable (>95%), less toxic, high flash and fire points (>300 °C) as compared to mineral oils [24]. Vegetable oils possess very high moisture and viscosity than MO. Vegetable oil is not stable like MO when subjected to temperature stresses.

The process of extracting vegetable oil starts with the process of seeding. The seeds are refined, bleached, and deodorized to form pure vegetable oils. The first step is to refine the oil using alkaline refinement to remove fatty acids, followed by bleaching to eliminate the coloring materials and the deodorization. The volatile and odoriferous materials are removed by using vacuum steam distillation at high temperatures [25]. Vegetable oils such as sunflower, rice bran, rapeseed, palm, coconut, and soybean, have been studied extensively on their dielectric nature for transformer insulation and cooling. Natural oil age monitoring in the transformers should be based on various dielectric properties such as dielectric strength, dissipation factors, acidity, dissipation factor [5,17]. The degradation rate of vegetable oil can be compared to that of mineral oil. The relationship between hottest spot temperature and long-term operations for both mineral oil and vegetable oil-based transformer are determined under medium voltages.

The work discusses statistical approach to understand the performance of the insulating fluid that is normally required when the field testing is complicated. Such analysis is meaningful in performing reliability analysis to determine the time to failure. Moreover, the statistical analysis helps to realize the temperament of insulating fluid under accelerated heating conditions. This approach used in the research is simple and it derives mathematical equations using interpolation function, which helps the power engineer to recognize the actual condition of the insulating oil without the need to test the oil at field conditions. The functions are used to derive the bath-tub curves used in the reliability analysis, but in our case, we focus on the response of oils to nanoparticle and their stability to accelerated ageing.

2. Key Materials of Research

2.1. Marula Oil (MRO)

Marula or Sclerocarya birrea is a drought-resistant plant species, found in Southern Africa, Botswana. The tree can grow a diameter of 2 m and produces fruits [26]. The nuts are crushed to produce MRO, which has alcohol content range from 1% to 7% [27]. The oil contains amino acids, fatty acids, oleic and linoleic, flavonoids, catechins, and procyanidin and high amounts of antioxidants [27]. Oleic acid in MRO has a melting point of 360 °C, which helps the antioxidants to resist high-temperature stress. MRO naturally contains oleic acid, which shows high temperament against temperature stress. In addition, the oil contains vitamin-E (tocopherol), which naturally inhibits free electrons during the discharge phenomena [24]. Natural esters containing antioxidants helps to prevent formation of valence electrons during electron avalanche. Thus, natural ester retards ionization delays breakdown in oil. On the other hand, the procyanidin in the oil acts as an excellent antioxidant and inhibits reactive oxygen, thereby preventing the oil from ionization.

2.2. Conductive Nanoparticle (CNP)

The theory of nanoscience has its application as a drug carrier in various fields of engineering and health. These nano-sized particles (10 nm) are highly stable, and it is complicated to predict their physio-chemical characteristics. The application of CNP in the transformer liquid dielectric starts with the nanoparticle's ability to capture free electrons during the discharge or streamer application [28]. Discharge or streamer is initiated by the ionization; a process by which valence electrons are separated from shells and start bridging a conductive path between electrodes. This establishes a short-circuited path in liquid dielectric and results in dielectric breakdown [21,24]. The CNP traps the fast-moving electrons during the discharge mechanism and slows the process either by neutralizing or slowing down the ionization. This increases the breakdown strength of the liquid dielectric and reduces ageing. Unlike magnetic and semi-conductive nanoparticles, CNP is influenced by the direction of the electric field. This leads to CNP coming into action whenever there exists a strong field between the two electrodes [29]. Al_2O_3 shows very low charge relaxation time, metal chelation, and electron scavenging properties. This retards free electrons that are responsible for electron avalanche, reducing energy density of avalanche chain. It is also known that CNP improves the breakdown voltage at a concentration of 0.5 g/L in the host fluid as mentioned by Raymon et al. [28]. Al_2O_3 possesses amphoteric structure and it does not react with water in the presence of Natural Esters. Since Al_2O_3 is slower to saturate in water, it is used for adsorption purification of oils as adsorbent to capture hydrocarbon impurities from the air. The effect of Al_2O_3 has been well utilized in this study to understand the behavior in MRO and MO, and its suitability as an additive in liquid dielectric coolant for the transformer.

3. Dielectric Behavior of Host Fluids

MRO and inhibited MO are collected from Botswana and India, respectively. Both MRO and MO are considered as host fluids, and their dielectric characteristics are measured as per IEC (International Electrotechnical Commission) [30] and ASTM (American Society for Testing and Materials) International [31–34] standards and presented in Table 1.

The dielectric characteristics like breakdown voltage, kinematic viscosity and flash/fire point are considered in the statistical study. These parameters are much likely to influence other parameters of the liquid dielectric; hence, it is necessary to investigate the behavior analytically. The analysis involves the estimation of survival rate and hazard rate by populating the data points of the dielectric characteristics of base fluids as density and cumulative probability distributions.

Table 1. Dielectric characteristics of host fluids.

Host Fluids	Breakdown Voltage (KV)	Kinematic Viscosity (cSt)	Pour Point (°C)	Flash Point (°C)	Fire Point (°C)	Thermal Conductivity (W/m-K)	Moisture Content (ppm)	Acidity (mg KOH/g)
MRO	47	52	−23	225	237	0.182	84.237	0.020
MO	32	23	−17	153	158	0.130	<9.561	0.010

Experimental Procedure of the Investigated Dielectric Properties

The study makes use of the dielectric parameters such as breakdown voltage, kinematic viscosity, flash point, fire point to validate them with hazard function and survival function. Breakdown voltage of the liquid dielectric is defined as the maximum withstand voltage capacity of the liquid dielectric under standard room temperature and pressure. Usually, the breakdown voltage is measured by filling the measuring cup (test cup) with 500 mL of the liquid, spherical electrodes fully submerged in the liquid. Gap spacing between electrodes is 2.5 mm and 50 Hz ac test voltage is increased using the control knob at a rate of 2 kV/s until breakdown. The voltage is recorded as breakdown voltage of the sample. Between the successive measurements, a 2 min relaxation time is given, and the test cup is inspected for the carbon formation, then the oil is slowly stirred to avoid formation of bubbles. The test is repeated according to IEC 60156 standards and measurements are recorded [30]. Similarly, the kinematic viscosity is measured using a Redwood viscometer. Kinematic viscosity is defined as the shear stress of the liquid against its flow. The test cup of the Redwood viscometer is filled with the liquid and placed on the water bath, which is heated using the electric heater. The liquid temperature is measured using the thermometer setup and the orifice is controlled manually to drain the oil to a vessel placed under the viscometer. The time of flow is measured using the stopwatch and kinematic viscosity is calculated by considering the pipette and burette constants according to ASTM D445 standards [32]. On the other hand, the fire point and flash point are measured using the Pensky–Martens closed cup apparatus according to ASTM D93 standards [33]. The test cup is filled with the oil and placed in a water bath then heated using an electric heater. The temperature of the liquid is measured using a thermometer and flame is introduced to a small opening in the top surface of the test cup. At some temperature, the liquid begins producing vapour that can be a combustible source when oxygen is introduced to the source in the presence of a fire source. This phenomenon is termed as flash point (smoke point). Likewise, when the combustion becomes continuous for at least 5 s, it is called as fire point (ignition point). Although there are no direct connections between the dielectric properties, they can be individually affected by factors like concentration of the nanoparticles or temperature, which is used to describe the changes in the dielectric properties.

4. Assessment of Breakdown Voltage of Host Fluids

The dielectric breakdown voltage of MRO and MO is distributed as a density plot shown in Figure 1. Readings were taken two times a day for 30 successive days. During this period, the samples are stored in a dark container to protect the fluid from photo oxidation and dust particles. Density plot is a continuous plot of variables that are random in space. The breakdown strength of MRO is higher than that of MO, with MRO showing a wider range of breakdown voltage as compared to MO. MRO's breakdown voltage limits are from 36 kV to 47 kV while MO falls between 22 kV and 33 kV. The maximum strength of MRO breakdown voltage lies between 45 kV to 47 kV. On the other hand, MO shows endurance between 22 kV to 25 kV and 30 kV to 33 kV. This indicates that most of the breakdown strength's measured values lie within these indicated regions for both MRO and MO. Breakdown endurance is due to the presence of CNP that retards free electrons detached from valence shell and then reduces the energy density of the electron chain between electrodes. Figure 2 illustrates the cumulative probability of random breakdown voltage of MRO and MO that occur at different voltage instances (sequential events). The sequential events are independent of each other, and it is critical not to have two events occur simultaneously [35].

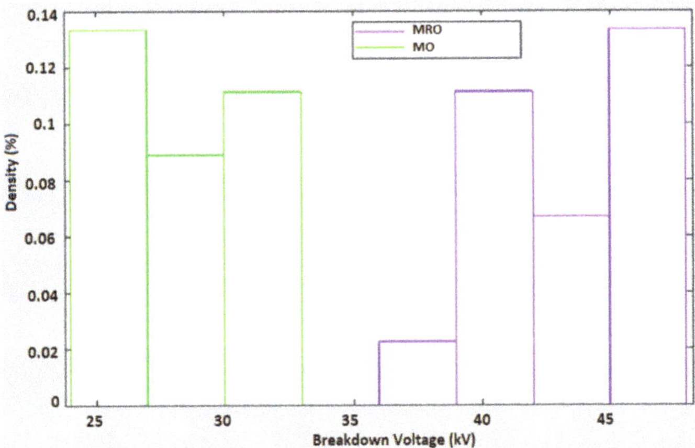

Figure 1. Showing the different ranges of breakdown voltage at which, the MRO and MO have its % density.

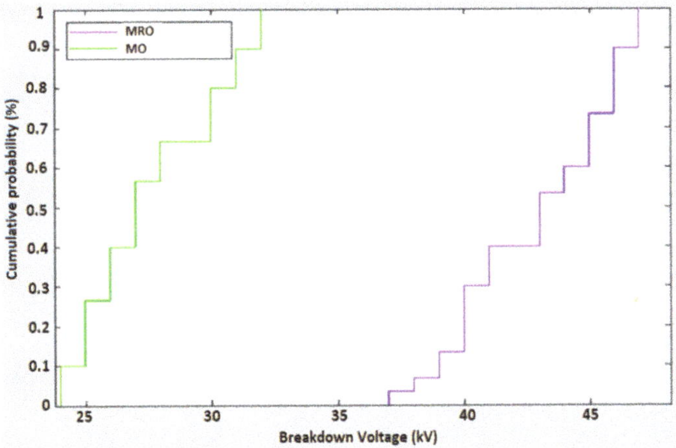

Figure 2. Cumulative probability of breakdown instances of MRO showing wider range and MO showing linear range.

A wider range of breakdown voltage of the MRO indicates that it experiences a lesser effect on the successive breakdown events. The density plot also shows a quick recovery between breakdown events. Such recovery is due to the presence of natural antioxidants in MRO. Hence, it is hypothetical and practical to conclude that MRO has a lesser propensity to electrical stress. Reliability analysis such as survival function and hazard function are used in this study to understand and compare the endurance of the liquid dielectrics. Survival function determines the number of data points that survive over time or in other words the data points that fail in the expected duration of time. Survival function is determined using the Kaplan–Meier function. A simple way to calculate the distribution of the survival function is through a hypothesis which evaluates the risk over a constant time i.e., $\lambda(t) = \lambda$. From the expression of distribution function; $F(t) = 1 - \exp(-\lambda t)$, the survival function is; $S(t) = \exp(-\lambda t)$, can be expressed as $S(t) = 1 - F(t)$. Likewise, hazard function indicates the rate at which the data points experience hazard over time or in other words it is the likelihood or frequency of failure per unit time. The above survival and hazard functions are used to determine the reliability of the liquid dielectric. Such mathematical analysis is therefore essential to realize the bathtub function of the liquid dielectric. When the distribution becomes exponential, the hazard function becomes; $\lambda(t) = F(t)/S(t)$ [36]. According to Figure 3, the proportion of MRO surviving 40 kV is 70%. On the other hand, 80% of the time MO survives only 26 kV and 20% of the time it survives 30 kV. MRO shows 50% survivalat 44 kVwhile MO is at 28 kV. This indicates that MRO has a superior survival rate than MO. From Figure 4, at 50% cumulative hazard, MRO is less likely to fail until it reaches a breakdown voltage of 45 kV. It is also evident that MO experiences a 50% hazard at 32 kV and above voltages. Both $\lambda(t)$ and $H(x)$ from Figures 3 and 4 shows some data exceeding the confidence bounds of MO. This clearly indicates that MRO is likely to follow the upper survival bounds. Most of the time, it is less likely to fail above 40 kV. Here, the Weibull parameters are obtained from the generated Figures 3 and 4 with a shape parameter "K > 1". This shows the rate of failure increases with time as ageing goes on with time. Similarly, the scale parameter "λ" is assumed to be greater than "0" to spread the distribution evenly.

Figure 5 shows the failure probability of MRO and MO for 30 days from the breakdown voltage tests as per IEC standard [30]. The probability of failure of the dielectric fluid is the ability of the fluid to show a constant breakdown strength during successive testing. The probability of failure is measured at the breakdown voltages of MRO (47 kV) and MO (32 kV). Typically, MRO shows no changes upto 5 days, while MO shows 40% probability to breakdown from day 1 of testing. By day 15, MO has 87% breakdown probability which is a 47% rise from day 1 while MRO has a 60% rise for the same period. There is an inflexion point on day 23, where MRO shows the same rate of failure as MO. MO has shown slower growth in failure rate even though its probability of failure was initially 40% higher than that of MRO. This indicates that electrical stress has a greater impact on MRO than MO from the inflexion point on day 23. In this case, MRO needs enhancement by adding a CNP such as Al_2O_3.

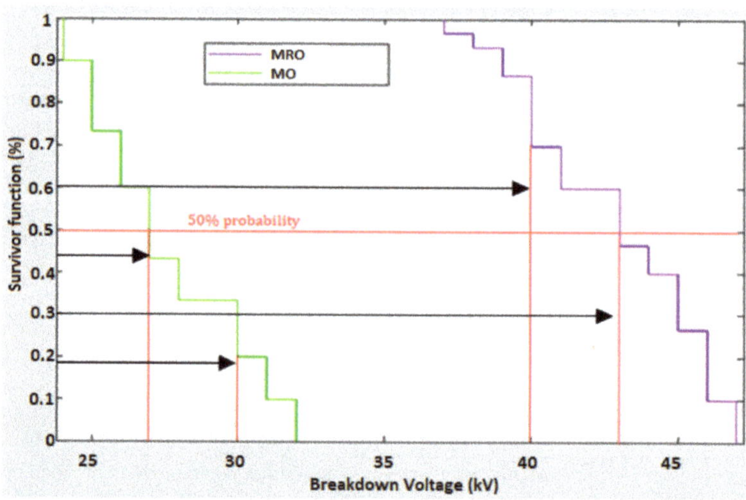

Figure 3. Survival function showing the cumulative distribution of MRO and MO, for any positive number "T" surviving over the time "t".

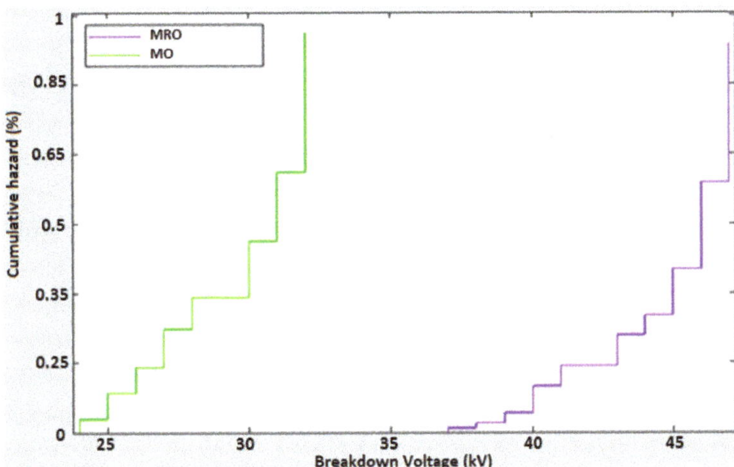

Figure 4. Cumulative hazard of MRO and MO showing the integral of hazard function $H(x) = -\ln(1 - f(x))$ which shows the probability of failure at time x has given survival until time x.

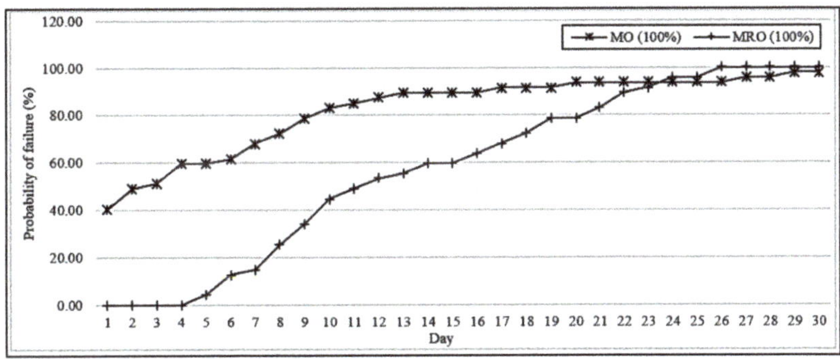

Figure 5. Showing the endurance of breakdown voltage and probability of failure of MRO and MO for 30 days.

5. Preparation of the Nanofluids

The nanofluids are prepared by amalgamation of MRO and MO with nano sized Al_2O_3 in different concentrations from 0.1 g/L to 2 g/L with concentration increase interval of 0.25 g/L. They are heated up to 100 °C and treated in an ultrasonication bath with 30 kHz as mixing frequency. The oil moisture is removed during the heating process. The nanofluid is collected in a closed container after the procedure and kept at room temperature out of reach of sunlight to avoid ultraviolet and photo oxidation. The nanofluid prepared using MRO shows a fine dispersion of Al_2O_3 since it contains oleic acid which naturally allows the additive to float in the host fluid. Comparatively, the MO-based nanofluid shows agglomeration after 5 h and needs repeated dispersion using the ultrasonication bath.

5.1. Nanoparticle Impact on the Breakdown Voltage

The impact of nanoparticle on the breakdown voltage of MRO and MO is presented for varying concentrations starting from 0.1 g/L, and 0.25 g/L increment upto 2 g/L. The addition of Al_2O_3 has resulted in improved breakdown voltage of the host fluids as seen from Table 2.

Table 2. Enhancement of Host Fluids using Al_2O_3 Nanoparticle.

Concentration of Al_2O_3 (g/L)	Breakdown Voltage of MRO (kV)	% Enhanced	Breakdown Voltage of MO (kV)	% Enhanced
0	47	0	32	0
0.1	51	7.84	34	5.88
0.25	54	12.96	37	13.51
0.5	59	20.34	39	17.95
0.75	64	26.56	42	31.91
1	60	21.67	40	27.27
1.25	58	18.97	38	21.95
1.5	54	12.96	33	20.00
1.75	45	−4.44	31	11.11
2	39	−20.51	30	−6.67

There is a constant rise in the breakdown voltage when Al_2O_3 is added from the concentration of 0.1 g/L to 0.75 g/L and it begins to decline after further additions of Al_2O_3. This variation is seen from Table 2 where the highest enhancement of breakdown voltage is measured at 0.75 g/L with 26.6% for MRO and with 31.2% for MO. A composite cross-section of liquid-nanoparticle helps to reduce the random nature of breakdown of fluid by energy reduction achieved by field grading. This implies that breakdown is an extreme-value process. For further additions of Al_2O_3 at 1.75 g/L, enhancement is lost for MRO, showing a decline of 4.4% from the host's original value. A similar behaviour is measured for MO at 2 g/L with a 6.7% decline. Both oils do not experience a major stochastic change in the breakdown voltage with the addition of CNP. % enhancement from Table 2 indicates that addition of nanoparticles increases the electron trapping density of the liquid when an ac breakdown voltage is applied between the electrodes. The increase in the kinematic viscosity by the addition of CNP above 0.75 g/L distracts the capture cross-section of the liquid molecules. As a result, the mean free path is very short for breakdown to occur, which gradually decreases breakdown voltage after 0.75 g/L concentration of CNP. A polynomial interpolation function is developed by considering the breakdown voltage of MRO and MO with the concentration of the Al_2O_3. The Lagrange interpolation polynomial is found for a given set of different breakdown voltage readings. The polynomial function $P(x)$ is calculated for points 'x_j' which is given in Equations (1) and (2).

$$P(x) = \sum_{j=1}^{n} P_j(x) \tag{1}$$

$$P_j(x) = y_j \prod_{k=1, k \neq j}^{n} \frac{x - x_k}{x_j - x_k} \tag{2}$$

The Lagrange Interpolation helps to find the polynomial which takes certain values at arbitrary points. When more data points are used for developing a polynomial function, greater data turbulence is observed between data points. The breakdown voltage data is used to develop a mathematical function by concentrating on four limits for reducing the polynomial and data turbulence order. Equations (3) and (4) represent the Lagrange polynomial interpolation functions of MRO and MO, respectively, as calculated for the breakdown voltage variations for the addition of Al_2O_3. Here, $U(x)$ is breakdown voltage at the instant and U is the maximum withstand capacity of oil. The generated generalized form of polynomial function for MRO and MO is given in Equation (5). Equation (5) is then used to estimate arbitrary values of the breakdown voltage of the host fluids transformed using the nanoparticle.

$$\frac{U(x)}{U} = 0.3977x^4 - 1.567x^3 + 1.4324x^2 + 0.1364x^1 + 1 \tag{3}$$

$$\frac{V(x)}{U} = 0.1873x^4 - 0.8746x^3 + 0.8906x^2 + 0.1874x^1 + 1 \tag{4}$$

$$\frac{U(x)}{U} = \int_{x=0}^{x=2} K1x^4 + K2x^3 + K3x^2 + K4x^1 + C \, dx \tag{5}$$

$$f(x) = a1 * exp\left(-\left(\frac{x - b1}{c1}\right)^2\right) \tag{6}$$

A Gaussian function shown in Equation (6) is a function for predicting arbitrary real constants. It is a probability density function that calculates the value of the distribution function for the specified concentrations between a range. Here, the concentration is a random value "x", which is usually distributed as seen from Figure 6a,b. Though the data points are typically distributed, from Table 3, the measured maximum breakdown voltage is higher than the predicted voltages. The predicted values face deviation of 0.82 kV and 0.24 kV for MRO and MO, respectively, with low root mean square error (RMSE). This shows that there is no data disorder for the confidence interval of 95%. Lagrange interpolation is useful in finding arbitrary points and the development of the mathematical functions. At 95% confidence, a maximum conformity (R^2) is achieved in Lagrange interpolation polynomial function. Another reason for using the Lagrange interpolation polynomial is that for a given set of points (with no two values equal), the function assumes the lowest degree at each value the corresponding value. This helps the functions to coincide with each other at each point. Other interpolation functions seem to be not fitting well with the arbitrary points.

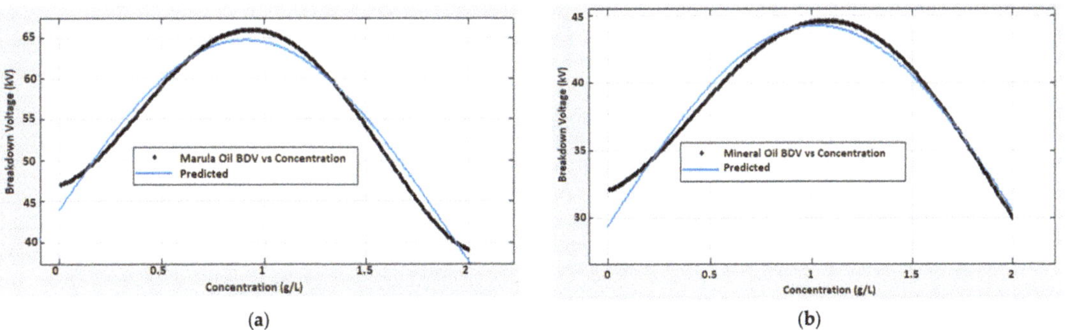

Figure 6. Measured breakdown voltage distribution using Lagrange polynomial function is compared with the predicted breakdown voltage distribution using Gaussian function. (a) MRO, (b) MO.

Table 3. Variation of Gaussian prediction of the breakdown voltage of MRO and MO with 95% confidence bound with conformity more than 98%.

Oil	Confidence	R^2	RMSE	Max Measured (kV)	Max Predicted (kV)	Deviation (kV)
MRO	95%	0.98	0.97	65.94	65.12	0.82
MO	95%	0.98	0.61	44.52	44.28	0.24

Assessment of Survival and Hazard Functions of Breakdown Voltage

The density plot of the breakdown voltage for the fluids with varying concentrations of Al_2O_3 is presented in Figure 7a,b. From Figure 7a, the frequency of breakdown voltage for MRO is highest when 0.75 g/L of the nanoparticle is added, and density (%) begins to rise for the breakdown voltage from 59 to 67 kV.

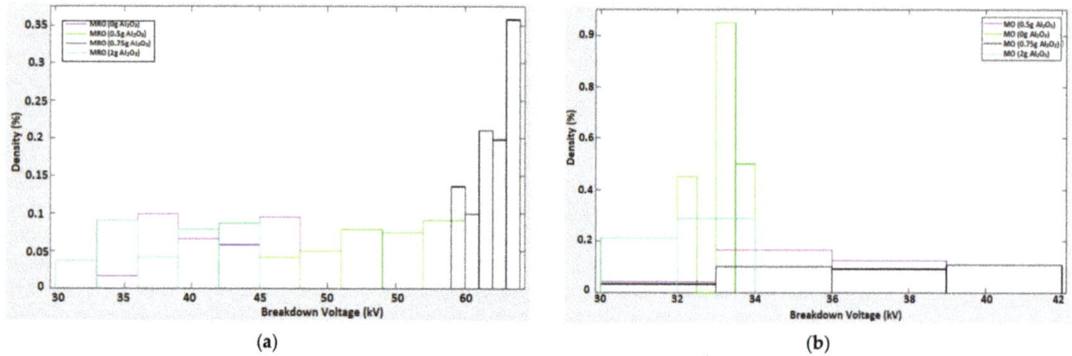

Figure 7. % Density function of the breakdown voltage in different concentration of the Al2O3, (**a**) MRO, (**b**) MO.

Similarly, without the addition of nanoparticles MRO has the lowest density of 10% with a range of 33 to 47 kV while a maximum of 6% density is observed for the 0.5 g/L addition with a range of 45 to 60 kV. The least density is obtained with 2 g/L addition for the range of 32 to 45 kV. This explains that MRO's breakdown voltage with the addition of 0.75 g/L Al_2O_3 has the highest density. The MO's highest density from Figure 7b is observed at 0 g/L, showing the breakdown voltage range of 32 to 34 kV. Unlike MRO, MO shows a wider density range for the concentration of 0.75 g/L of Al_2O_3, covering the range 30 to 42 kV with less than 15% probability that can be seen from Figure 8a,b. Like in MRO, the addition of 0.75 g/L of Al_2O_3 in MO shows a reduction in the dielectric breakdown voltage. At 50% of cumulative probability, MRO's data points with 0 g/L Al_2O_3 are likely to fail at 40 kV, showing a wider range from Figure 8a compared to MO observed from Figure 8b. Similarly, MO's breakdown voltage is 32 kV for 2 g/L, which is very close to the 30 kV for the 0 g/L concentration. At 0.5 g of Al_2O_3, MRO shows higher strength. A maximum of 62 kV obtained within the range of 59 to 62 kV, breakdown is evident with the addition of 0.75 g/L Al_2O_3, which shows a narrow probability for MRO. For the 0.75 g concentration, MO shows 37 kV to 42 kV with 50% cumulative probability as seen from Figure 8b, below which the MO is very likely to fail. At 0.75 g/L, MRO shows a 26.56% breakdown voltage rise from Table 3 compared to the MO's breakdown voltage. Above 0.75 g/L concentration of Al_2O_3, the fluids' breakdown voltage begins to decrease.

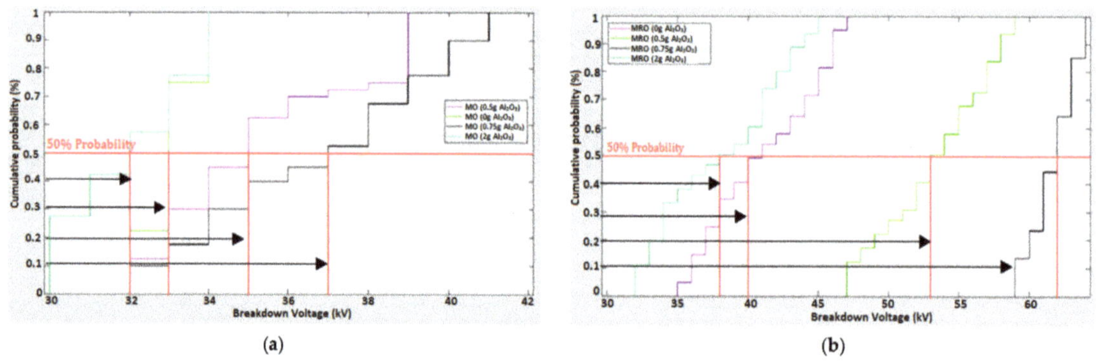

Figure 8. Breakdown voltage density function in different concentration of the Al_2O_3, (**a**) MO, (**b**) MRO.

The cumulative hazard function for the breakdown voltage of MRO and MO under varying concentrations of Al_2O_3 is presented in Figure 9a,b. There are two to three outbounds of the cumulative hazard as seen in Figure 9a for Al_2O_3 concentrations of 0 g/L, 0.5 g/L, and 0.75 g/L. For the 2 g/L concentration as seen from Figure 9a,b, with both overreaching upper and lower limits of Weibull distribution, the hazard rate is 70%. The data for MRO at 0.5 and 0.75 g/L concentrations has closely followed the Weibull distribution as shown in Figure 9a.

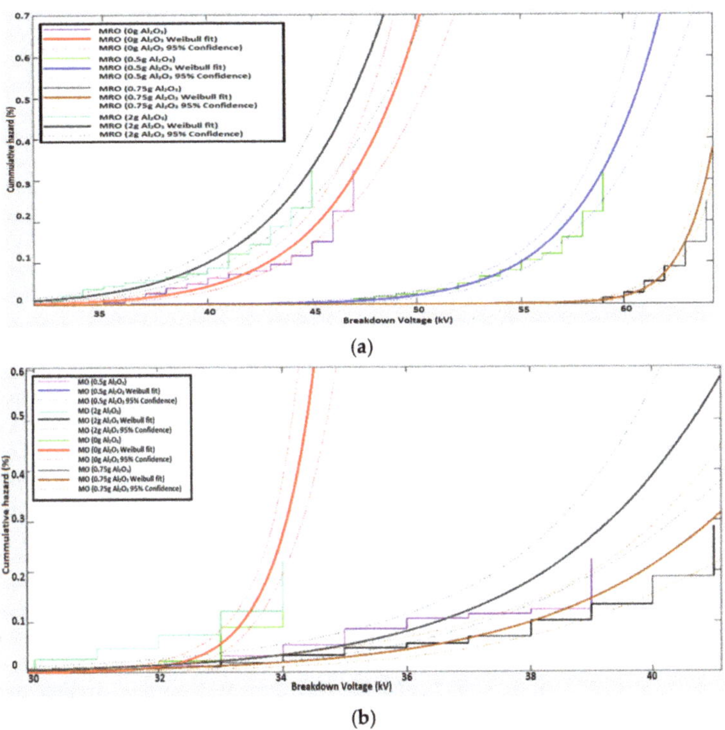

Figure 9. Cumulative breakdown voltage hazard function in different concentration of the Al_2O_3 with Weibull fit showing the 95% confidence, (**a**) MRO, (**b**) MO.

Moreover, 0.75 g/L addition shows a narrow distribution which is less likely to survive than the other concentrations. In the event of cumulative breakdown voltage, the hazard rate for MO is higher as seen in Figure 9b for the 2 g/L Al_2O_3 concentration. Only in the 0.5 g/L and 0.75 g/L concentrations of Al_2O_3 the hazard functions follow the Weibull distribution with wider extremities out of the 95% confidence limit. It is also observed that the survivors of MRO are higher than those of MO from Figure 10a,b. The data of MO is completely out of the 95% confidence limits for 0 g/L and 2 g/L concentrations of Al_2O_3. From observation, it can be concluded that the wider the range of the survivor breakdown voltage, the lesser the probability of survival as can be seen from Figures 9b and 10b. However, the trend is narrow for MRO, especially for the 0.75 g concentration, which shows the highest data survival from Figures 9a and 10a.

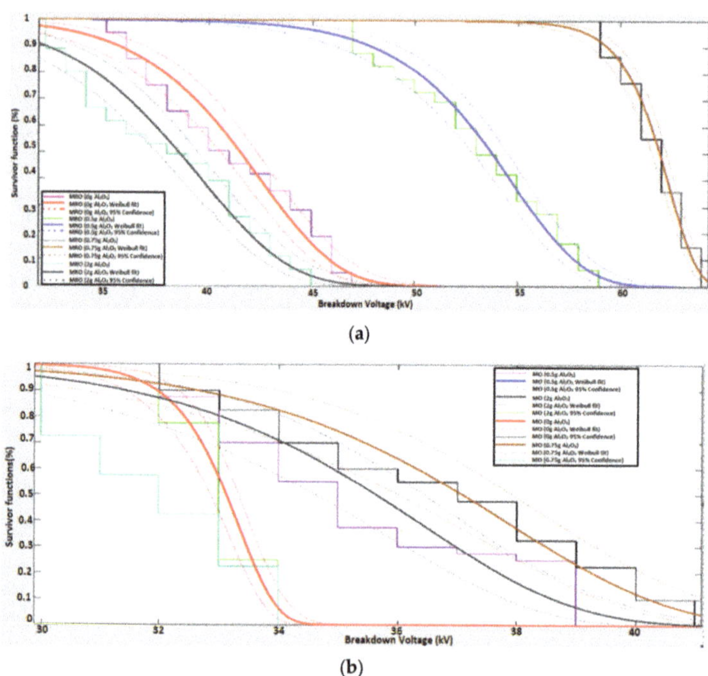

(a)

(b)

Figure 10. Survivor breakdown voltage function in different concentration of the Al_2O_3 with Weibull fit showing the 95% confidence, (**a**) MRO, (**b**) MO.

5.2. Nanoparticle Impact on Kinematic Viscosity

The kinematic viscosity of MRO and MO for the various concentrations of Al_2O_3 and temperature is studied and presented in Table 4. The kinematic viscosity of MRO and MO is measured at temperatures of 0 °C, 30 °C, 60 °C, 90 °C, and 110 °C while increasing the concentration of Al_2O_3 to assess the molecular relaxation of the fluid at the same time. The results show a promising reduction in kinematic viscosity at 0.75 g/L for all temperatures with a reduction of 40% at 0 °C, 66% at 30 °C, 54% at 60 °C, 66% 90 °C, and 62% at 110 °C for MRO. The interpolation polynomial function for the kinematic viscosity under varying temperatures is shown in Equations (7)–(11). Here, $V(x)$ is kinematic viscosity at the instant and V is the actual measured kinematic viscosity. The generalized interpolation function for the kinematic viscosity under various temperatures is presented in Equation (12).

$$\frac{V(x)}{V} = -4.805x^5 + 14.7782x^4 - 15.3986x^3 + 6.4599x^2 - 1.2938x^1 + 1 \tag{7}$$

$$\frac{V(x)}{V} = -14.4195x^5 + 45.106x^4 - 48.207x^3 + 20.385x^2 - 3.196x^1 + 1 \tag{8}$$

$$\frac{V(x)}{V} = -9.8904x^5 + 30.303x^4 - 31.3381x^3 + 12.7129x^2 - 2.0508x^1 + 1 \tag{9}$$

$$\frac{V(x)}{V} = -9.8904x^5 + 30.303x^4 - 31.3381x^3 + 12.7129x^2 - 2.0508x^1 + 1 \tag{10}$$

$$\frac{V(x)}{V} = -12.3566x^5 + 39.168x^4 - 41.3506x^3 + 17.6939x^2 - 3.0946x^1 + 1 \tag{11}$$

$$\frac{V(x)}{V} = \int_{x=0}^{x=2} K1x^5 + K2x^4 + K3x^3 + K4x^2 + K5x^1 + C \, dx \tag{12}$$

$$f(x) = a_0 + a_1 \cos(xw) + b_1 \sin(xw) \tag{13}$$

Table 4. Variation of kinematic viscosity (cSt) of MRO under the influence of Al_2O_3 concentration and Temperature with 95% confidence.

Conc (g/L)	Temperature										Goodness of Fit	
	0 °C	% Decrement	30 °C	% Decrement	60 °C	% Decrement	90 °C	% Decrement	110 °C	% Decrement	R^2	RMSE
0.1	52	0	45	0	37	0	20	0	13	0	0.99	2.227
0.25	47	−10.64	41	−9.76	34	−8.82	19	−5.26	11	−18.18	0.98	2.166
0.5	43	−20.93	39	−15.38	31	−19.35	15	−33.33	10	−30.00	0.99	1.839
0.75	37	−40.54	27	−66.67	24	−54.17	12	−66.67	8	−62.50	0.96	2.964
2	39	−3317.33	31	−45.16	28	−32.14	21	4.76	13	0	0.92	2.685

The concentration at which the kinematic viscosity changes with temperature is estimated using Fourier distribution with Equation (13) and its parameters are presented in Table 5. The interpolation function is used to determine the arbitrary points between the concentrations of Al_2O_3. The data is used to determine the Fourier distribution through which the predicted curve is generated. The measured and predicted distribution of MRO's kinematic viscosity is presented in Figure 11a–e.

Table 5. Factors associated with Fourier distribution for MRO with 95% confidence.

Oil and Temperature	a0	a1	b1	w	R^2
MRO 0 °C	43.22	5.499	2.257	4.291	0.9628
MRO 30 °C	37.21	−2.851	8.326	6.61	0.7669
MRO 60 °C	30.03	1.862	5.246	5.615	0.9176
MRO 90 °C	5.391×10^8	-5.391×10^8	1.113×10^5	−0.0003764	0.8396
MRO 110 °C	6.027×10^7	-6.027×10^7	2.871×10^4	−0.0009218	0.8421

The dispersion of the nanoparticle is well examined up to a concentration of 0.75 g/L Al_2O_3. Above 0.75 g/L, MRO begins to show viscosity rise. The dispersion of nanoparticles in the host fluid never agglomerate due to the presence of oleic acid [37,38] in MRO. The presence of oleic acid acts as surface coating for the CNP and reduces the density in liquid. Moreover, the surface coating protects the CNP from thermal stresses and releases CNP at the time of ionization process. Both low and high temperatures affect the viscosity of MRO. The negative sign (from Table 5) in the decrement clearly shows that viscosity reduction is significant when compared to the host fluid at 0 or 0.1 g/L concentration of Al_2O_3. The positive sign indicates the increment or saturation effect compared to the host viscosity at 0 or 0.1 g/L concentration of Al_2O_3. No change can be seen in MRO's sample with 2 g/L concentration of Al_2O_3 where the viscosity is already saturated. At 0.75 g/L of Al_2O_3, the curve shows good conformity with the host fluid.

A similar effect from Table 6 can be seen in MO with an impressive reduction at 0.75 g/L at all temperatures with the drop of 46.67% at 0 °C, 53.85% at 30 °C, 50% at 60 °C, 42.86% at 90 °C, and 75% at 110 °C. Unlike MRO, MO shows saturation at all temperatures above 0.75 g/L concentration of Al_2O_3. The interpolation polynomial function for various temperatures is presented in Equations (14)–(18). Here the trend shows better conformity (R^2) than in MRO, with a low RMSE. The generalized interpolation function for the kinematic viscosity under various temperatures is presented in Equation (19). The predicted curve is generated using the Fourier distribution given in Equation (20), and its parameters are presented in Table 7. Comparison of measured and predicted curves for MO is presented in Figure 12a–e.

$$\frac{V(x)}{V} = 2.433x^4 - 2.6938x^3 + 0.2383x^2 - 0.11537x^1 + 1 \tag{14}$$

$$\frac{V(x)}{V} = -5.694x^4 + 14.7804x^3 - 11.6376x^2 + 2.3692x^1 + 1 \tag{15}$$

$$\frac{V(x)}{V} = -10.0209x^4 + 25.052x^3 - 19.15003x^2 + 4.0884x^1 + 1 \tag{16}$$

$$\frac{V(x)}{V} = -2.4335x^4 + 10.4507x^3 - 10.2361x^2 + 2.44708x^1 + 1 \tag{17}$$

$$\frac{V(x)}{V} = 6.0244x^4 - 4.2292x^3 - 3.0693x^2 + 1.5797x^1 + 1 \tag{18}$$

$$\frac{V(x)}{V} = \int_{x=0}^{x=2} K1x^4 + K2x^3 + K3x^2 + K4x^1 + C \, dx \tag{19}$$

$$f(x) = a_0 + a_1 \cos(xw) + b_1 \sin(xw) \tag{20}$$

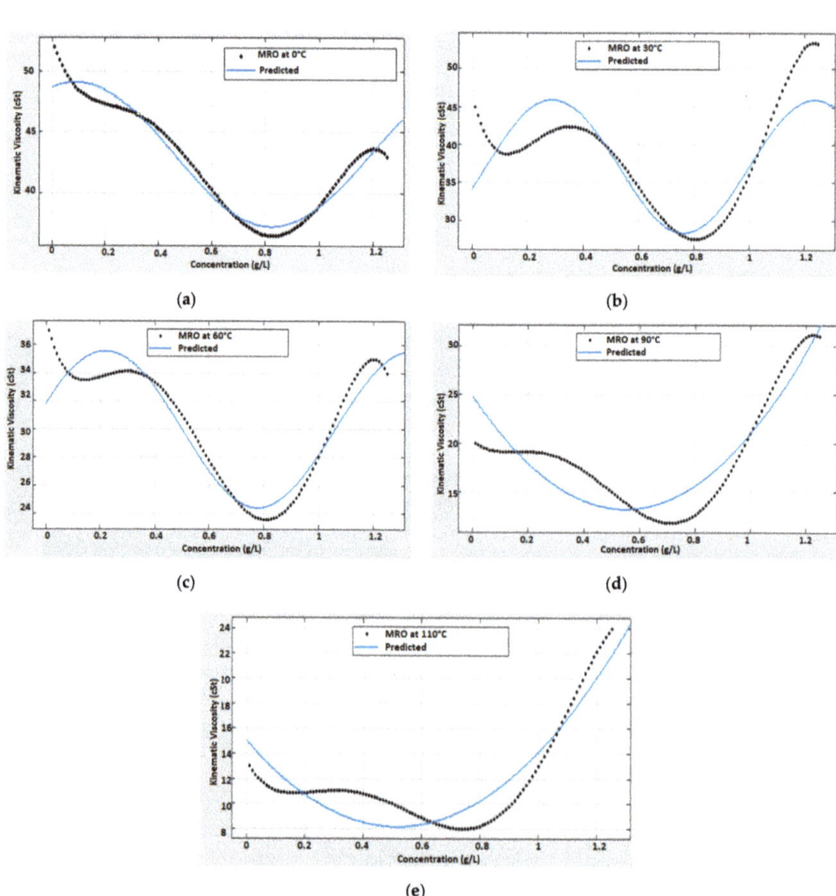

Figure 11. Comparison of measured and predicted kinematic viscosity of MRO using Lagrange polynomial interpolation function and Fourier distribution function. The predicted value showing deviation of ±2 cSt and 0.6 ± 0.2 g/L from the measured value, (**a**) at 0 °C, (**b**) 30 °C, (**c**) 60 °C, (**d**) 90 °C, (**e**) 110 °C.

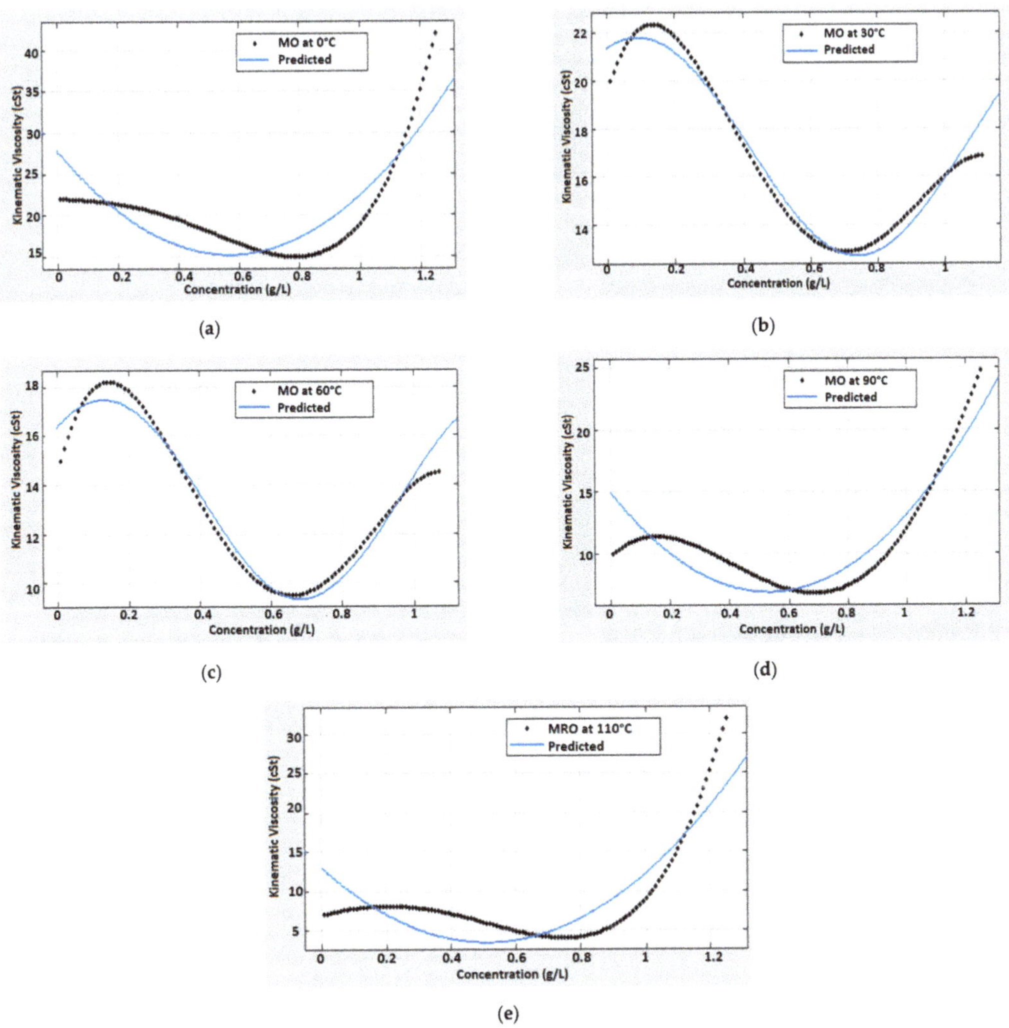

Figure 12. Comparison of measured and predicted kinematic viscosity of MO using Lagrange polynomial interpolation function and Fourier distribution function. The predicted value showing deviation of ± 1 cSt and 0.6 ± 0.2 g/L from the measured value, (**a**) at 0 °C, (**b**) 30 °C, (**c**) 60 °C, (**d**) 90 °C, (**e**) 110 °C.

Table 6. Variation of kinematic viscosity (cSt) of MO under the influence of Al_2O_3 concentration and temperature with 95% confidence.

Conc (g/L)	0 °C	% Decrement	30 °C	% Decrement	60 °C	% Decrement	90 °C	% Decrement	110 °C	% Decrement	R^2	RMSE
0.1	22	0	20	0	15	0	10	0	7	0	0.99	0.21
0.25	21	−4.76	21	4.76	17	11.76	11	9.09	8	12.50	0.99	0.30
0.5	18	−22.22	15	−33.33	11	−36.36	8	−25.00	6	−16.67	0.99	0.23
0.75	15	−46.67	13	−53.85	10	−50.00	7	−42.86	4	−75.00	0.99	0.51
2	19	−15.79	16	−25.00	14	−7.14	12	16.67	9	22.22	0.97	0.76

Table 7. Factors associated with Fourier distribution for MO with 95% confidence.

Oil and Temperature	a0	a1	b1	w	R^2
MO 0 °C	3.6×10^8	-3.6×10^8	9.4×10^4	-0.0	0.73
MO 30 °C	17.2	4.1	1.9	4.8	0.97
MO 60 °C	13.4	3.0	2.7	5.6	0.97
MO 90 °C	1.1×10^6	-1.1×10^6	4.3	-0.0	0.85
MO 110 °C	3.5×10^7	-3.5×10^7	-2.5×10^4	0.0	0.77

Assessment of Survival and Hazard Functions of Kinematic Viscosity

The density plots from Figures 13 and 14 show the temperature effect on MRO and MO's kinematic viscosity at various concentrations of Al_2O_3. The concentration of data points of viscosity at high temperatures for MRO is populated within the range of less than 25 cSt while MO has a wider range up to 32 cSt. It can be seen that at concentration of 1 g/L of Al_2O_3, the viscosity of MO increases by 22% from the host fluid viscosity value. From Figures 15 and 16, the cumulative probability of MRO and MO at various concentrations of Al_2O_3 indicates that above 50% probability the temperature has reduced the viscosity of the fluids. There exists inverse relationship between temperature and oil's viscosity according to the above results and results of [21,24]. According to that, when temperature increases, the viscosity begins to reduce. Moreover, temperature rise increases the volume of the oil and reduces consumption of oil in transformers. It can also be observed that at 50% probability, MRO and MO at 110 °C have a viscosity less than 11 cSt and 7 cSt, respectively.

It is evident from Figures 15 and 16 that molecular relaxation is highly effective in MRO than MO. At a temperature of 90 °C from Figure 15, the kinematic viscosity begins to shift from 23 cSt at 60 °C to 12 cSt, which is not realized in MO. Moreover, both Figures 15 and 16 show saturation of viscosity above the 70% probability with a much steeper increase than the previous curve or inflexion points. The MO's saturation in Figure 16 supports the argument that irrespective of temperature, the concentration of Al_2O_3 begins to increase the viscosity followed by agglomeration.

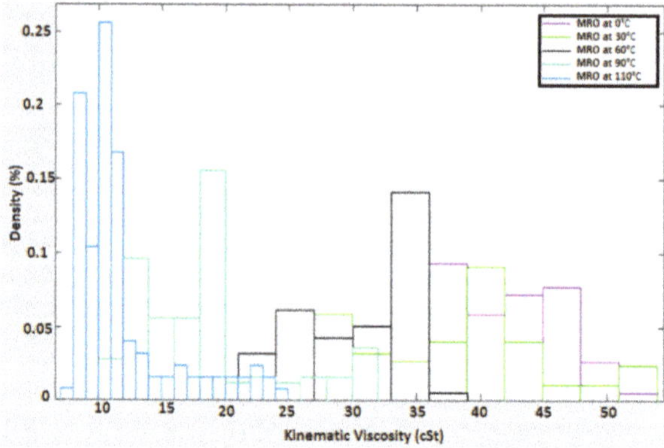

Figure 13. Distribution of kinematic viscosity of MRO as % density function.

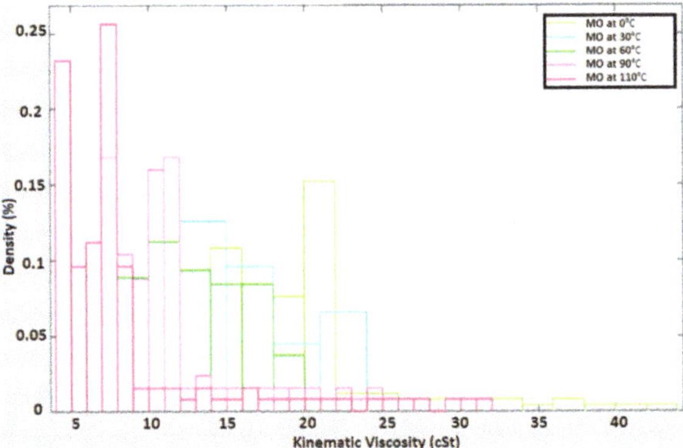

Figure 14. Distribution of kinematic viscosity of MO as % density function.

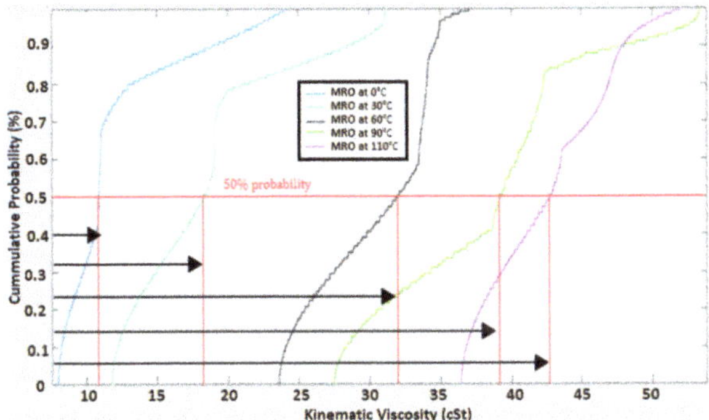

Figure 15. Cumulative probability density function of MRO showing variation of kinematic viscosity in different temperature level.

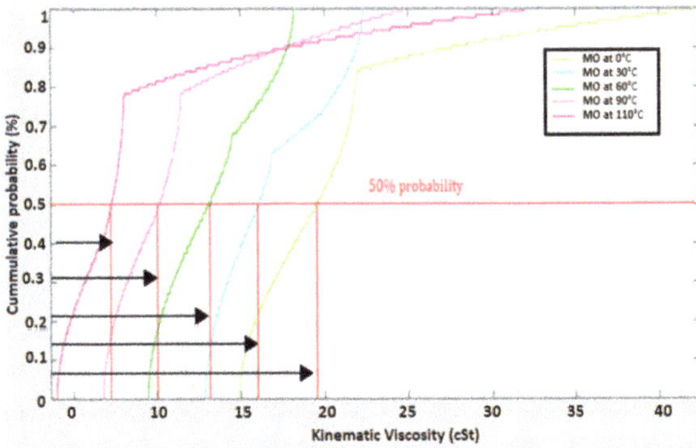

Figure 16. Cumulative probability density function of MO showing variation of kinematic viscosity in different temperature level.

However, viscosity of MRO from Figure 15 shows similar behavior to that in MO, with slower saturation above the 0.75 g/L concentration of Al_2O_3. The survival function of MRO and MO's kinematic viscosity is shown by the data quantile or inverse cumulative survival rate as seen from Figures 17–22. MRO and MO have a better survival rate at different temperatures within an 80% probability from Figures 19 and 20.

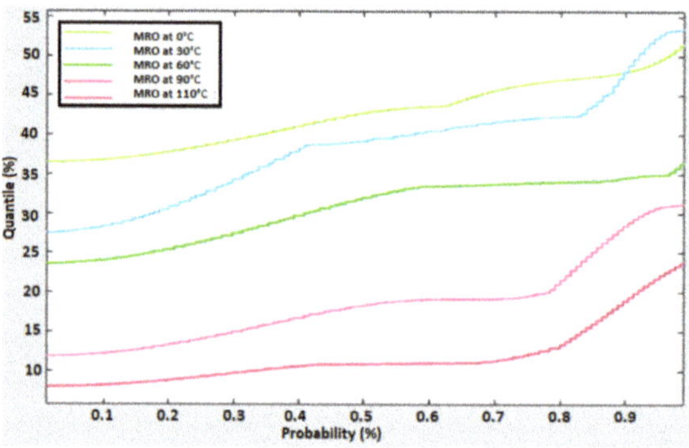

Figure 17. Inverse cumulative probability of MRO showing the probability of data items with different survival rate.

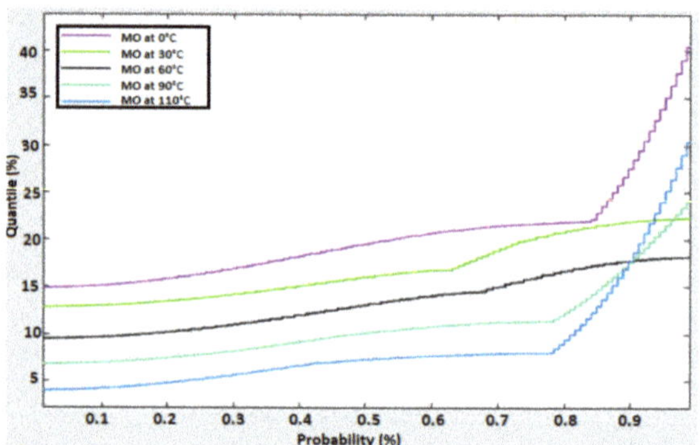

Figure 18. Inverse cumulative probability of MO showing the probability of data items with different survival rate.

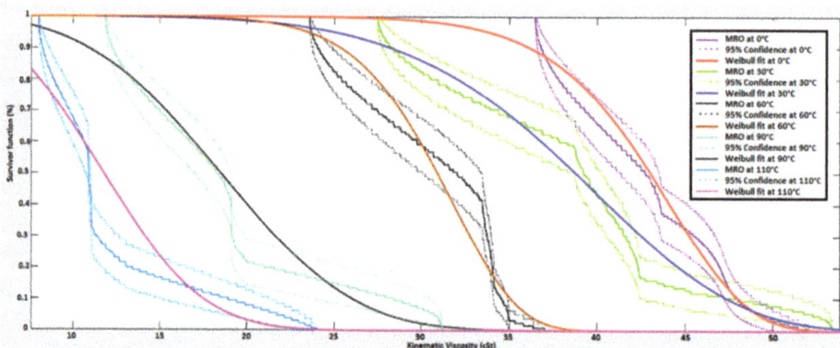

Figure 19. Kinematic viscosity survival function of MRO showing the confidence of 95% fitted to the Weibull.

The nanoparticle concentration effect on kinematic viscosity reverses after 0.75 g/L in MO as seen from Figure 20 with a 20% probability. However, the impact on MRO from Figure 18 is slower and shows that at 30 °C MRO experiences saturation at 0 °C with 20% probability. Whenever a steeper rate is observed from Figures 19 and 20, there is less survival within the 95% confidence bounds, which is a risk for both MRO and MO. There is more significant data turbulence in the steeper curve with insufficient proximity to the Weibull curve, which is higher for MO as seen in Figure 22 than MRO in Figure 23. The cumulative hazard, on the other hand, supports the conclusion devised from the survival rate of MRO and MO. From Figures 21 and 22, 20% of the data points experience hazards whenever the saturation starts. It is also desirable to maintain a minimum concentration of Al_2O_3 to avoid agglomeration and thermal conductivity alteration that results in local hot spots.

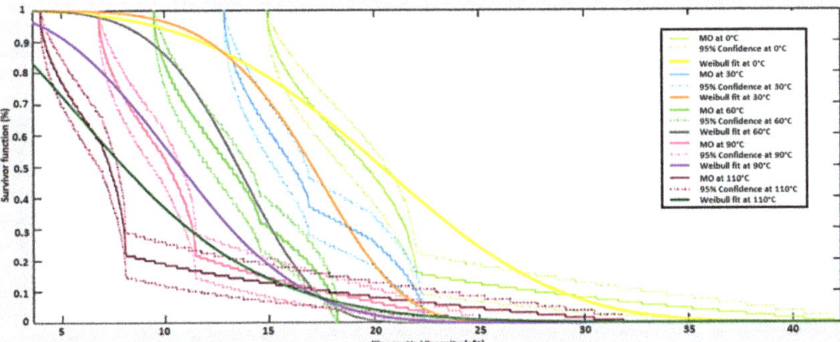

Figure 20. Kinematic viscosity survival function of MO showing the confidence of 95% fitted to the Weibull.

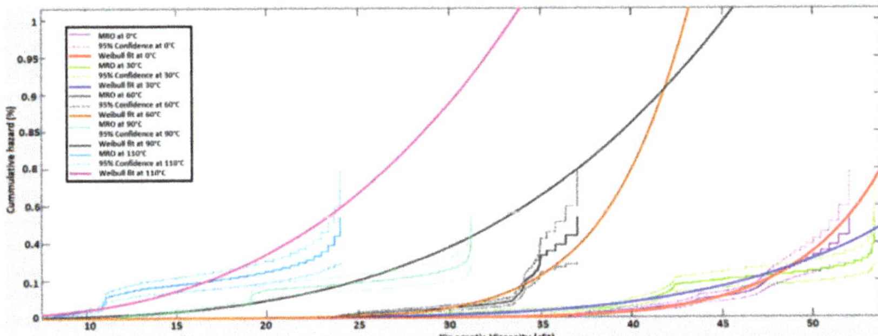

Figure 21. Cumulative hazard of MRO showing the data points of kinematic viscosity under varying concentration of Al_2O_3 in different temperature levels.

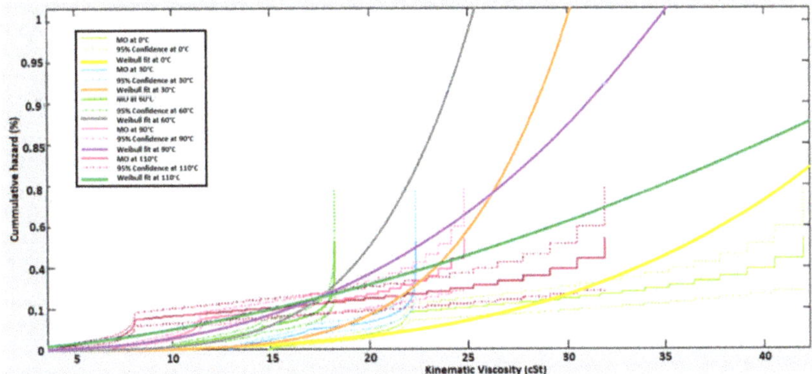

Figure 22. Cumulative hazard of MO showing the data points of kinematic viscosity under varying concentration of Al_2O_3 in different temperature levels.

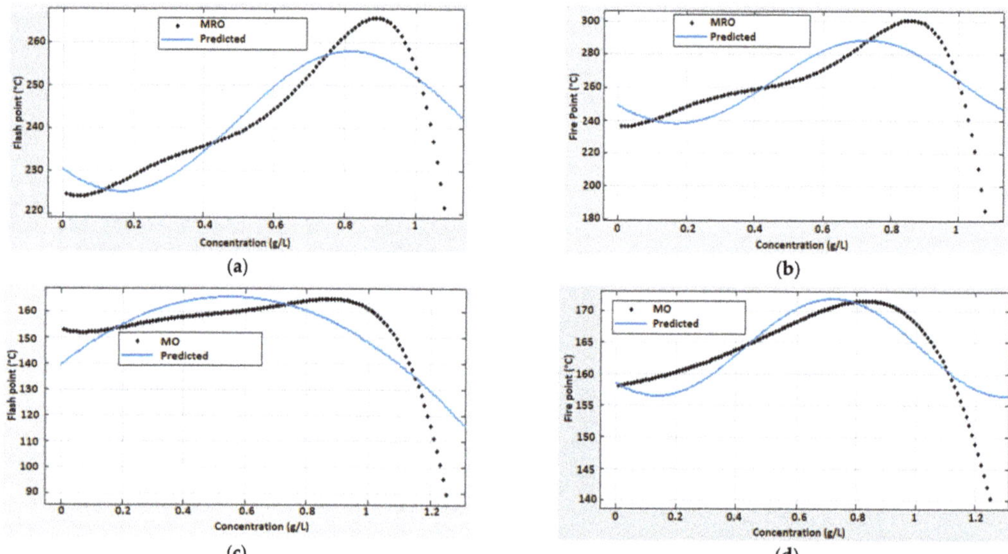

Figure 23. Comparison of measured and predicted flash point and fire point of MRO and MO using Lagrange polynomial interpolation function and Fourier distribution function. The predicted value showing deviation of ±3 °C and ±0.2 g from the measured value, (**a**) flash point of MRO, (**b**) fire point of MRO, (**c**) flash point of MO, (**d**) fire point of MO.

5.3. Nanoparticle Impact on Flash Point and Fire Point

MRO and MO's thermal characteristics are well discussed by calculating the absolute difference and relative difference between the flash and fire points. At various concentrations of Al_2O_3 in MRO and MO, the flash and fire points are modified as seen from Table 8. Al_2O_3 is an excellent heat conduction additive that significantly alters the fluids thermal characteristics. Although Al_2O_3 increases the viscosity of the fluids above 0.75 g/L concentration, it also affects the flash and fire points. The addition of nanoparticles increases the viscosity and reduces thermal conduction. Even though viscosity decreases with an increase in temperature, it increases above 0.75 g/L addition, resulting in a rise in thermal conduction. Effectively, this leads to a drop-in flash and fire points above 0.75 g/L concentration. Comparing the absolute and relative difference of MRO and MO's flash and fire points from Table 8, there is a significant absolute and relative difference of 34 °C and 12.41% observed for MRO. MRO absolute difference is almost four times higher than MO. On the other hand, MO has shown 8 °C and 4.79%, which need rapid extinction whenever smoke is identified. The drop is well predictable in MO as it shows clear evidence of the temperature drop below the base values of flash point and fire point. It is also evident to have a minimum of 7 to 10% relative difference between the flash and fire points reported by Raymon et al. [5]. The Lagrange interpolation polynomial function for flash point and fire point for MRO and MO is presented in Equations (21)–(24). Here, $FLP(x)$ and $FRP(x)$ is fire point and flash point measured at the instant. The generalized Lagrange function and Fourier function (predicted) for the varying concentration of Al_2O_3 is given in Equations (25) and (26). The parameter for Fourier function is presented in Table 9. The graphs obtained using Lagrange and Fourier functions shown in Figure 23a–d show lower data deviation for MRO and higher data deviation for MO, which is a clear indication that the actual parameter is closely predicted with a deviation of $\pm 3\,^\circ$C and ± 0.2 g Al$_2$O$_3$ concentration.

$$\frac{FLP(x)}{FLP} = -5.43768x^5 + 11.7452x^4 - 8.7423x^3 + 2.7637x^2 - 0.200152x^1 + 1 \tag{21}$$

$$\frac{FRP(x)}{FRP} = -10.4497x^5 + 14.7782x^4 - 15.6351x^3 + 4.4423x^2 - 0.18219x^1 + \tag{22}$$

$$\frac{FLP(x)}{FLP} = -2.9763x^5 + 7.1620x^4 - 6.2319x^3 + 2.3427x^2 - 0.2441x^1 + 1 \tag{23}$$

$$\frac{FRP(x)}{FRP} = -0.14713x^5 - 0.0372x^4 - 0.1507x^3 + 0.0390x^2 + 0.0579x^1 + 1 \tag{24}$$

$$\frac{f(x)}{f} = \int_{x=0}^{x=2} K1x^4 + K2x^3 + K3x^2 + K4x^1 + C\,dx \tag{25}$$

$$f(x) = a_0 + a_1\cos(xw) + b_1\sin(xw) \tag{26}$$

Table 8. Flash and fire points of MRO and MO under various concentration of Al$_2$O$_3$.

Concentration of Al$_2$O$_3$ (g/L)	MRO				MO			
	Flash Point (°C)	Fire Point (°C)	Absolute Difference	Relative Difference	Flash Point (°C)	Fire Point (°C)	Absolute Difference	Relative Difference
0	225	237	12	5.19	153	158	5	3.22
0.1	225	240	15	6.45	152	159	7	4.50
0.25	231	252	21	8.70	155	161	6	3.80
0.5	239	263	24	9.56	159	166	7	4.31
0.75	257	291	34	12.41	163	171	8	4.79
1	254	264	10	3.86	161	168	7	4.26
1.25	246	258	12	4.76	154	161	7	4.44
1.5	241	248	7	2.86	148	154	6	3.97
1.75	237	241	4	1.67	143	148	5	3.44
2	231	234	3	1.29	136	140	4	2.90

Table 9. Factors associated with the predicted Fourier distribution function of MRO with 95% confidence.

Parameter	a0	a1	b1	w	R2
MRO Flash Point	241.6	-11.2	-11.9	4.8	0.82
MRO Fire Point	263.7	-14.2	-20.7	5.6	0.60
MO Flash Point	-2.226×10^9	2.26×10^9	3.421×10^5	0	0.63
MO Fire Point	164.2	-5.6	5.242	5.3	0.70

Assessment of Survival and Hazard Functions of Flash Point and Fire Point

The density (%) and cumulative curves for the flash point and fire point of MRO and MO are presented in Figure 24a,b and Figure 25a,b. From Figure 25a, MRO has the highest data traces covering 230 to 291 °C, which is 58% higher than MO as seen in Figure 25a. The density of MO shows the highest residues in the range 150 to 163 °C from Figure 24a. Figure 25b shows that its probability to maintain 160 °C flash points is only 40%. MRO is likely to keep 240 °C with a probability of 40%, which is almost an 80 °C difference to ensure the safety of solid transformer insulation and core components. Figure 25b shows that MRO's fire point within the 40 to 60% probability range is nearly 258 to 268 °C. Similarly, for MO at 40 to 60% probability, the range is between 158 to 160 °C. This in particular shows that the temperature difference between the flash point and fire point of MRO and MO is 10 °C and 2 °C, respectively, which indicates that MRO is comparatively safer than MO.

Therefore, the determination of survival and hazard for the flash point is adequate to discuss because the relative percentage between the flash point and fire point is already discussed. The survival rate and cumulative hazard rate of MRO and MO's flash point characteristics are presented in Figure 26a,b. At 50% probability, the data points of fire point of MO leading to survival is 159 °C and MRO is 242 °C. At 90% probability, MO begins to show a poor survival rate and increased hazard, while MRO shows less hazard rate. Moving from the 50% to 90% probability, MO can survive at a maximum of 140 °C. On the other hand, MRO shows survival of up to 222 °C. There is a difference in the temperature drop from 50% probability to 90% probability observed for MO and MRO which is 19 °C and 20 °C, respectively. The decline is mainly due to the concentration rise of Al_2O_3 above 0.75 g/L, thereby affecting saturation in the fluid's dielectric properties.

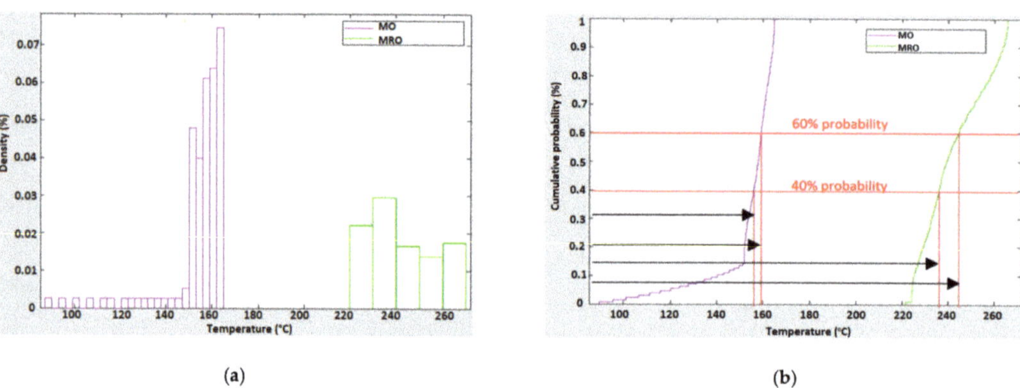

(a) (b)

Figure 24. Distribution of data point of flash points of MRO and MO, (**a**) showing the % density plot, (**b**) showing the cumulative probability with survival rate between the range of 40–60% probability.

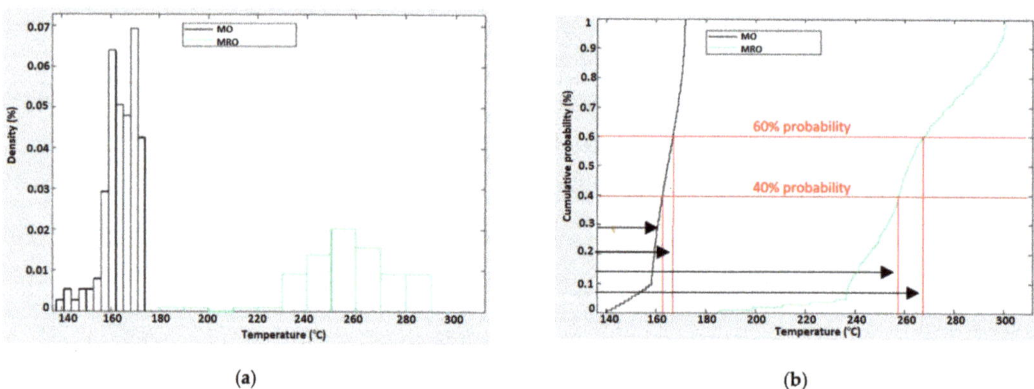

(a) (b)

Figure 25. Flash point of MRO and MO with (**a**) survival, (**b**) hazard function with 95% confidence interval fitted to the Weibull.

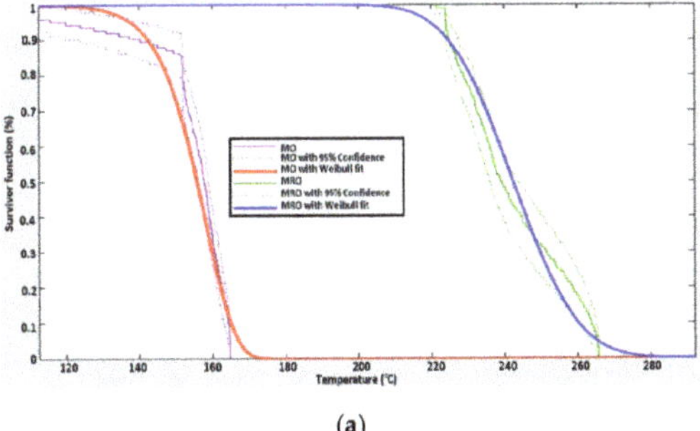 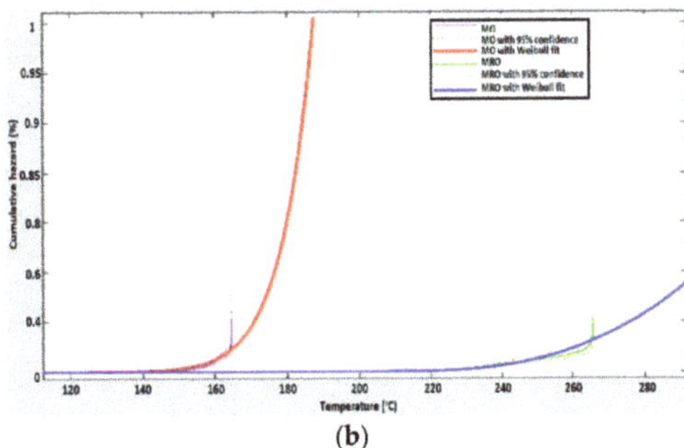

(a)   (b)

Figure 26. Flash point of MRO and MO with (**a**) survival, (**b**) hazard function with 95% confidence interval fitted to the Weibull.

6. Conclusions

The results of statistical analysis shows the treatment of natural esters like Marula oil with Al_2O_3 enhances the oil's electrical, physical, and thermal characteristics. The survival function and hazard function enable analysis of the host fluids' dielectric properties treated with Al_2O_3, as well as the electron scavenging properties of Al_2O_3. An oil's stability is accessed by subjecting the oil to accelerated heating, it also shows the temperament of the oil after accelerated ageing. At the same time, it helps to derive mathematical function that actually fits the linear relationship of the parameter, which are then used for prediction of the future events using the supervised and unsupervised methods. The rate of increase in breakdown voltage at 0.75 g/L concentration of Al_2O_3 shows the Al_2O_3 nature helps to capture the free electrons generated during the inception of ionization. The addition of CNP is effective at electron trapping, reducing the mean free path between the two electrons. The increase in the valence electron trap density particularly with the CNP is moreover reduced with a small concentration (0.75 g/L) of Al_2O_3 than the higher concentration. Thus, the addition of Al_2O_3 delays ionization and improves the breakdown voltage of the MRO. The presence of oleic acid in MRO ensures a fine dispersion without the need to coat the CNP with oleic acid and at the same time improves the dielectric breakdown voltage of the insulating fluid. MRO treated with CNP is thermally stable and show endurance for an extended time. A higher absolute and relative difference is seen in MRO than MO. This indicates that the fluid possesses high heat resistance and immunity to temperature stress. The statistical reliability analysis helps to do reliability analysis (time to failure) and enables one to evaluate the lifetime behavior of the insulating oil when subjected to ageing. According to the observations recorded for the different samples, MO become no longer useful and loses its endurance. Such modeling is normally required when actual testing is complicated or impractical. The addition of Al_2O_3 to MRO acts as a superior nano-ester insulating fluid and it will perform well in actual transformer conditions. The superior nano-ester insulating fluid can serve various power apparatus like circuit breakers, reactors, capacitors, and cables. Thus, maintaining the life of the equipment for an extended period without the risk of failure.

Author Contributions: Conceptualization, R.A.R., A.Y.; methodology, R.A.R.; data curation, R.A.R.; writing—original draft, R.A.R.; writing-review and editing, R.A.R., A.Y., M.M.; validation. R.A.R., A.Y., M.M.; data curation, R.S.; writing—review and editing, R.S. All authors have read and agreed to the published version of the manuscript.

Funding: The APC was funded by Botswana International University of Science and Technology (BIUST), Private Bag 16, Palapye, Botswana through the GRAND REF: DVC/RDI/2/1/7 V(57) and PROJECT CODE: S00299.

Conflicts of Interest: The authors declare no conflict of interest.

Abbreviations

MRO—Marula Oil; MO—Mineral Oil; CNP—Conductive Nanoparticle; PT—Power Transformer; APAC—Asia-Pacific, CAGR—Compound Annual Growth Rate; NEC—National Electric Code; PAO—Polyalphaolefins; BHT—Butylated Hydroxy Toluene; BHA—Butylated Hydroxy Anisole, P.G.—Propyl Gallate; αT—Alpha Tocopherol; β-T—Beta Tocopherol; C.A.—Citric Acid; A.A.—Ascorbic Acid; R.A.—Rosemary extracts; Al_2O_3—Aluminium-III-Oxide; g/L—grams per Litre; IEC—International Electrotechnical Commission; ASTM—American Society for Testing and Materials; kV/s—kilo Volts per Second; mm—millimeter; Hz—Hertz; mL—milli Litre; kV—kilo Volts; W/m-K—Watts per meter Kelvin; ppm—parts per million; mgKOH//g—milligrams of Potassium Hydroxide per gram; %—Percentage; R^2—Conformity; RMSE—Root Mean Square Error; cSt—Centistokes; °C—degree Celsius; Temp—Temperature.

References

1. Bakrutheen, M.; Raymon, A.; Pakianathan, P.S.; Rajamani, M.P.E.; Karthik, R. Enhancement of Critical Characteristics of Aged Transformer Oil using Regenerative Additives. *Aust. J. Electron. Electron. Eng.* **2014**, *11*, 77–86. [CrossRef]
2. IEA. 2019. Available online: https://www.iea.org/reports/world-energy-outlook-2019 (accessed on 26 November 2019).
3. WICZ. 2020. Available online: http://www.wicz.com/story/42539724/global-transformers-market-size-and-share-by-industry-demand-worldwide-research-leading-players-updates-emerging-trends-investment-opportunities-andrevenueexpectationtill2025 (accessed on 25 August 2020).
4. ndutr gmnt utlook, arkt Assment. Available online: https://www.alliedmarketresearch.com/transformer-oil-market (accessed on 25 August 2020).
5. Raymon, A.; Ravi, S.; Abid, Y.; Modisa, M. Enhancement of dielectric properties of Baobab Oil and Mongongo Oil using cost-effective additive for power transformer insulating fluids. *Environ. Technol. Innov.* **2020**, *20*, 1–13.
6. Rafiq, M.; Shafique, M.; Azam, A.; Ateeq, M.; Ahmad Khan, I.; Hussain, A. Sustainable, renewable and environmental-friendly insulation systems for high voltages applications. *Molecules* **2020**, *25*, 3901. [CrossRef] [PubMed]
7. Raymon, A.; Ravi, S.; Abid, Y.; Modisa, M. An overview of potential liquid insulation in power transformer. *J. Energy Convers.* **2020**, *8*, 126–140.
8. Kumar, S.S.; Iruthayarajan, M.W.; Bakrutheen, M. Analysis of vegetable liquid insulating medium for applications in high voltage transformers. In Proceedings of the International Conference on Science Engineering and Management Research (ICSEMR), Chennai, India, 27–27 November 2015.
9. Suhaimi, S.N.; Rahman, A.R.; Din, M.F.M.; Hassan, M.Z.; Ishak, M.T.; Jusoh MT, B. A review on oil-based nanofluid as next generation insulation for transformer application. *J. Nanomater.* **2020**, *2020*, 1–17. [CrossRef]
10. Harlow, J.H. *Electric Power Transformer Engineering*, 3rd ed.; CRC Press: London, UK, 2012; pp. 1–693.
11. Boris, H.; Gockenbach, E.; Dolata, B. Ester fluids as alternative for mineral based transformer. In Proceedings of the IEEE International Conference on Dielectric Liquids, Virginia Beach, VA, USA, 18–21 October 2009.
12. Raymon, A.; Ravi, S.; Abid, Y.; Modisa, M. Performance evaluation of natural esters and dielectric correlation assessment using artificial neural network (ANN). *J. Adv. Dielectr.* **2020**, *10*, 1–10.
13. Hernandez-Herrera, H.; Jorge, I.S.-O.; Mejia-Taboada, M.; Jacome, A.D.; Torregroza-Rosas, M. Natural Ester Fluids Applications in Transformers as a Sustainable Dielectric and Coolant. *Proc. AIP Conf.* **2019**, *2123*, 020049. [CrossRef]
14. Ab Ghani, S.; Muhamad, N.A.; Noorden, Z.A.; Zainuddin, H.; Abu Bakar, N.; Talib, M.A. Methods for improving the workability of natural ester insulating oils in power transformer applications: A review. *Electr. Power Syst. Res.* **2018**, *163*, 655–667. [CrossRef]
15. Cristian, O.; Cristina, M.; Félix, O.; Fernando, D.; Alfredo, O. Titania nanofluids based on natural ester: Cooling and insulation properties assessment. *Nanomaterial* **2020**, *10*, 603.
16. Ayalew, Z.; Kobayashi, K.; Matsumoto, S.; Kato, M. Dissolved Gas Analysis (DGA) of Arc Discharge Fault in Transformer Insulation Oils (Ester and Mineral Oils). In Proceedings of the 2018 IEEE Electrical Insulation Conference, San Antonio, TX, USA, 17–20 June 2018.
17. Rouabeh, J.; M'barki, L.; Hammami, A.; Jallouli, I.; Driss, A. Studies of different types of insulating oils and their mixtures as an alternative to mineral oil for cooling power transformers. *Heliyon* **2019**, *5*, 1–15. [CrossRef] [PubMed]
18. Chandrasekar, S.; Montanari, G. Analysis of partial discharge characteristics of natural esters as dielectric fluid for electric power apparatus applications. *IEEE Trans. Dielectr. Electr. Insul.* **2014**, 1251–1259. [CrossRef]
19. Rafiq, M.; Lv, Y.Z.; Ma, K.B.; Wang, W.; Li, C.R.; Wang, Q. Use of vegetable oils as transformer oils—A review. *Renew. Sustain. Energy Rev.* **2015**, *52*, 308–324. [CrossRef]
20. Abdelmalik, A.A.; Fothergill, J.C.; Dodd, S.J.; Abbott, A.P.; Harris, R.C. Effect of Side Chains on the Dielectric Properties of Alkyl Esters Derived from Palm Kernel Oil. In Proceedings of the 2011 IEEE International Conference on Dielectric Liquids, Trondheim, Norway, 26–30 June 2011.
21. Raymon, A.; Samuel Packianathan, P.; Rajamani, M.P.E.; Karthik, R. Enhancing the critical characteristics of natural esters with antioxidants for power transformer applications. *IEEE Trans. Dielectr. Electr. Insul.* **2013**, *20*, 899–912. [CrossRef]

22. Jeong, J.I.; An, J.S.; Huh, C.S. Accelerated aging effects of mineral and vegetable transformer oils on medium voltage power transformers. *IEEE Trans. Dielectr. Electr. Insul.* **2012**, *19*, 156–161. [CrossRef]
23. Raymon, A.; Ravi, S.; Abid, Y.; Modisa, M. Comparison of ageing characteristics of superior insulating fluids with mineral oil for power transformer application. *IEEE Access* **2020**, *8*, 141111–141122.
24. Raymon, A.; Karthik, R. Reclaiming aged transformer oil with activated bentonite and enhancing reclaimed and fresh transformer oils with antioxidants. *IEEE Trans. Dielectr. Electr. Insul.* **2015**, *22*, 548–555. [CrossRef]
25. Dudrow, F.A. Deodorization of edible oil. *J. Am. Oil Chem. Soc.* **1983**, *60*, 272–274. [CrossRef]
26. Marula Production Guidelines. Available online: https://www.ethicalsuppliers.co.za/wp-content/uploads/2017/09/Marula-production-guidelines.pdf (accessed on 8 September 2020).
27. Sclerocarya_Birrea. Available online: https://en.wikipedia.org/wiki/Sclerocarya_birrea (accessed on 10 September 2020).
28. Raymon, A.; Sakthibalan, S.; Cinthal, C.; Subramaniaraja, R.; Yuvaraj, M. Enhancement and comparison of nano-ester insulating fluids. *IEEE Trans. Dielectr. Electr. Insul.* **2016**, *23*, 892–900. [CrossRef]
29. Nerusu, N.; Narasimha Rao, K.V. Performance of nanoparticles (Al_2O_3) combined with mineral based lubricating oil in refrigeration and air-conditioning systems: A review. *J. Adv. Res. Dyn. Control Syst.* **2018**, *10*, 1741–1760.
30. IEC. *Insulating Liquids—Determination of the Breakdown Voltage at Power Frequency–Test Method*, 3rd ed.; Standard IEC60156; IEC: Geneva, Switzerland, 2003.
31. IEC. *Standard Test Method for Determination of Water by Coulometric Karl Fischer Titration*; Standard IEC60814; IEC: Geneva, Switzerland, 1994.
32. IEC. *Standard Test Method for Kinematic Viscosity of Transparent and Opaque Liquids (and Calculation of Dynamic Viscosity)*; Standard ASTM D445; IEC: Geneva, Switzerland, 2011.
33. IEC. *Standard Test Methods for Flash Point by Pensky-Martens Closed Cup Tester*; Standard ASTM D93; IEC: Geneva, Switzerland, 2012.
34. IEC. *Standard Test Method for Acid and Base Number by Color Indicator Titration*; Standard ASTM D974; IEC: Geneva, Switzerland, 2014.
35. The Idea of a Probability Density Function. Available online: http://mathinsight.org/probability_density_function_idea (accessed on 15 November 2020).
36. Sheldon Ross, M. *Introduction to Probability and Statistics for Engineers and Scientists*, 6th ed.; Academic Press: London, UK, 2020; pp. 1–687.
37. Rajňák, M.; Kurimský, J.; Cimbala, R.; Čonka, Z.; Bartko, P.; Šuga, M.; Paulovičová, K.; Tóthová, J.; Karpets, M.; Kopčanský, P.; et al. Statistical analysis of AC dielectric breakdown in transformer oil-based magnetic nanofluids. *J. Mol. Liq.* **2020**, *309*. [CrossRef]
38. Víctor, A.P.; García, B.; Carlos Burgos, J.; Ricardo, A. Enhancing transformer liquid insulation with nanodielectric fluids: State of the art and future trends. In Proceedings of the Advanced Research Workshop on Transformers, La Toja Island, Spain, 3–5 October 2016.

Article

Energy Distribution of Optical Radiation Emitted by Electrical Discharges in Insulating Liquids

Michał Kozioł 

Faculty of Electrical Engineering, Automatic Control and Informatics, Opole University of Technology, Proszkowska 76, 45-758 Opole, Poland; m.koziol@po.edu.pl

Received: 26 March 2020; Accepted: 29 April 2020; Published: 1 May 2020

Abstract: This article presents the results of the analysis of energy distribution of optical radiation emitted by electrical discharges in insulating liquids, such as synthetic ester, natural ester, and mineral oil. The measurements of optical radiation were carried out on a system of needle–needle type electrodes and on a system for surface discharges, which were immersed in brand new insulating liquids. Optical radiation was recorded using optical spectrophotometry method. On the basis of the obtained results, potential possibilities of using the analysis of the energy distribution of optical radiation as an additional descriptor for the recognition of individual sources of electric discharges were indicated. The results can also be used in the design of various types of detectors, as well as high-voltage diagnostic systems and arc protection systems.

Keywords: optical radiation; electrical discharges; insulating liquids; energy distribution

---

1. Introduction

One of the characteristic features of electrical discharges is the emission to the space in which they occur, an electromagnetic wave with a very wide range. Such typical ranges of emitted radiation include ionizing radiation, such as X-rays, optical radiation, acoustic emission, and radio wave emission. Based on most of these emitted ranges, diagnostic methods were developed, which enables the detection and location of the source of electrical discharges, which is a great achievement in the diagnostics of high-voltage electrical insulating devices [1–4]. These methods are constantly being improved and modified in terms of increasing their effectiveness and speed of operation. Parallel to these activities, research was also carried out in the field of basic studies aimed at learning new possibilities of using the physicochemical properties of electrical discharge forms [5–10]. Examples of not fully understood areas are X-ray radiation and optical radiation emitted by electrical discharges [11–14].

The research topic discussed in this article is focused in particular on the analysis of optical radiation emitted by electrical discharges, which is usually interpreted using a designated spectrum. For this study, the optical radiation range from 200 nm to 1100 nm was assumed. The radiation spectrum represents the visual form of electromagnetic radiation distributed over the individual components of the wavelengths. Using the radiation spectrum, information about the range of waves that are involved in the analyzed radiation is presented, but their quantitative values were not determined. The dependence of the quantitative size on the occurring wavelength component was represented by the spectral distribution. Spectral distribution, in addition to the range of wavelengths of occurring radiation, most often shows the intensity value of individual components of wavelengths.

Registration of optical radiation is a particularly difficult task in the case of emissions in insulating liquids where there is a large attenuation of the optical signal [15–17]. In addition, there was also an environment with high electric field strength. Therefore, to record radiation in such conditions, it required the use of advanced measuring devices that enabled transmission and processing of optical signals without interference. An additional problem was the correct positioning of the measuring probe (optical fiber) in the expected location of the electrical discharge, so that the emitted optical radiation can be introduced and transmitted by means of an optical fiber. Currently, effective measuring probes have not yet been developed, and all measurements carried out in this area are of an experimental nature.

Conducted and published studies were mainly focused on the possibility of recording discharges and determining spectral distributions on their basis [18–21]. However, there is much less work devoted to the development of useful descriptors which, determined on the basis of the obtained spectral distributions, could be used to identify the forms of electrical discharges in various insulation systems (both gas and liquid). Such an approach was presented in the work [22], where a group of descriptors for identifying forms of electrical discharges in insulating oil were developed.

With regard to the already conducted research related to the registration and analysis of optical radiation emitted by electric discharges, in terms of the possibility of using their results in high-voltage diagnostics, the author proposed a new approach to the interpretation of recorded spectral distributions. This solution is based on the analysis of the optical spectrum in terms of the share of individual ranges of optical radiation and their use as a descriptor to recognize single-source forms of electrical discharges.

2. Method of Measuring Optical Spectra

The tests were carried out on two electrode systems, on which single-source forms of electrical discharges were generated. The first system consisted of needle–needle electrodes, where a high voltage was applied to one of the electrodes and the other was earthed. The second system consisted of a needle electrode, and a solid dielectric was used to generate surface discharges. Both systems can be used as models of potential damage in the high power insulating liquid filled transformers, where the needle–needle electrode system was a model of damage to neighboring transformer windings, while the surface discharge system was a model of damage at the contact between the solid and liquid dielectric. The electrode systems were subsequently immersed in insulating liquids, and the measurements were carried out in identical metrological conditions for each variant. Figure 1 shows the general scheme of the measuring system.

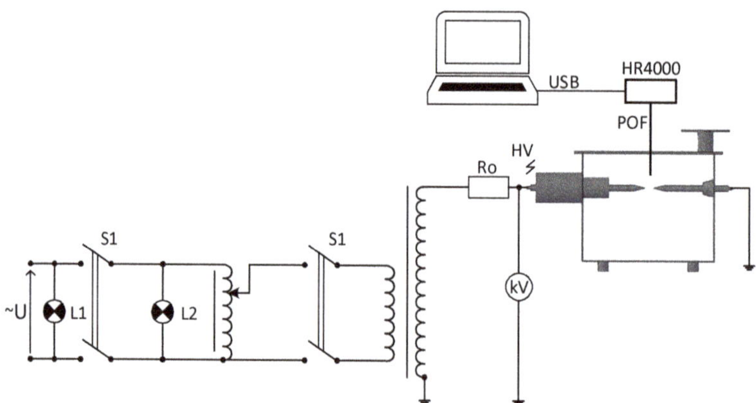

Figure 1. Diagram of the measuring system: Ro—protective water resistor (1.1 MΩ); POF—Polymer Optical Fiber; HR4000—optical spectrophotometer; L1 and L2—control signalling; S1—voltage switch; kV—voltmeter; and U—mains voltage 230 V.

The schematic diagram and general view of the spark gap for generating electric discharges in a needle–needle system is shown in Figure 2. Two identical electrodes with the following dimensions were used: total length, 35 mm; base diameter, 20 mm; apex angle, 32°; and needle head diameter, 0.8 mm. Distance between the electrodes in the needle–needle system was constant for all cases and was 2 cm. The electrodes were made of copper, and their surfaces were electroplated with nickel. The galvanic coating of the copper electrode with nickel improves its mechanical resistance and thermal strength, which also allows multiple uses of the same electrode.

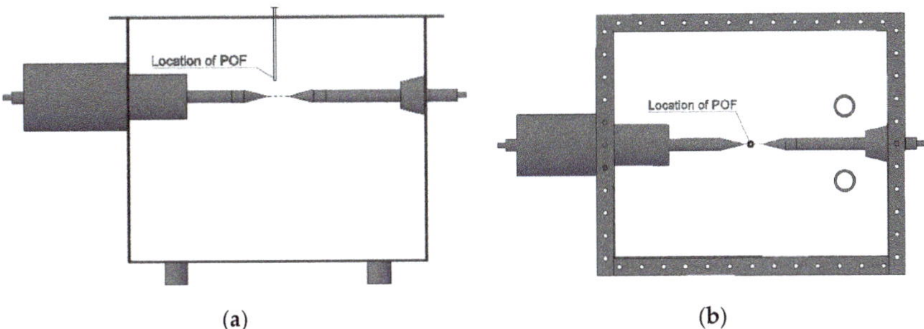

Figure 2. Needle–needle electrode system: schematic diagram side view (**a**) and view from above (**b**).

A system of two metal electrodes with a solid dielectric between them was used to generate electrical discharges in the surface system. A schematic diagram and general view of the spark gap in the surface system is shown in Figure 3. The supplying electrode was a needle electrode, and the grounded electrode was a flat cylindrical plate with a base diameter of 69 mm and thickness of 9 mm, which was made of metal. The spark gap electrodes were separated by a solid dielectric, which was a plate made of sodium glass, with external dimensions of 90 mm × 90 mm and a thickness of 10 mm.

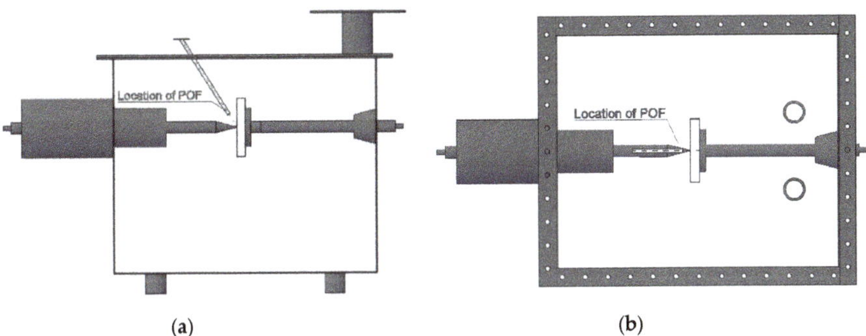

Figure 3. Surface discharge system: schematic diagram side view (**a**) and view from above (**b**).

The three most frequently used electroinsulating liquids in power engineering were used for testing natural ester Midel 1204, synthetic ester Midel 7131, and mineral oil Orlen Trafo EN. All liquids were brand new and free of any contamination. The temperature of the insulating liquids were the same in all examined cases and was 20 °C. Due to the experimental nature of basic research, the influence of liquid temperature on the obtained measurement results was not analyzed.

The optical spectrophotometer, HR4000 from Ocean Optics (Dunedin, FL, USA) was used to record the optical radiation emitted by electrical discharges. The applied spectrophotometer recorded optical radiation in the ultraviolet, visible, and near-infrared range (UV–VIS–NIR spectral range from 200 nm to 1100 nm). The device is equipped with a 3648-element linear silicon CCD array and an optical resolution of 0.47 nm FWHM (Full Width at Half Maximum). This enabled the detection of 3648 components of the recorded optical spectrum in the range of 200 nm to 1100 nm.

Polymer optical fiber 600SR (POF) manufactured by Ocean Optics was used as the measuring head, and one of its endpoints was placed near the electrode system. The basic parameters of the optical fiber were presented in Table 1. During the emission of optical radiation by electrical discharge, the light beam was introduced into the optical fiber and sent to spectrophotometer. The spectrophotometer converts the light beam into a parallel stream with a spectral range of 200 nm to 1100 nm and counts the number of emitted photons for each wavelength. The integration time of the spectrophotometer (matrix exposure time) was the same in all cases and was set to 1 s. Obtained data were presented in the form of spectral characteristics, where the intensity corresponded to the number of counts for wavelengths in the analyzed range.

Table 1. Basic parameters of the optical fiber.

Parameter	Value
spectral range	200 nm–1100 nm
fiber core type	polymer
core diameter	$600 \pm 10\ \mu m$
operating temperature range	$-65\ ^\circ C \ldots +300\ ^\circ C$
fiber bend radius	12 cm
acceptance angle	25°

Figure 1 shows how the fiber was placed in the electrode system area. The optical fiber head was placed at a distance of 2.5 cm from the expected source of optical radiation emission. This distance was determined on the basis of the fiber acceptance cone parameter and due to the metal elements of the fiber head.

All measurements were made in a darkened laboratory room, separated from external sources of optical radiation. Before each measurement series, the spectrophotometer was calibrated to determine the minimum background level. This operation aimed to eliminate interference resulting from the process of converting the optical signal to digital form. The background calibration function is available in the device software. Spectral calibration of the spectrophotometer with a dedicated POF was performed by the manufacturer. The supply voltage of electrode systems was from 25 to 50 kV (RMS voltage) of 50 Hz alternating current (AC), with gradation every 5 kV. In order to limit the discharge current, a water resistor (Ro) of 1.1 MΩ was used, which limited the current to mA range (about 100 mA for this system). For each supply voltage, 10 measurement tests were performed.

3. Measurement Results

Figure 4 presents examples of registered optical spectra emitted by electrical discharges generated on the system of needle–needle electrodes for each of the tested insulating liquids.

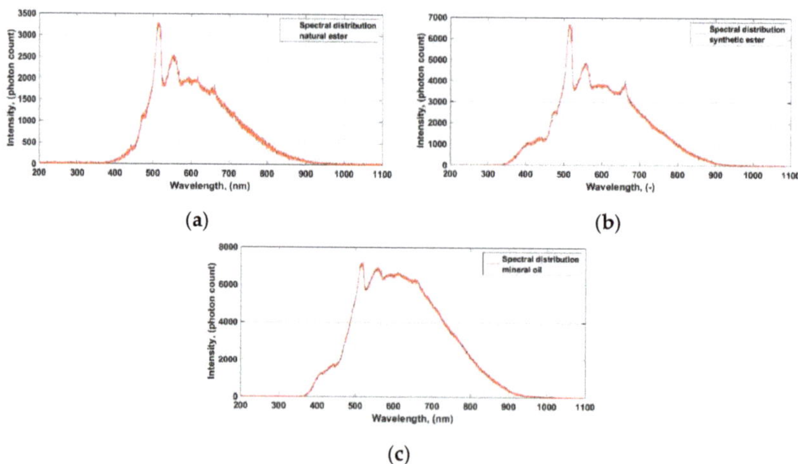

Figure 4. Example results of measurements generated on a needle–needle spark gap at 35 kV supply voltage for the following insulating liquids: Midel 1204 (**a**); Midel 7131 (**b**); and Mineral oil (**c**).

Obtained optical spectra in the needle–electrode system shows some similarity in the shape of the spectral characteristics in all three analyzed liquids. The spectral range of the characteristics mainly includes visible light and, to a small extent, near infrared and ultraviolet. Figure 5 presents examples of registered optical spectra emitted by electrical discharges generated on the surface discharge system for each of the tested insulating liquids.

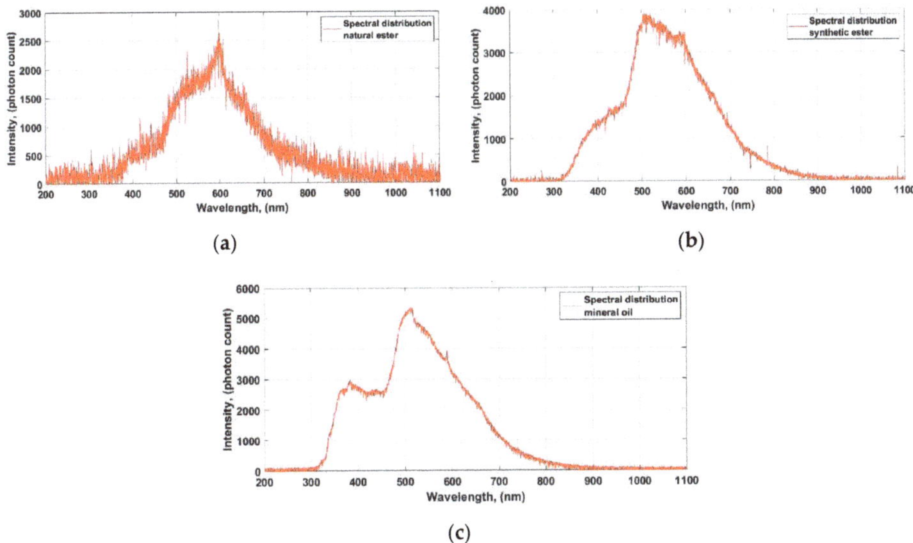

Figure 5. Example results of measurements generated on a surface discharge system at 35 kV supply voltage for the following insulating liquids: Midel 1204 (**a**); Midel 7131 (**b**); and Mineral oil (**c**).

By comparing the obtained spectral characteristics in both analyzed electrode systems and for the same supply voltage level, significant differences in their shapes and spectral ranges can be observed. For the needle–needle electrode system, the spectral range is mainly in visible light, and a small extent in the near-infrared and ultraviolet range. In turn, the spectra obtained for the surface discharge system contained a higher proportion of ultraviolet radiation. This showed the potential possibility of using optical spectra analysis for the recognition of single-source forms of electrical discharges.

4. Optical Radiation Energy

Based on the obtained characteristics of spectral distributions and using the quantum description where optical radiation is described as a photon stream, the share of emitted energy can be estimated. Each wavelength of emitted radiation corresponds to a specific photon energy which can be determined from the relation:

$$E = n{\cdot}h{\cdot}v \tag{1}$$

where: E—total energy of the photon stream (J), n—number of photons counted (-), h—Planck constant 6.626×10^{-34} (J·s), and v—wave frequency (1/s).

The frequency of the waveform is expressed by the formula:

$$v = \frac{c}{\lambda} \tag{2}$$

where: v—wave frequency (1/s); c—phase speed of the wave, speed of light in vacuum 2.998×10^8 (m/s); and λ—wavelength (nm).

Table 2 presents examples of the calculated energy values of optical radiation emitted by electric discharges generated in the tested electrode systems. In order to better present the determined total energy values, the physical unit of energy description in the form of electronvolts (eV) was used. They were calculated from a simple relationship resulting from the definition of eV, where $1\,J \approx 6.241509126(38) \times 10^{18}$ eV. The calculated energy was not the total energy radiated by electrical discharges. The presented values were estimated based on the registered optical radiation by the spectrophotometer. This stage of research does not include an attempt to prepare energy balance, but only presents the possibility of applying energy distribution analysis to recognize the form of electrical discharges.

Table 2. Examples of optical radiation energy values.

Type of Liquid	Energy in UV Range (E_{UV}) 200 nm–380 nm	Energy in VIS Range (E_{VIS}) 380 nm–780 nm	Energy in NIR Range (E_{NIR}) 780 nm–1100 nm	Total Energy ($E_c = E_{UV} + E_{VIS} + E_{NIR}$)	
	(J)	(J)	(J)	(J)	GeV
needle–needle system					
natural ester	6.56×10^{-13}	9.07×10^{-11}	2.21×10^{-12}	9.35×10^{-11}	0.58
synthetic ester	2.39×10^{-12}	1.96×10^{-10}	4.50×10^{-12}	2.03×10^{-10}	1.27
mineral oil	1.08×10^{-12}	3.07×10^{-10}	1.02×10^{-11}	3.17×10^{-10}	1.98
surface discharge system					
natural ester	9.04×10^{-12}	8.15×10^{-11}	3.34×10^{-12}	9.38×10^{-11}	0.59
synthetic ester	1.07×10^{-11}	1.57×10^{-10}	1.63×10^{-12}	1.69×10^{-10}	1.05
mineral oil	3.09×10^{-11}	2.05×10^{-10}	1.74×10^{-12}	2.37×10^{-10}	1.48

Analysis of the Variability of Optical Energy Shares in the UV–VIS–NIR Range

The analysis of optical radiation variability in the UV–VIS–NIR range was carried out using traditional statistical methods. The main aim of the variability analysis was to determine the share of the energy of optical radiation, in particular spectral ranges for recorded optical spectra emitted by electric discharges generated in the analyzed electrode configurations. The results of the variability analysis are presented in Table 3. The results presented in Table 3 were determined on the basis of 10 registered characteristics for each variant of the measuring system.

Table 3. Comparison of the share of optical radiation emitted by electric discharges.

Type of Liquid	Range of Optical Radiation	Average Energy (%E_c)	Variance	Standard Deviation	Coefficient of Variation	Range of Typical Values of Radiation Energy Share
		$\bar{x}$	s^2	s	V_s	X_{typ}
needle–needle system						
Natural ester	UV	0.70	0.01	0.06	8.57	$0.64 < X_{typ} < 0.76$
	VIS	97.60	0.62	0.79	0.81	$96.81 < X_{typ} < 98.39$
	NIR	1.70	0.08	0.29	17.06	$1.41 < X_{typ} < 1.99$
Synthetic ester	UV	0.50	0.02	0.15	30.00	$0.35 < X_{typ} < 0.65$
	VIS	97.90	1.13	1.06	1.08	$96.84 < X_{typ} < 98.96$
	NIR	1.60	0.13	0.36	22.50	$1.24 < X_{typ} < 1.96$
Mineral oil	UV	0.40	0.07	0.26	65.0	$0.64 < X_{typ} < 0.76$
	VIS	97.80	0.51	0.71	0.73	$97.09 < X_{typ} < 98.51$
	NIR	1.80	0.03	0.16	8.89	$1.64 < X_{typ} < 1.96$

Table 3. Cont.

Type of Liquid	Range of Optical Radiation	Average Energy (%E_c)	Variance	Standard Deviation	Coefficient of Variation	Range of Typical Values of Radiation Energy Share
		$\bar{x}$	s^2	s	V_s	X_{typ}
		surface discharge system				
Natural ester	UV	6.70	0.30	0.55	7.46	$6.15 < X_{typ} < 7.25$
	VIS	90.20	1.65	1.28	1.42	$88.92 < X_{typ} < 91.48$
	NIR	3.10	0.10	0.32	10.32	$2.78 < X_{typ} < 3.42$
Synthetic ester	UV	12.10	0.48	0.69	5.70	$11.41 < X_{typ} < 12.79$
	VIS	86.00	0.38	0.62	0.72	$85.38 < X_{typ} < 86.62$
	NIR	1.90	0.08	0.28	14.74	$1.62 < X_{typ} < 2.18$
Mineral oil	UV	13.00	0.10	0.32	2.46	$12.68 < X_{typ} < 13.32$
	VIS	86.30	0.10	0.31	0.36	$85.99 < X_{typ} < 86.61$
	NIR	0.70	0.01	0.06	8.57	$0.64 < X_{typ} < 0.76$

Figures 6–8 present the percentage share of energy in optical radiation ranges emitted by electrical discharges generated on the system of needle–needle electrodes and surface discharge systems in the insulating liquids adopted for testing. The UV radiation was less than 1% of the total energy radiation and poorly detectable in all tested insulating liquids. However, it was different in the case of electrical discharges generated in a surface discharge system, where the proportion of ultraviolet radiation was much higher.

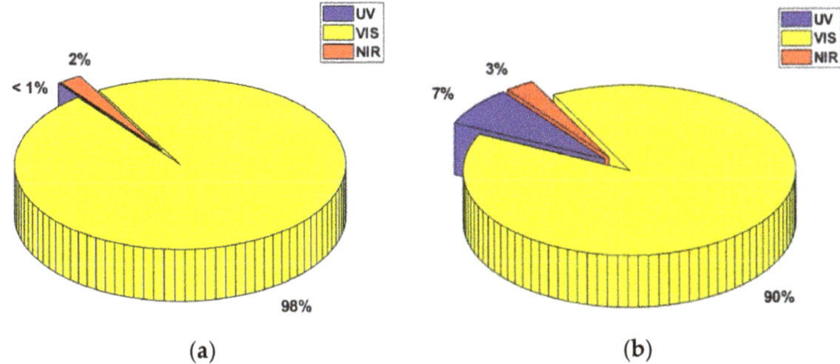

Figure 6. Percentage of optical radiation energy for electrical discharges generated in natural ester, Midel 1204, on electrode systems: needle–needle (**a**) and for surface discharges (**b**).

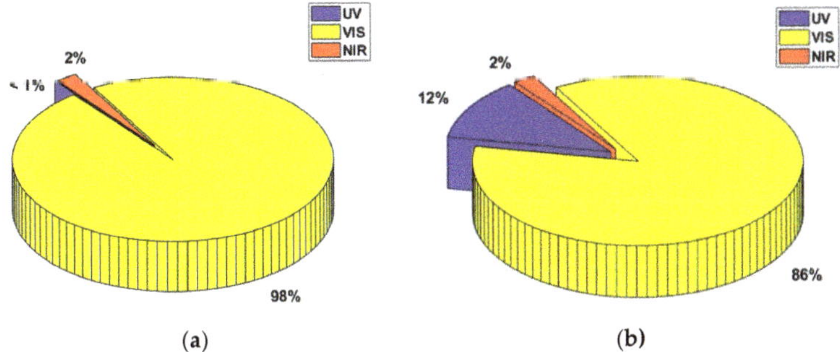

Figure 7. Percentage of optical radiation energy for electrical discharges generated in natural ester, Midel 7131, on electrode systems: needle–needle (**a**) and for surface discharges (**b**).

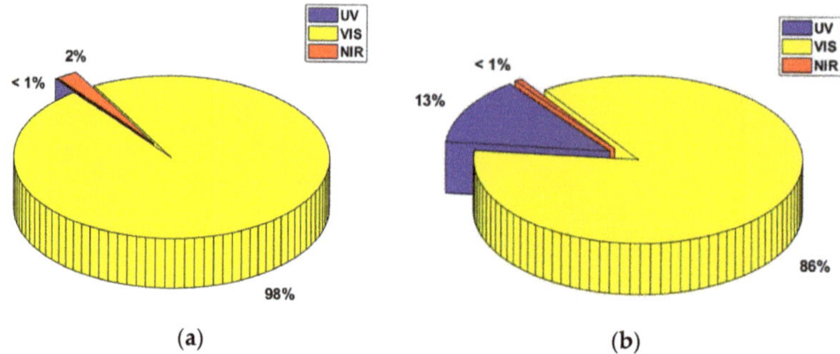

Figure 8. Percentage of optical radiation energy for electrical discharges generated in mineral oil, Orlen Trafo EN, on electrode systems: needle–needle (**a**) and for surface discharges (**b**).

This is probably due to various energy transformations accompanying the phenomenon of electrical discharges, which occurs in the used electrode systems. In the electrode system of the needle–needle type, electric discharges mostly occur in the form of a spark, while in the system for surface discharges, the phenomena occur continuously. Radiation in the ultraviolet range, in the case of discharges in the needle–needle system, is short-lived and effectively suppressed.

5. Conclusions

The spectral distribution of the optical radiation emitted by electrical discharges in insulating liquids differed according to the electrode geometry. Needle–needle electrodes had < 1% UV radiation in all analyzed cases. In contrast, surface discharges had 7% or more UV radiation, depending on the type of electrical insulating liquid. This result might allow identification of the discharge type from the radiation spectrum and might be incorporated in expert diagnostic systems used in various technical areas. The results also justify further research, in terms of the applicability of the proposed indicator, for recognizing forms of electrical discharges in high-voltage insulation systems.

Funding: This research was co-funded by the National Science Centre, Poland (NCN) as a part of the Preludium Research Project No. 2017/25/N/ST8/00590.

Conflicts of Interest: The author declares no conflict of interest. The funders had no role in the design of the study; in the collection, analyses, or interpretation of data; in the writing of the manuscript; or in the decision to publish the results.

References

1. Kunicki, M. Variability of the UHF signals generated by partial discharges in mineral oil. *Sensors* **2019**, *19*, 1392. [CrossRef] [PubMed]
2. Wotzka, D.; Koziol, M.; Nagi, L.; Urbaniec, I. Experimental analysis of acoustic emission signals emitted by surface partial discharges in various dielectric liquids. In Proceedings of the IEEE 2nd International Conference on Dielectrics (ICD 2018), Budapest, Hungary, 1–5 July 2018.
3. Mukhtaruddin, A.; Isa, M.; Noor, M.M.; Adzman, M.R.; Ain, M.F. Review of UHF detection of partial discharge experimentation in oil-filled power transformer: Objectives, methodologies and results. In *AIP Conference Proceedings*; AIP Publishing LLC: Melville, NY, USA, 2018; Volume 2013.
4. Meitei, S.N.; Borah, K.; Chatterjee, S. Modelling of Acoustic Wave Propagation Due to Partial Discharge and Its Detection and Localization in an Oil-Filled Distribution Transformer. *Frequenz* **2019**, *74*, 73–81. [CrossRef]
5. Dincer, S.; Duzkaya, H.; Tezcan, S.S.; Dincer, M.S. Analysis of Insulation and Environmental Properties of Decomposition Products in SF6-N2 Mixtures in the Presence of H2O. In Proceedings of the IEEE International Conference on Environment and Electrical Engineering and IEEE Industrial and Commercial Power Systems Europe, EEEIC/I and CPS Europe, Genova, Italy, 11–14 June 2019.

6. Graczkowski, A.; Gielniak, J.; Przybyłek, P.; Walczak, K.; Morańda, H. Study of the Dielectric Response of Ester Impregnated Cellulose for Moisture Content Evaluation. In Proceedings of the 23rd Nordic Insulation Symposium, Trondheim, Norway, 9–12 June 2013.

7. Mukhtaruddin, A.; Isa, M.; Adzman, M.R.; Hasan, S.I.S.; Rohani, M.N.K.H.; Yii, C.C. Techniques on partial discharge detection and location determination in power transformer. In Proceedings of the 3rd International Conference on Electronic Design (ICED), Phuket, Thailand, 11–12 August 2016; pp. 537–542.

8. Jiang, J.; Wang, K.; Wu, X.; Ma, G.; Zhang, C. Characteristics of the propagation of partial discharge ultrasonic signals on a transformer wall based on Sagnac interference. *Plasma Sci. Technol.* **2019**, *22*, 24002. [CrossRef]

9. Locke, B.R.; Thagard, S.M. Analysis and review of chemical reactions and transport processes in pulsed electrical discharge plasma formed directly in liquid water. *Plasma Chem. Plasma Process.* **2012**, *32*, 875–917. [CrossRef]

10. Piotrowski, T.; Rozga, P.; Kozak, R. Comparative analysis of the results of diagnostic measurements with an internal inspection of oil-filled power transformers. *Energies* **2019**, *12*, 2155. [CrossRef]

11. Nagi, Ł.; Kozioł, M.; Kunicki, M.; Wotzka, D. Using a scintillation detector to detect partial discharges. *Sensors* **2019**, *19*, 4936. [CrossRef] [PubMed]

12. Fracz, P.; Urbaniec, I.; Turba, T.; Krzewiński, S. Diagnosis of High Voltage Insulators Made of Ceramic Using Spectrophotometry. *J. Spectrosc.* **2016**, *2016*. [CrossRef]

13. Schultz, T.; Pfeiffer, M.; Franck, C.M. Optical investigation methods for determining the impact of rain drops on HVDC corona. *J. Electrostat.* **2015**, *77*, 13–20. [CrossRef]

14. Nagi, Ł. Detection of ionizing radiation generated by electrical discharges in the air using sphere-sphere system. *E3S Web Conf.* **2017**, *19*, 01045. [CrossRef]

15. Biswas, S.; Koley, C.; Chatterjee, B.; Chakravorti, S. A methodology for identification and localization of Partial Discharge sources using optical sensors. *IEEE Trans. Dielectr. Electr. Insul.* **2012**, *19*, 18–28. [CrossRef]

16. Kebbabi, L.; Beroual, A. Optical and electrical characterization of creeping discharges over solid/liquid interfaces under lightning impulse voltage. *IEEE Trans. Dielectr. Electr. Insul.* **2006**, *13*, 565–571. [CrossRef]

17. Shih, K.Y.; Locke, B.R. Optical and electrical diagnostics of the effects of conductivity on liquid phase electrical discharge. *IEEE Trans. Plasma Sci.* **2011**, *39*, 883–892. [CrossRef]

18. Shcherbanev, S.A.; Nadinov, I.U.; Auvray, P.; Starikovskaia, S.M.; Pancheshnyi, S.; Herrmann, L.G. Emission Spectroscopy of Partial Discharges in Air-Filled Voids in Unfilled Epoxy. *IEEE Trans. Plasma Sci.* **2016**, *44*, 1219–1227. [CrossRef]

19. Rosolem, J.B.; Tomiyama, E.K.; Dini, D.C.; Bassan, F.R.; Penze, R.S.; Ariovaldo, A.; Floridia, C.; Fracarolli, J.P. A Fiber Optic Powered Sensor Designed for Partial Discharges Monitoring on High Voltage Bushings. In Proceedings of the 2015 SBMO/IEEE MTT-S International Microwave and Optoelectronics Conference (IMOC), Porto de Galinhas, Brazil, 3–6 November 2015; pp. 1–5.

20. Dilecce, G. Optical spectroscopy diagnostics of discharges at atmospheric pressure. *Plasma Sour. Sci. Technol.* **2014**, *23*, 015011. [CrossRef]

21. Rózga, P.; Tabaka, P. Spectroscopic Measurements of Electrical Breakdown in Various Dielectric Liquids. In Proceedings of the IEEE 11th International Conference on the Properties and Applications of Dielectric Materials (ICPADM), Sydney, NSW, Australia, 19–22 July 2015; pp. 524–527.

22. Kozioł, M.; Boczar, T.; Nagi, Ł. Identification of electrical discharge forms, generated in insulating oil, using the optical spectrophotometry method. *IET Sci. Meas. Technol.* **2019**, *13*, 416–425. [CrossRef]

 *energies*

Article

Comparative Analysis of Optical Radiation Emitted by Electric Arc Generated at AC and DC Voltage

Łukasz Nagi *, Michał Kozioł and Jarosław Zygarlicki

Faculty of Electrical Engineering, Automatic Control and Informatics, Opole University of Technology, Proszkowska 76, 45-758 Opole, Poland; m.koziol@po.edu.pl (M.K.); j.zygarlicki@po.edu.pl (J.Z.)
* Correspondence: l.nagi@po.edu.pl

Received: 7 June 2020; Accepted: 25 September 2020; Published: 2 October 2020

Abstract: The article presents a comparison of the spectra of electromagnetic radiation emitted by an electric arc. The spectrum ranges from ultraviolet through visible light to near infrared. Spectra from electric arcs were compared for different frequencies of generating current and for direct current. Characteristic peaks for each measurement were described, and the percentage of individual components of light emitted through the arc was presented. An electric arc is an undesirable phenomenon in many areas, and its detection and control depends largely on its source. There are also areas where an electric arc is used. A better understanding of the physical phenomena involved in different arcs can help optimize the use of the electric arc. Safety and economy through the elimination of parasitic energy shares i.e., in the welding arc can be based on the control of the arc by controlling its optical spectrum. The optical method used in this study is one of the methods of electrical discharge detection in electrical devices and systems.

Keywords: electric arc; gas insulation; arc welding; optical method; spectrophotometer; electromagnetic radiation; arc lamps

1. Introduction

The main aim of the research presented in this article was to determine and analyze the spectrum of optical radiation emitted by an electric arc in air for DC and different AC voltage frequencies. The shape of the spectrum and range of the radiation correspond with the applied voltage. Electric arc is an undesirable phenomenon in many areas. Uncontrolled electrical discharges from Corona Discharges (CD) to Partial Discharges (PD), breakdowns to the arc, or rapid changes in conditions allow the insulation medium to ionize and create a plasma channel for the arc. This is particularly dangerous for electrical equipment or the aviation industry. The aerospace industry is developing toward optimizing fuel consumption for this purpose by reducing the weight of aircraft. The transition from pneumatic or hydraulic systems to electrical systems is associated with a number of electrical failure problems. The next generation of aircraft that are to be fully electric (more electric aircrafts, MEA) will probably have to deal with the problems of choosing insulating materials for cables or protecting equipment, modules, and electric conductors against damage associated with electrical discharges and the detection of related damage and electrical phenomena. In aeronautics, as in other industries, the early detection of electrical discharges improves safety. In the paper [1], the authors presented a method of CD detection with a cheap CMOS (Complementary Metal-Oxide-Semiconductor) camera recording the spectrum of electromagnetic radiation in the UV (Ultraviolet) and VIS (Visual light) ranges. For measurements of this range of radiation, devices such as spectrophotometers with a larger range of radiation recording can also be used: not only the wider UV range, but also the near infrared (NIR) light frequencies [2]. This is a well-known method for the detection and measurement of partial discharges in power devices [3]. The methods of apparent charge measurement, acoustic emission method [4],

X-ray recording in Gas Insulated Switchgear (GIS) [5], or measurement of UHF signals [6] are also used. However, in aircraft conditions at reduced pressure, the methods based on electromagnetic radiation detection—UHF or Optical methods—have the best efficiency. The atmospheric conditions are important in such measurements, as well as the arrangement of elements at the test site, which may constitute obstacles to the signals [7]. The electric arc, which is a consequence of CD–PD breakdown, causes not only damages in electrical equipment, but also damages to the insulation, which increases the probability of occurrence of more such events and, consequently, failures. Along with the increase of voltages, which is required with the increased consumption of electric energy, the electric stresses in the insulations of wires also increase. The risk of electric arcs also increases. A comprehensive review of the knowledge and needs concerning arc tracking is presented in the article [8]. In the paper [9], the authors analyzed the electric arc spectrum for alternating current (AC). In the earlier mentioned article [1], the research on CD spectra were related to direct current (DC). This article presents research on electric arc spectra coming from both DC and AC for selected frequencies of current generating an arc in the air under normal pressure.

There are also areas in industry where the arc is used in many processes. Generated in appropriate conditions and controlled by the selection of many parameters, the electric arc is a phenomenon without which many industries could not work. One of them is welding. The electric arc connects the workpiece with a suitable admixture and produces a weld. Transformer (classic) welding machines usually supply the electrodes (so-called lagging) with alternate voltage–network voltage. They operate at a low frequency, which is unfavorable, because the arc near the sinusoidal transition is extinguished by 0V, and as a result, an uneven weld is obtained on the welded material, and the material is easily melted [10]. Recent methods of current inversion technology and the use of microprocessor controller technologies have resulted in new alternatives for old-fashioned methods: manual metal arc (Manual Metal Arc, MMA) welding and TIG (Tungsten Inert Gas) welding techniques [11]. The TIG method is associated with a non-fusible wolfram electrode and a negative voltage power supply. In MMA-type inverter welders, there is a lagging electrode usually supplied with positive voltage. The use of constant voltage improves the quality of the weld and enables a more accurate control of the arc, using electronic circuits that precisely regulate the voltage, current, and circuit induction of the electrode. In TIG welding, it is recommended to limit the energy released during the arc fracture, as the increased arc discharge energy is accompanied by the erosion of the non-flammable electrode [12]. In all types of welders, a significant part of energy is radiated to the environment in the form of electromagnetic radiation as a result of uncontrolled stimulation of the arc shielding gas atoms. It is important for these processes to limit this radiation by controlling the arc supply. This will limit the impact of this radiation on the health of workers and reduce losses in a broad sense from the economic point of view [13]. The second of the hypotheses put forward in these studies is the assumption that it is possible to increase the efficiency of welding equipment by modifying the parameters of the current supplying the electric arc. The supply of a high-frequency voltage to the arc allows for more precise control of the arc formation process and the reduction of excitations of the shielding gas atoms, thus reducing the parasitic emission of electromagnetic waves. Moreover, the differentiation of electric arcs based on spectra can be applied in arc furnaces. Diagnosis of the insulation of such devices is based on PD detection. The optical method is one of the diagnostic methods. Efficient differentiation of whether the signal comes from a generated arc or from a damaged insulation is important to ensure operational safety. This problem was partially addressed in [14].

Recording of the spectrum of electromagnetic radiation generated by electric arcs has so far focused on DC arcs and was performed by means of UV cameras, visible light cameras, and spectrophotometers. Arcs with currents from 50 to 200 A were studied [15] as well as arcs with current peak levels from 10 to 100 kA with short peak duration: about 15 μs [16]. The first of them presents the results recorded for the whole process related to the arc formation and its development until its expiry. In the article by Martins et al. [16], the camera was synchronized with the arc generator trigger. The development of this research is presented in the article [17]. The authors investigated arcs generated by the current with peak values from 100 to 250 kA. In the study, a high-speed camera from the visible light range was used. The use of the spectrophotometer extends the spectrum of electromagnetic radiation by UV and NIR. The article [18] presents the use of a system with a spectrophotometer to record the spectrum of an electric arc generated on a Yacob ladder powered by direct current. The color of the DC arc changes from blue to violet and then yellow, creating a flame as the current increases, which is all in the visible light range [19]. Spectra from outside the visible light range are presented in the literature as residual, mainly for direct current supply. There is no work that would analyze the spectra for AC and DC in such a wide range. Arc power supply for different frequencies gives surprising results and is presented in our previous article [9]. The main aim of the research presented in this article was to determine and analyze the spectrum of optical radiation emitted by an electric arc in air for DC and different AC voltage frequencies. The shape of the spectrum and range of the radiation correspond with the applied voltage.

The article is an extension of comparative tests of DC and AC electric arcs. The aim of the authors was to present generalized experimental studies on the DC electric arc, in relation to the AC electric arc, with a wide spectrum of supply voltage frequencies. The research was of laboratory and cognitive character, and the parameters of electric arc supply were not optimized for any of the applications proposed in the article.

2. Materials and Methods

In this study, two systems were used to generate an electric arc. Block diagrams of the systems are shown in Figures 1 and 2. The first system (Figure 1) generated a high-frequency electric arc of a given frequency. The second system (Figure 2) was prepared to generate an arc with a DC-powered spark gap.

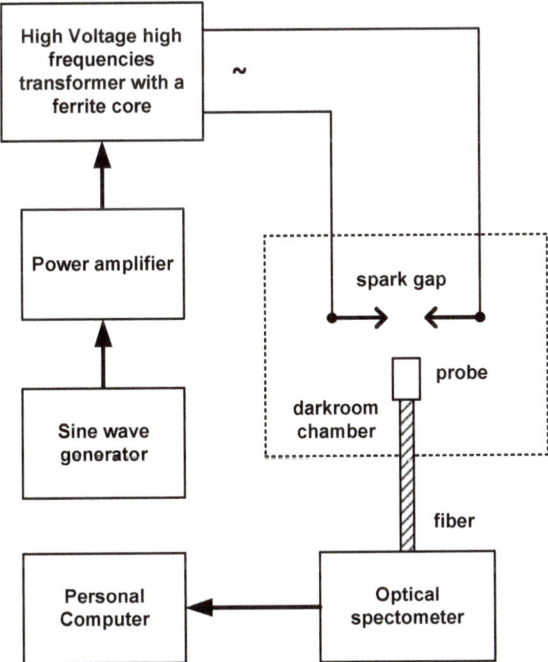

Figure 1. Block diagram of a system to generate an electric arc at a given frequency.

The circuit from Figure 1 consisted of a functional generator (Tektronix AFG1022), whose signal output was connected to the input of a power amplifier using an integrated amplifier on the Texas Instruments OPA541 circuit. The output of the power amplifier supplied the primary winding of the high-voltage transformer adapted to work with high frequency by using a ferrite core. The transformer's gear ratio was 1:500. The transformer's secondary winding fed directly to the spark plug terminals located inside the shielded chamber, which also accommodates the spectrometer probe to record the electromagnetic radiation emitted by the electric arc. The optical input surface of the spectrometer was positioned parallel to the generated electric arc at a distance of 1 cm. The optical radiation emitted by the electric arc was recorded with an optical spectrophotometer type HR4000 from Ocean Optics, which allowed registration in the range from 200 to 1100 nm. The transmission of the optical signal from the emission source to the spectrophotometer took place using a polymer optical fiber (POF) with the same spectral range as the spectrophotometer. The optical system was calibrated by the instrument manufacturer and included compensation for the spectral transmission characteristics of the POF optical fiber. The spectrometer data output was connected to a PC.

The system in Figure 2 was created by modifying the first system in which the output of the secondary winding of the high-voltage transformer was connected to a single-half rectifier with a ripple filter. The first end of the high-voltage transformer secondary winding was connected to the anode of a high-voltage rectifier diode (BY16 Diotec Semiconductor), whose cathode was connected to the first electrode of a 1nF high-voltage ceramic capacitor and the first spark plug. The second end of the secondary winding of the high-voltage transformer was connected to the second electrode of the 1nF high-voltage ceramic capacitor and the second spark plug.

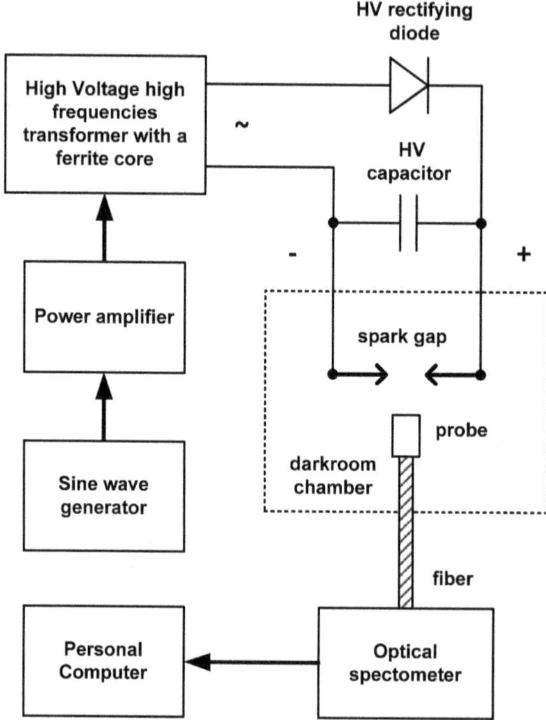

Figure 2. Block diagram of a system for generating an electric arc with a DC-powered spark gap.

The sinusoidal voltage signal from the function generator was fed to the power amplifier, in which it was amplified with current and from the amplifier output fed the primary winding of the high-voltage transformer. The stimulated high-voltage transformer induced in its secondary winding a sinusoidal voltage of adjustable frequency and 4 kV amplitude. In the first circuit, the alternating high voltage directly supplied the spark gap. In the second circuit, the high alternating voltage was regulated to DC and filtered from the ripple and then fed to the spark plug inputs. An electric arc was formed in the spark gap, the light spectrum of which was recorded by a probe and optical spectrometer system. Data from the spectrometer were sent to a PC, where they were archived and further analyzed in the Matlab calculation program.

The measurements were taken in a laboratory with constant ambient conditions, using the systems described above. Arc initialization took place by shortening the electrodes to each other and then extending them over a distance of several millimeters. The measurements were made at an ambient temperature of 22 °C, humidity of 45%, and air pressure of 1000 hPa. The constant voltage of the spark supply from the second system was 4 kV, and the same voltage amplitude was also set for alternating voltages in the first system. During the measurements, none of the electrodes of the spark plug was earthed: the spark plug worked on floating reference potential.

For the purpose of interpretation and analysis of recorded optical signals, the following division of the spectral range was assumed: ultraviolet radiation (UV, from 200 to 380 nm), visible light (VIS, from 380 to 780 nm) and near-infrared (NIR, from 780 to 1100 nm). Each measurement cycle was preceded by a background light calibration procedure.

3. Results and Discussion

Figure 3 shows the registered spectra of optical radiation in the UV-VIS-NIR range for an electric arc generated at 4 kV AC at different frequencies. The obtained optical spectra of electric arcs for different frequencies of current presented in Figure 3 are mostly similar. The same arc release voltage, 4 kV, was used in the test. The characteristic peaks for each frequency are mainly in the ultraviolet range. However, some differences appear for very different frequency values of the arc generating current. For low frequencies, the spectrum range can be seen in the near-infrared part and partially in the visible light range (Figure 3a). For increasingly higher frequencies (Figure 3b,c), the visible light and NIR components disappear. For frequencies of 100 and 130 kHz (Figure 3d,e), this range is already marginally recorded. It can also be noted that the component in the VIS range for 380–440 nm from a clear peak at low frequencies clearly becomes a Pareto and may not be an important descriptor for this frequency range. Characteristic wavelengths for the presented graphs are collected and presented in Table 1.

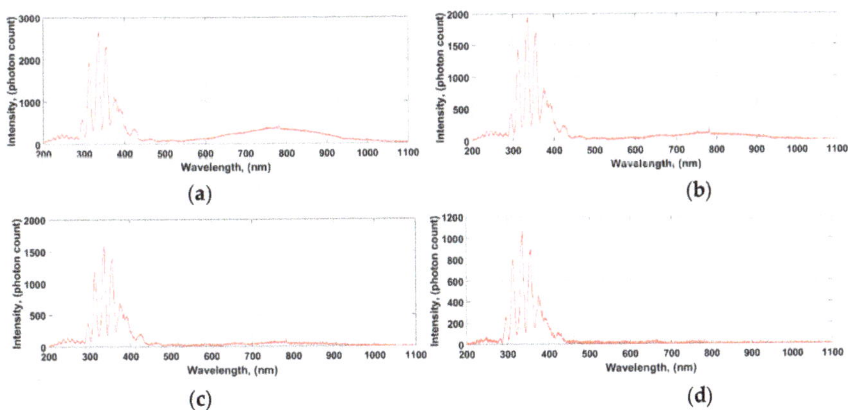

Figure 3. *Cont.*

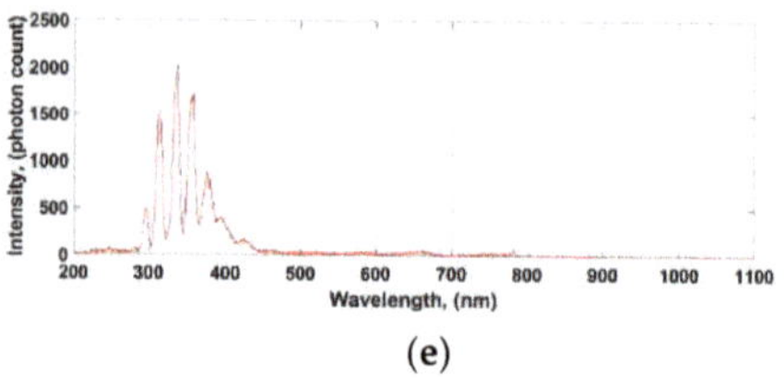

(**e**)

Figure 3. Examples of spectral characteristics for an electric arc generated in the air at a voltage of 4 kV AC with a frequency: 5 kHz (**a**); 30 kHz (**b**); 40 kHz (**c**); 100 kHz (**d**); 130 kHz (**e**).

Figure 4 shows the radiation spectrum in the UV-VIS-NIR range for the electric arc generated by DC. The whole range of spectral lines with characteristic wavelengths of higher intensity than the continuous spectrum can be observed. Table 1 shows the characteristic line lengths for this spectrum. Compared to the spectra obtained from arcs generated by alternating voltage, the main part of the electromagnetic wave being recorded is in the visible light range. In addition, a greater share of NIR can be observed than it was visible for arcs generated by AC. The comparison of the spectra to better illustrate the differences is shown in Figure 5.

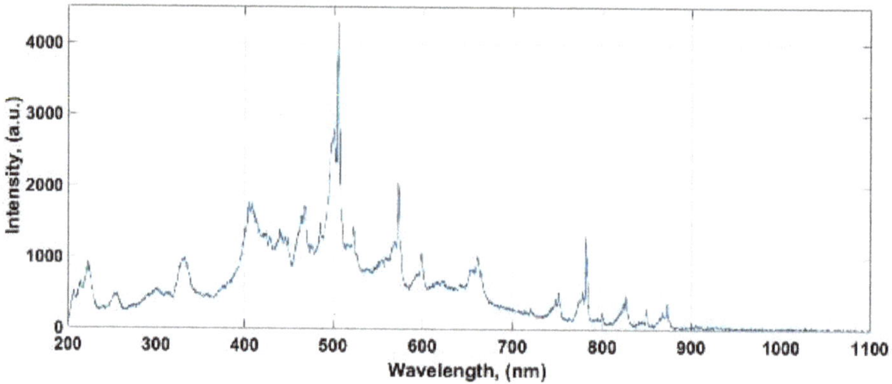

Figure 4. Examples of spectral characteristics for an electric arc generated in the air at a voltage of 4 kV DC.

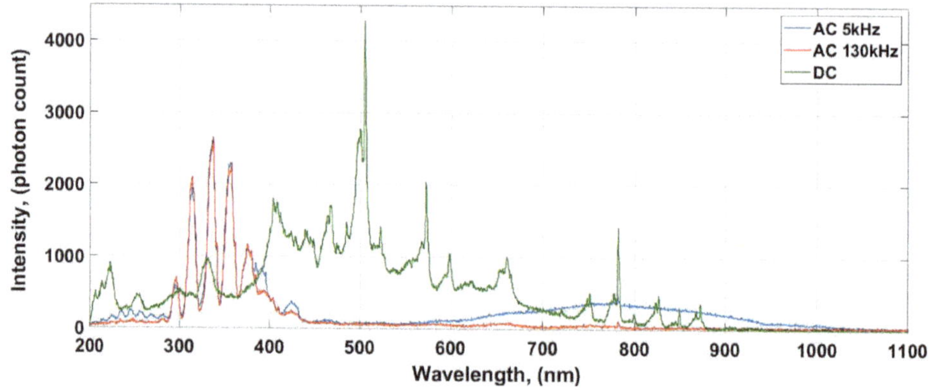

Figure 5. Comparison of spectral characteristics for an electric arc generated in the air at a voltage 4 kV of: 5 kHz AC, 130 kHz AC and DC.

Using the assumptions of quantum mechanics, we can describe optical radiation as a stream of photons, in which each elementary particle is a carrier of energy. The amount of energy that a single photon has can be determined by the following equation:

$$E = h\upsilon \tag{1}$$

where E—energy of a single photon (J), h—Planck constant $6.626 \times 10{-}34$ (J·s), υ— wave frequency (1/s). The optical radiation wave frequency is expressed by the formula:

$$\upsilon = c\lambda \tag{2}$$

where υ—wave frequency (1/s); c—speed of light in vacuum 2.998×108 (m/s); λ—wavelength (nm).

From the number of photons recorded by the spectrophotometer for each wavelength, it is possible to estimate the relative amount of energy that this component emits (E_w). For this purpose, the number of photons (n) is included in Equation (1):

$$E_w = nh\upsilon \tag{3}$$

The sum of the energies of all components makes it possible to determine the energy for the whole analyzed spectral range (UV-VIS-NIR). Thus, the energy calculated is not the total energy of the optical radiation emitted by the electric arc. It is a relative value estimated on the basis of recorded spectral characteristics, which, for better illustration, is presented as a percentage of individual spectral ranges (Figure 6).

When analyzing the spectral ranges for the energy emitted, it can be concluded that for an electric arc generated at alternating voltage, regardless of its frequency, ionizing radiation dominates in the ultraviolet range. Its share is about 60–70% of the emitted energy in the form of optical radiation. The remaining part is the energy of visible light photons, whose share is about 20–30% and a small share, about 2–6% of infrared radiation energy.

Slightly different energy distribution was observed for the optical radiation emitted by the electric arc generated at DC voltage. In this case, the energy of ionizing radiation is only 30% of the energy emitted as optical radiation. However, the dominant share is for the energy of visible light photons. Infrared radiation energy is very low and represents only 1% of the energy of the whole spectral range.

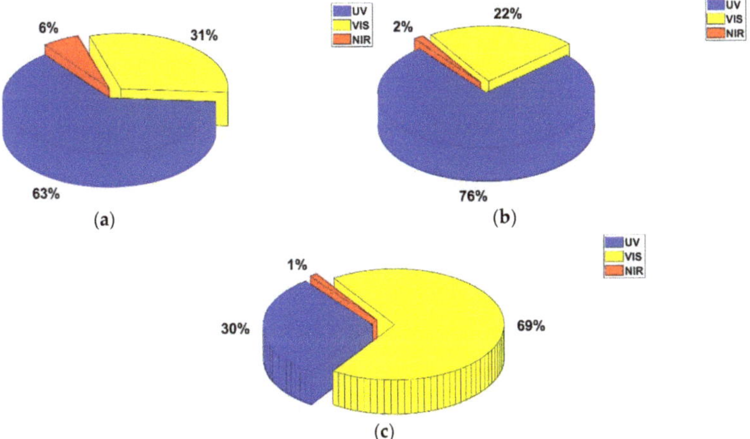

Figure 6. Percentage of optical radiation energy for the analyzed spectral ranges for an electric arc generated in the air at a voltage of 4 kV AC with a frequency 5 kHz (**a**); 130 kHz (**b**); and DC voltage (**c**).

Table 1. Characteristic peaks from electromagnetic spectrum from arc.

Spectral Range	5 kHz AC	130 kHz AC	DC
		Wavelength	
UV 200–380	224.943	-	223.610
	236.666	-	-
	245.714	-	-
	255.817	-	253.957
	268.033	-	-
	282.619	-	-
	295.595	295.595	-
	313.572	313.572	-
	337.050	337.050	329.143
	355.999	355.999	-
	375.959	375.959	-
VIS 380–780	391.686	391.686	-
	-	-	403.724
	424.099	424.099	-
	-	-	467.565
	-	-	500.472
	-	-	504.868
	-	-	522.425
	-	-	572.312
	-	-	598.673
	-	-	660.771
	-	-	751.471
NIR 780–1100	781.955	781.955	781.955
	-	-	826.197
	-	-	849.200
	-	-	872.617

Differences in the energy shares of spectral ranges as well as in spectral characteristics obtained for the arc generated at alternating and DC voltage result, among others, from physicochemical phenomena occurring at the atomic level. The high frequency of alternating voltage significantly reduces the transmission of energy needed to stimulate gas atoms (in this case, it is air). It causes a rapid change in polarization, which causes an increase in electron acceleration and at the same time slowing down the electrons. This has a significant impact on the kinetic energy transmitted by them, which, as a result of the rapid polarization, cannot reach a sufficiently high value to interact with air particles. Only part of the electrons reaches a sufficiently high level of energy at which, as a result of the collision, ionizing radiation is emitted and registered in the ultraviolet range. This is clearly visible on alternating voltage spectra, where a small number of band spectra, characteristic for the activity of high-energy particles, are observed (Figure 3).

In the case of the arc generated at DC voltage, the transmitted energy is supplied continuously. This maintains a constant and stable electric field in which moving electrons increase their kinetic energy. When the electron collides with gas particles, it transfers most of its energy to their excitation, which can be observed in the form of numerous band spectra occurring in the spectral characteristics (Figure 4).

4. Conclusions

On the basis of the results obtained, significant differences in the recorded spectra of optical radiation emitted by the electric arc generated at alternating and direct voltage were shown. The percentage share of optical radiation energy was estimated for particular spectral ranges. For example, the results obtained may be useful in the design of arc-protective systems. They can also be a starting point for designing electronic and power-electronic systems of arc generation with given physical and physicochemical properties. In turn, this can translate into new designs, prototypes, and constructions of different devices. The generalized studies presented in the article may be a guide for scientists dealing with welding arcs, arc lamps, or chemical catalyst systems, improving their energy efficiency and functional parameters. Research on the physical properties of the electric arc, correlated with the frequency parameters of the voltage generating the electric arc, can also be applied to the design of electric actuators, their electrical bundles, and power supply systems, which are used in challenging environmental conditions conducive to the formation of an undesirable electric arc, e.g., at reduced atmospheric pressure in aircraft.

Author Contributions: Conceptualization, Ł.N., M.K. and J.Z.; methodology, M.K. and J.Z.; software, M.K. and J.Z.; validation, J.Z. and Ł.N.; formal analysis, Ł.N. and M.K.; investigation, J.Z.; resources, Ł.N., M.K. and J.Z.; data curation, Ł.N., M.K. and J.Z.; writing—original draft preparation, Ł.N., M.K. and J.Z.; visualization, J.Z. and M.K.; supervision, Ł.N.; project administration, Ł.N.; funding acquisition, Ł.N and M.K. All authors have read and agreed to the published version of the manuscript.

Funding: This work was co-financed by funds of the National Science Centre Poland (NCS) as part of the PRELUDIUM research project No. 2014/15/N/ST8/03680 and the PRELUDIUM Research Project No. 2017/25/N/ST8/00590.

Conflicts of Interest: The authors declare no conflict of interest.

References

1. Riba, J.R.; Gómez-Pau, Á.; Moreno-Eguilaz, M. Experimental study of visual corona under aeronautic pressure conditions using low-cost imaging sensors. *Sensors* **2020**, *20*, 411. [CrossRef] [PubMed]
2. Kozioł, M. Energy Distribution of Optical Radiation Emitted by Electrical Discharges in Insulating Liquids. *Energies* **2020**, *13*, 2172. [CrossRef]
3. Rozga, P.; Tabaka, P. Comparative analysis of breakdown spectra registered using optical spectrometry technique in biodegradable ester liquids and mineral oil. *IET Sci. Meas. Technol.* **2018**, *12*, 684–690. [CrossRef]
4. Pierzga, R.; Boczar, T.; Wotzka, D.; Zmarzły, D. Studies on infrasound noise generated by operation of low-power wind turbine. *Acta Physica Polonica A* **2013**, *124*, 542–545. [CrossRef]
5. Li, T.; Pang, X.; Jia, B.; Xia, Y.; Zeng, S.; Liu, H.; Tian, H.; Lin, F.; Wang, D. Detection and Diagnosis of Defect in GIS Based on X-ray Digital Imaging Technology. *Energies* **2020**, *13*, 661. [CrossRef]
6. Beura, C.P.; Beltle, M.; Tenbohlen, S. Positioning of UHF PD Sensors on Power Transformers Based on the Attenuation of UHF Signals. *IEEE Trans. Power Deliv.* **2019**, *34*, 1520–1529. [CrossRef]
7. Prakash Beura, C.; Beltle, M.; Tenbohlen, S.; Siegel, M. Quantitative analysis of the sensitivity of UHF sensor positions on a 420 kV power transformer based on electromagnetic simulation. *Energies* **2019**, *13*, 3. [CrossRef]
8. Riba, J.R.; Gómez-Pau, Á.; Moreno-Eguilaz, M.; Bogarra, S. Arc tracking control in insulation systems for aeronautic applications: Challenges, opportunities, and research needs. *Sensors* **2020**, *20*, 1654. [CrossRef] [PubMed]
9. Nagi, Ł.; Kozioł, M.; Zygarlicki, J. Optical radiation from an electric arc at different frequencies. *Energies* **2020**, *13*, 1676. [CrossRef]
10. Choi, S.W.; Lee, J.M.; Lee, J.Y. High-Efficiency Portable Welding Machine Based on Full-Bridge Converter with ISOP-Connected Single Transformer and Active Snubber. *IEEE Trans. Ind. Electron.* **2016**, *63*, 4868–4877.
11. Pereira, A.B.; De Melo, F.J.M.Q. Quality assessment and process management of welded joints in metal construction—A review. *Metals* **2020**, *10*, 115. [CrossRef]
12. Burlaka, V.; Lavrova, E. Impoving energy characteristics of the welding power sources for TIG-AC welding. *Eastern-Eur. J. Enterp. Technol.* **2019**, *5*, 38–43. [CrossRef]
13. Freschi, F.; Giaccone, L.; Mitolo, M. Arc Welding Processes: An Electrical Safety Analysis. *IEEE Trans. Ind. Appl.* **2017**, *53*, 819–825. [CrossRef]
14. Karandaeva, O.I.; Yakimov, I.A.; Filimonova, A.A.; Gartlib, E.A.; Yachikov, I.M. Stating diagnosis of current state of electric furnace transformer on the basis of analysis of partial discharges. *Machines* **2019**, *7*, 77. [CrossRef]
15. Guan, R.; Jia, Z.; Fan, S.; Zhang, X.; Wang, T.; Deng, Y. DC arc self-extinction and dynamic arc model in open-space condition using a Yacob Ladder. *IEEE Trans. Plasma Sci.* **2019**, *47*, 4721–4728. [CrossRef]

16. Martins, R.S.; Zaepffel, C.; Chemartin, L.; Lalande, P.; Soufiani, A. Characterization of a high current pulsed arc using optical emission spectroscopy. *J. Phys. D. Appl. Phys.* **2016**, *49*, 415205. [CrossRef]

17. Martins, R.S.; Zaepffel, C.; Chemartin, L.; Lalande, P.; Lago, F. Characterization of high current pulsed arcs ranging from 100 kA to 250 kA peak. *J. Phys. D. Appl. Phys.* **2019**, *52*, 185203. [CrossRef]

18. Kozioł, M.; Wotzka, D.; Boczar, T.; Frącz, P. Application of Optical Spectrophotometry for Analysis of Radiation Spectrum Emitted by Electric Arc in the Air. *J. Spectrosc.* **2016**, *2016*, 1814754.

19. Guan, R.; Jia, Z.; Fan, S.; Deng, Y. Performance and Characteristics of a Small-Current DC Arc in a Short Air Gap. *IEEE Trans. Plasma Sci.* **2019**, *47*, 746–753. [CrossRef]

 *energies*

Article

Ageing Tests of Samples of Glass-Epoxy Core Rods in Composite Insulators Subjected to High Direct Current (DC) Voltage in a Thermal Chamber

Krzysztof Wieczorek [1,*], **Przemysław Ranachowski** [2], **Zbigniew Ranachowski** [2] and **Piotr Papliński** [3]

1   Department of Electrical Engineering Fundamentals, Wroclaw University of Science and Technology, 50-370 Wroclaw, Poland
2   Department of Experimental Mechanics, Institute of Fundamental Technological Research Polish Academy of Sciences, 02-106 Warsaw, Poland; pranach@ippt.gov.pl (P.R.); zranach@ippt.pan.pl (Z.R.)
3   Environmental Impact and Overvoltage Protection Laboratory, Institute of Power Engineering-Research Institute, 01-330 Warsaw, Poland; piotr.paplinski@ien.com.pl
*   Correspondence: Krzysztof.Wieczorek@pwr.edu.pl

Received: 1 November 2020; Accepted: 16 December 2020; Published: 20 December 2020

Abstract: In this article, we presented the results of the tests performed on three sets of samples of glass-reinforced epoxy (GRE) core rods used in alternating current (AC) composite insulators with silicone rubber housing. The objective of this examination was to test the aging resistance of the rod material when exposed to direct current (DC) high voltage. We hypothesized that the long-term effects of the electrostatic field on the GRE core rod material would lead to a gradual degradation of its mechanical properties caused by ionic current flow. Further, we hypothesized that reducing the mechanical strength of the GRE core rod would lead to the breakage of the insulator. The first group of samples was used for reference. The samples from the second group were subjected to a temperature of about 50 °C for 6000 h. The third group of samples were aged by temperature and DC high voltage for the same time. The samples were examined using the 3-point bending test, micro-hardness measurement and microscopic analysis. No recordable degradation effects were found. Long-term temperature impact and, above all, the combined action of temperature and DC high voltage did not reduce the mechanical parameters or change the microstructure of the GRE material.

Keywords: DC high voltage; composite insulator; glass-reinforced epoxy core; 3-point bending test; mechanical strength; micro-hardness

---

1. Introduction

Progress in the construction of high voltage power converter systems and the dynamic development of electricity generation systems from so-called renewable sources have resulted in an increasing interest in the transmission of electricity through high voltage direct current (HVDC) transmission lines [1,2]. The high cost of converter stations—from alternating current (AC) to direct current (DC) and vice versa—are compensated by a significant reduction in loss of energy transmitted over long distances via HVDC transmission lines, especially when compared to systems that operate at AC voltages and have lower construction costs [3].

Current high voltage lines are more often equipped with modern composite insulators. The main advantage of this is surface hydrophobicity. In polluted environments, this property makes it impossible to create water paths that conduct leakage currents for housing composite insulators. Silicone elastomer insulator housings have the ability to hydrophobize surface pollution and regenerate temporarily in lost surface properties. Compared to ceramic and glass insulators, they are significantly lighter and, in many countries, cheaper. However, the use of HVDC for the transmission of electricity unfortunately raises some technical problems. The constant electric field, forced by the HVDC line, has a significant impact on the integrity of dielectric materials usually produced for AC applications [4]. Compared to systems operating at alternating voltages, electrostatic phenomena may cause changes in degradation processes, electrical strength [5,6] and even a four-fold increase in the accumulation of surface soiling [7]. Under such conditions, when the leakage current flows through the polluted surface without passing through the zero curve of voltage and current, the ignition of non-extinguishing concentrated surface discharges may occur. This can lead to degradation of the composite housing. The ageing process in the presence of HVDC results in an increased accumulation of spatial charge in areas where material is more degraded, thus contributing to a stronger distortion of the electric field [8]. This phenomenon may cause intensification of partial discharges. In addition, if the glass-reinforced epoxy (GRE) core material is exposed to the long-term electrostatic field, then the ionic current flow may cause gradual degradation of mechanical properties [9]. This process could be activated at increased temperatures. Electrolysis of the carrying material (glass fibers) could then lead to the breakage of such insulators and, consequently, to serious failures.

This experiment aimed to test the mechanical strength of GRE core rod samples after 6000 h of aging at a temperature of about 50 °C and in the presence of HVDC.

2. Materials and Methods

In the ageing comparative tests, we used samples of GRE material cut from the carrying rod of a typical high voltage composite insulator for AC lines. Fibers in the rod were made of ECR-glass (ECR—type of glass electrical chemical reinforced) [10,11]. Apart from silicon (SiO_2), these types of fibers contain calcium from $CaCO_3$ (as flux and stabilizer) and aluminum from Al_2O_3 (to improve chemical resistance), which are used in the glass composition. Typical ECR-glass contains over 58% SiO_2, about 22% CaO and less than 12% Al_2O_3, as well as smaller amounts of other additives. The presence of mobile sodium cations (Na^+) was excluded.

The samples had a diameter of 24.0 mm and a length of 120.0 mm. The first group, marked with the letter A, were fresh reference samples and reflected the material's initial structure. The samples of the second series, marked with the letter B, were subjected to a temperature of about 50 °C ± 2 °C for 6000 h. This temperature was selected following the measurements of the insulator housing temperature made on a sunny, cloudless day with an ambient temperature of about 27 °C. Measured temperature values under actual insulator operation conditions reached about 46 °C, as shown in Figure 1.

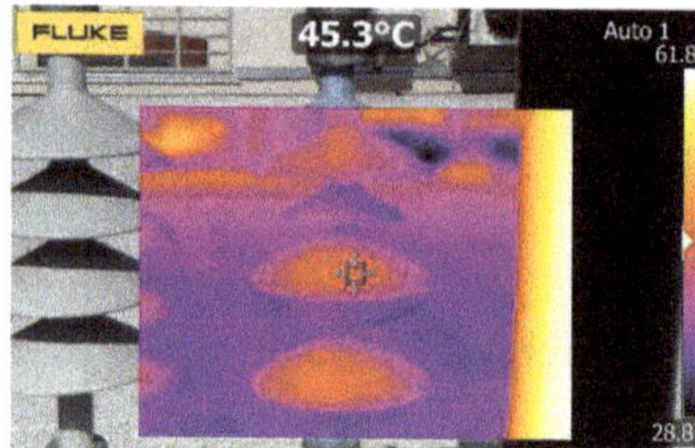

Figure 1. The surface temperature of the composite insulator housing measured with a thermal imaging camera.

The samples of the third series, marked with the letter C, with electrodes applied to their front surfaces, were subjected for 6000 h to a temperature of about 50 °C ± 2 °C and a DC voltage of 20 kV. The average value of the voltage distribution along the main axis of the typical insulator was about 1 kV/cm. In this research, it was applied twice as high as the electric field strength, i.e., 2 kV/cm. Increasing the voltage was supposed to accelerate the aging process. Figure 2 shows one sample from each of the three series.

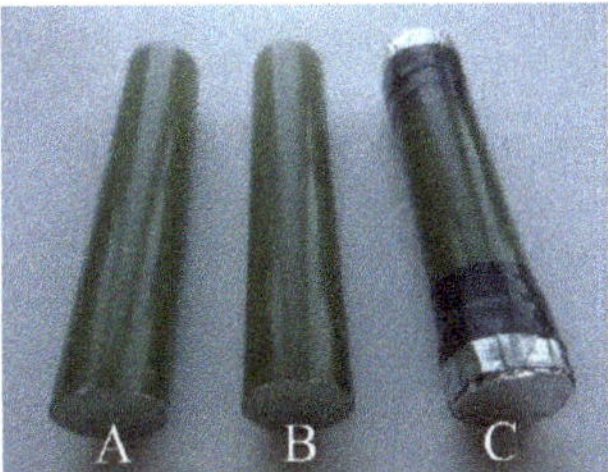

Figure 2. Glass-reinforced epoxy (GRE) material samples of the high voltage alternating current (HVAC) composite insulator carrying rod. From the left: reference sample—group A; thermally aged sample—group B; and DC and thermally aged sample—group C, with visible electrodes attached to the sample's front surfaces.

The samples were arranged in a special stand and placed in a heating chamber. Figure 3 shows the samples in the heating chamber (photograph taken with a fluke thermal imaging camera).

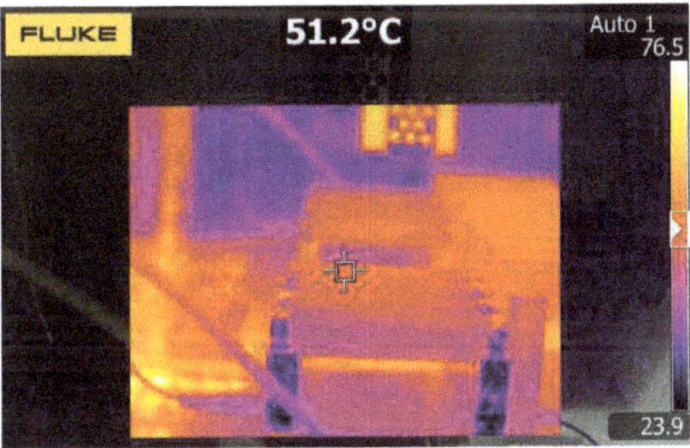

Figure 3. Samples placed in the heating chamber.

Samples from all three groups (both reference samples, thermally aged samples and those thermally and voltage aged) were tested using the 3-point bending test. For this purpose, a testing machine (INSTRON 1343) was used, which was extended with controllers and software by MTS. (MTS–Mathematisch Technische Software-Entwicklung GmbH, Berlin, Germany). Special adaptation of the support system was necessary as the tested samples had a cylindrical shape with a diameter of 24.0 mm. Therefore, in the steel rollers on which the samples were based, semi-circular notches with a radius of 12.0 mm and a depth of 10 mm were made at 100.0 mm of spacing. This is illustrated in Figure 4. A relatively low crosshead speed of 0.1 mm/min was set. This corresponded approximately to a force increase of 20 N/s.

The measuring system recorded the force acting on the sample, which was then converted to stress, according to the relation [12]:

$$\sigma_f = \frac{8 \times F \times l}{\pi \times d^3},$$

(1)

where:

σ_f—is the bending stress, in MPa;

F—force loading the sample, recorded by the measuring system, in N;

l—distance between supports in the measuring system, equal to 100 mm;

d—diameter of the cylindrical sample, equal to 24 mm.

Taking into account the fixed values occurring in relation (1):

$$\sigma_f = 0.0184 \times F \tag{2}$$

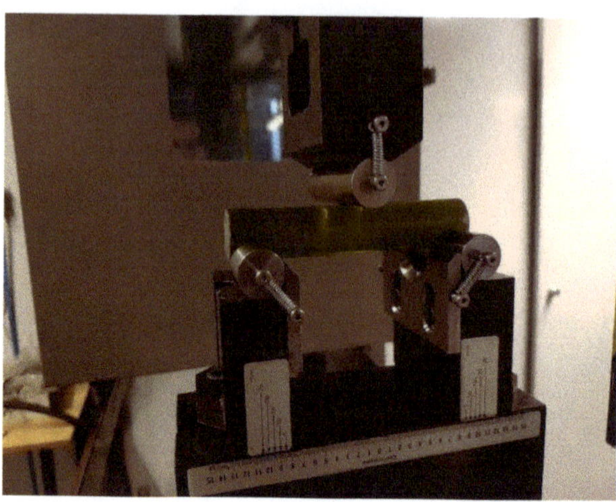

Figure 4. The sample of a group A from HVAC composite insulator carrying rod in a mechanical system for strength testing via the 3-point bending test. The notches in bottom supporting rollers are visible.

In addition to the 3-point bending test, we performed a micro-hardness examination of the sample material. It constituted an important supplement to the results of the optical method of the material testing. It also made it possible to independently assess the material homogeneity and cohesion. The measurements were made using the Vickers method (with a typical micro-hardness measurer) at 1 kG load of the indenter. We used a semi-automatic mode of measuring the imprint diameter. It should be emphasized that, apart from the obtained average values, the scatter of results (which proves the homogeneity of the microstructure of the material) provides important information.

3. Results

3.1. Mechanical Strength Tests

The mechanical characteristics of the 3-point bending test for all 13 tested samples showed a very high similarity (Figures 5–7). Small differences occurred individually for particular samples, but there were no differences in the characteristics of the samples that would be typical for groups A, B or C. For a stress that usually slightly exceeds 300 MPa (294–345 MPa for individual samples), the displacement linearly increased when stress increased. The slope of characteristics was identical for all tested samples. A slight non-linearity at the beginning of the characteristics resulted from the arrangement of the samples in the clamping system. When the stress reached about 300 MPa, the samples broke axially. Long cracks were formed that extended symmetrically from the middle of the samples and were present on the upper and lower side of the specimens, as illustrated in Figures 8 and 9. These cracks usually did not reach the ends of the samples. The formation of cracks, accompanied by a well audible crackle, was reflected by a clear fault, sometimes even two faults, on the samples' mechanical characteristics. This effect occurred for all tested samples and only the length of axial cracks differed. The formation of these cracks can be considered as a critical point. Sample stiffness was clearly reduced and, therefore, the slope of the further part of the characteristics already showed some differences for individual samples. There were also deviations from the straightforward course of the characteristics. Additional faults, visible on the characteristics of individual samples, corresponded to the formation of consecutive cracks and the enlargement of existing ones. This was largely random, hence the significant differences in the characteristics of the different shapes. However, it should be emphasized that there was no apparent link between these discrepancies and the group to which the samples belonged.

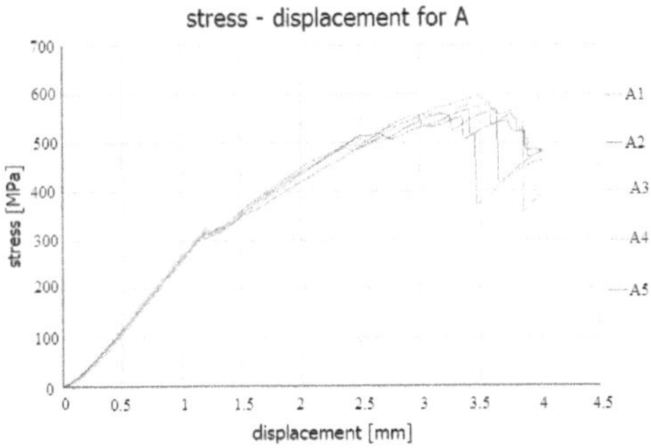

Figure 5. Mechanical characteristics of the 3-point bending test for reference samples–group A.

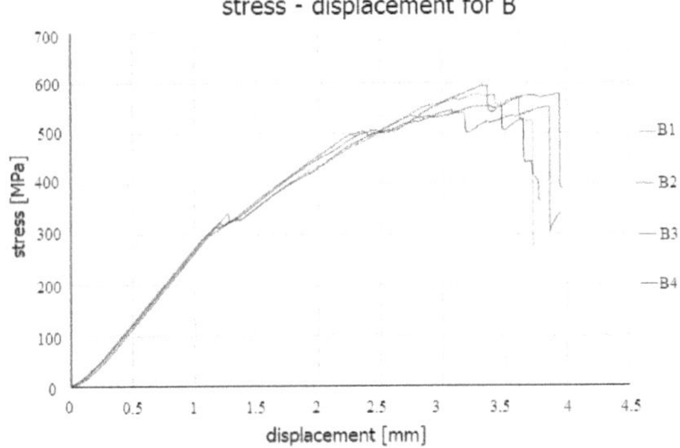

Figure 6. Mechanical characteristics of the 3-point bending test for samples subjected to temperature—group B.

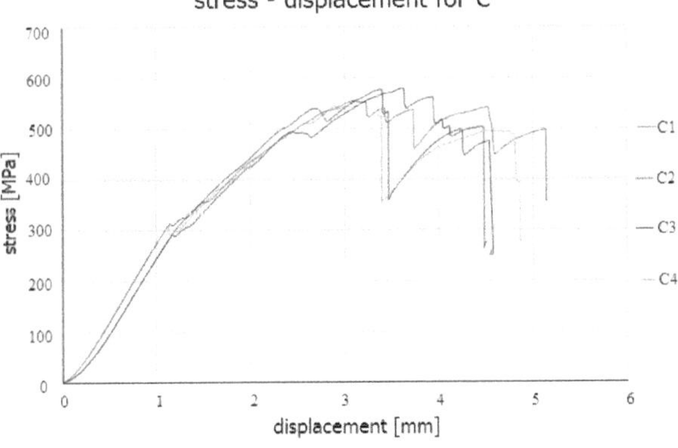

Figure 7. Mechanical characteristics of the 3-point bending test for samples subjected to high direct current (DC) voltage and temperature–group C.

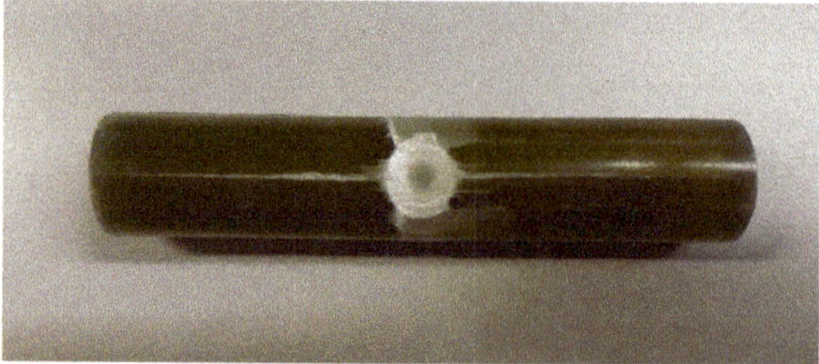

Figure 8. Long axial crack on top of sample A1, with an indentation caused by a crosshead of the strength testing machine.

Figure 9. Long axial crack in the lower part of sample C1.

Notwithstanding any differences in sample characteristics above the critical point (indicating the load where the breaking of the reinforcement initiates), samples exhibited high repeatability of the maximum stress value. When this value was reached, faults were produced, often with a large decrease in stress. In addition to audible cracklings, this indicates the formation of subsequent cracks that substantially reduced sample rigidity. It should also be noted that the samples did not deflect during the test. The recorded displacement of the crosshead (on the order of a few millimeters) caused the test samples to significantly indent, as illustrated in Figures 8 and 10. Additionally, at higher force values, steel components of the measuring system underwent deflection. After taking the samples out of the clamping system, they did not show the slightest bend. However, most of them had cracks on flat side surfaces, as seen in Figure 11. In Tables 1–3, the values of maximum force and stress were collected for all tested samples. Table 4 shows averaged values of maximum stress, together with standard deviation, for all three groups of samples.

Figure 10. Indentation at the point of operation of the crosshead of the strength testing machine, long axial crack and cracks on the side surface of sample C3.

Figure 11. Cracks on flat side surface of sample C4.

Table 1. Maximum force and stress values recorded for reference samples—group A.

Sample Designation	A1	A2	A3	A4	A5
Maximum force (kN)	32.33	30.85	30.82	31.38	30.00
Maximum stress (MPa)	595	568	567	577	552

Table 2. Maximum force and stress values recorded for samples subjected to temperature—group B.

Sample Designation	B1	B2	B3	B4
Maximum force (kN)	31.40	30.14	30.06	32.29
Maximum stress (MPa)	578	555	553	594

Table 3. Maximum force and stress values recorded for samples subjected to high DC voltage and temperature—group C.

Sample Designation	C1	C2	C3	C4
Maximum force (kN)	31.14	31.44	31.33	31.28
Maximum stress (MPa)	555	578	576	575

Table 4. Average values of maximum stress, including standard deviation, for the samples of all three groups.

Group of Samples	A	B	C
Average value maximum stress (MPa)	572 ± 14.1	570 ± 17.1	571 ± 9.7

3.2. Microscopic Examination of Samples

Microscopic examinations were carried out on the flat side surfaces of the samples, which were randomly selected from all three groups: A, B and C. For this purpose, fragments of the material were cut out of the selected samples. Further, we made so-called metallographic micro-sections, which were also used in micro-hardness measurements.

The metallographic micro-sections were prepared on a Struers LaboPol-2 polishing machine. The surface of the samples were grinded using SiC abrasive papers, and then polished using Struers DiaPro diamond suspension with the grain diameters of 3 μm and 1 μm. The final polishing was carried out on a colloidal SiO_2 suspension, with a grain size of 0.04 μm (Struers OP-S suspension). After each grinding and polishing step, the samples were washed in an ultrasonic washer in ethyl alcohol.

All recorded mechanical characteristics of the 3-point bending test were collected (Figure 12). The figures and tables illustrate the reproducibility of the results obtained from the 3-point bending test.

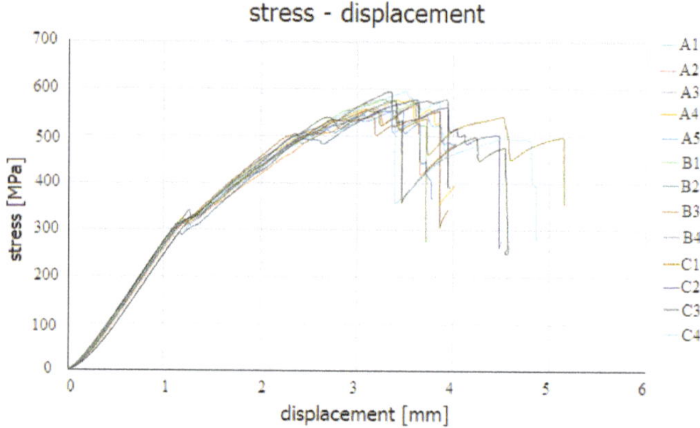

Figure 12. Mechanical characteristics of the 3-point bending test of all tested samples.

The analysis of microscopic images concluded that the microstructure of the tested material was characterized by high homogeneity. Images from different places on the tested surface, both from the same and different samples, showed no differences. The tight arrangement of fibers and binders (in the form of epoxy resin occupying about 1/3 surface) was analogous to all of the tested observation fields. The glass fiber diameter was also the same (several micrometers) with a small size dispersion. It should be emphasized that there were no differences for individual samples from groups A, B and C. Neither the long-term subjection to a temperature of about 50 °C nor the combined action of temperature and DC voltage of 20 kV caused any observable effects in the material's microstructure, which had been found in earlier studies [9]. Microstructure images of the sample material from all three groups are presented in Figures 13–15. There were few small dark chippings in the microstructure elements (i.e., fiber fragments and, less often, binder). They were created during the preparation of the surface of the samples.

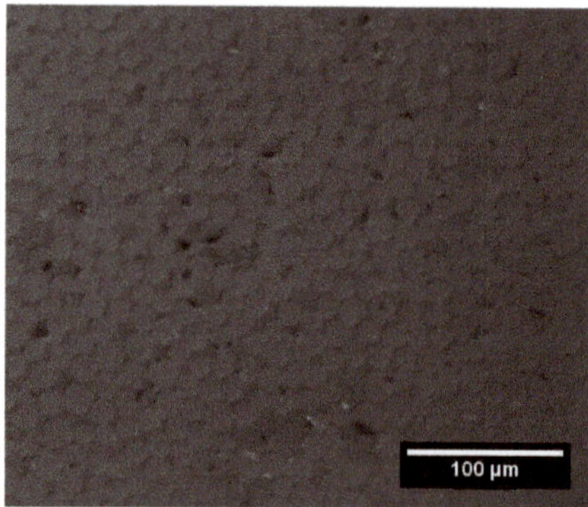

Figure 13. Microstructure image of sample A2 on its flat side surface, 200×. The ECR glass fibers in section and slightly darker epoxy resin are visible. Dark chipping of fiber fragments and less often of binder are few.

After appropriate reformatting of the 8-bit grey scale images and processing of these images with an optical microscope, the areas with the glass fibers and epoxy resin binder were depicted and distinguished more clearly, allowing for more accurate measurements. Image analysis was carried out with a Clemex computer-based analyzer. The blue phase was made up of glass fibers against a dark organic phase. The binary mask that was applied to the image, as shown in Figure 15, allowed for the quantitative measurement of glass fibers and resin content in the material and facilitated the determination of fiber diameters. Figure 16 shows the size distribution of glass fiber diameters obtained by averaging three microscopic images (A, B and C, one from each sample group).

Figure 14. Microstructure image of sample B2 on its flat side surface 100×. Few dark chipping of fiber fragments and binder are visible.

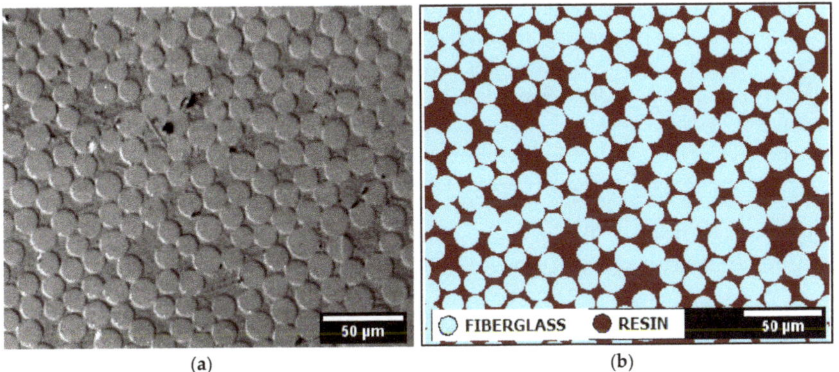

(a)   (b)

Figure 15. Microstructure image of sample C3 on its flat side surface, 200×. Few chipping of fiber fragments and binder are visible (**a**) on the right side and (**b**) the same area with a colored binary mask, which enables precise measurements to be taken.

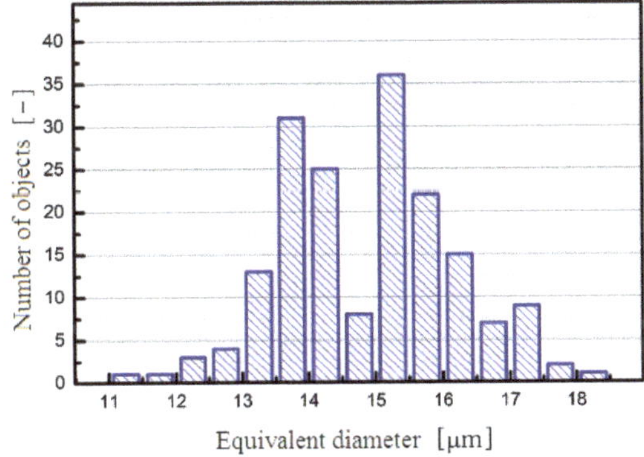

Figure 16. Size distribution of glass fiber diameters in the GRE material for the HVAC composite insulator carrying rod.

On all tested micro-sections of samples A2, B2 and C3, glass fibers represented, on average, 67.0%. The differences for individual observation fields did not exceed 2.1%. The average value of the glass fiber diameter was 14.9 mm. The whole distribution was between 11.0–18.5 mm. However, over 90% of the fibers had a diameter in a narrow range, from 13.0 to 16.5 mm. The distribution was clearly multimodal, with fibers with a diameter of 13.5–16.0 mm dominating.

The microstructure of the tested GRE material of the HVAC composite insulator carrying rod was clearly assessed as entirely appropriate, compact and homogeneous. Chipping created during the grinding and polishing process of test surfaces (mainly small fragments of glass fibers) did not exceed 3% of the surface. Tightly arranged glass fibers constituted 2/3 of the material by volume. The areas with only the resin visible in the micro-sections were not numerous and the size did not exceed several fibers. The ECR glass fibers used were uniform. Further, their diameter distribution was quite narrow and they did not raise any objections.

3.3. Micro-Hardness Measurements of Samples

Apart from the abovementioned microscopic tests, micro-hardness measurements were carried out for all samples. They allowed to access the cohesiveness and homogeneity of the material, being an important supplement to the results of other tests of the material. The measurements were carried out using the Vickers method, with a Struers Dura Scan universal micro-hardness meter with 1 kG of indenter load (HV1 – Hardness Vickers method, 1 means 1 kG). The measurements were carried out on the same side surfaces of the samples on which microscopic tests were performed. A semi-automatic mode of measuring the imprint diameter was used. If different diameters were obtained on the same imprint with generally small differences, their length was averaged (Figure 17). The following HV1 (International designation of Hardness Vickers method. HV1 means a hardness of 1 kG) values were obtained and averaged over 5 measurements on each sample:

- Reference sample A2–165 ± 9;
- Temperature-aged sample B2–169 ± 11;
- Temperature- and voltage-aged sample C3–155 ± 8.

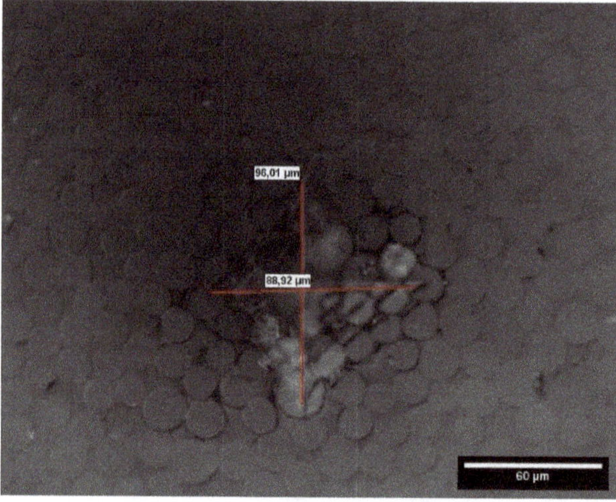

Figure 17. Image of an indenter imprint on the flat side surface of sample A2, 500×. In order to measure it correctly, the position of the markers was manually corrected. The measured values of the imprint diameter were averaged. There are no common microcracks running from the apex of the imprint, but there are visible fiber cracks as a result of the indenter operation.

The obtained micro-hardness values were high and clearly proved the quality of the tested GRE material of the HVAC composite insulator carrying rod. This confirmed the generally high evaluation of homogeneity and cohesiveness of the tested material. The imprints were often of a regular shape, allowing for automatic diameter measurements. In the case of less regular imprints, it was necessary to correct the marker setting manually, as illustrated in Figure 17. The typical phenomenon of relaxing the energy of load interaction through microcracks, especially running from the apex of the imprint, was not observed. However, the fibers located inside the area of the indenter operation often cracked. Images of typical imprints obtained on samples from three series are shown in Figures 17–19.

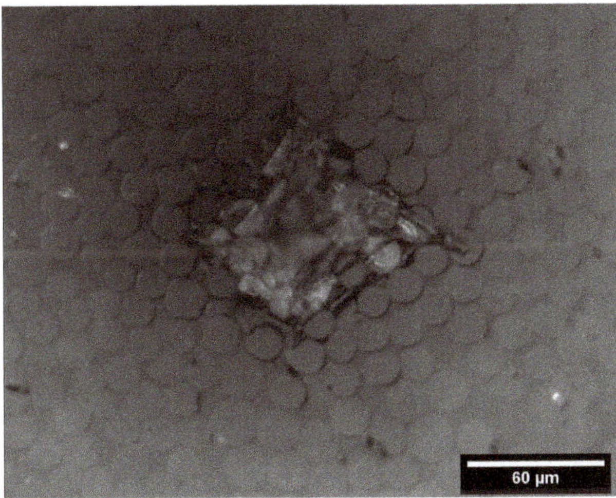

Figure 18. Image of an indenter imprint on the flat side surface of sample B2, 500×. Cracks and chipping of fiber fragments and resin damage as a result of the indenter operation are clearly visible.

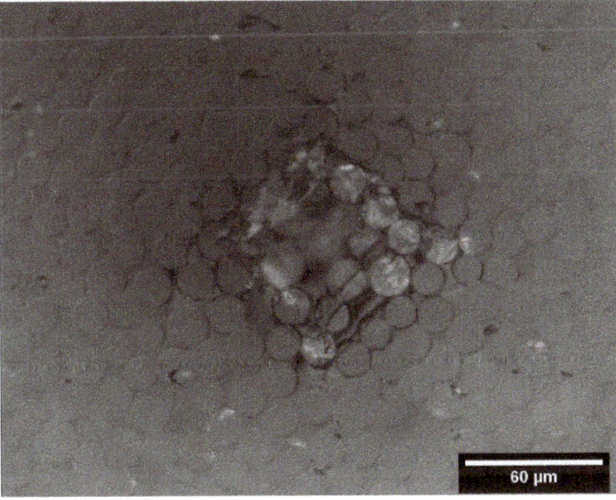

Figure 19. Image of an indenter imprint on the flat side surface of sample C3, 500×. Cracks and chipping of fiber fragments and resin damage in the area of the indenter operation are visible.

Compared to the average hardness of the reference sample, the thermally aged sample showed a slightly higher hardness (2.4%). The temperature and voltage-aged sample had a reduced hardness

(6.1%). The differences were small and comparable to the standard deviation values. Therefore, we can conclude that the material hardness of the tested samples from all three series remain at a similar level. There is no clear impact of ageing, neither as a result of temperature or the combined action of temperature and voltage on the material hardness, especially given the fact that the mechanical strength of all three series of tested samples remained at a similar level.

4. Discussion

Numerous publications that have focused on the use of composite insulators in high DC voltage lines have mainly examined their surface properties. Meanwhile, an important research issue that has not been mentioned in the literature is the long-term mechanical strength of the GRE cores exposed to high DC voltage. Long-term exposure to high DC voltage can lead to the development of an ionic current in the GRE core. This process may reduce the mechanical strength and, consequently, the insulator can break and the line would fall to the ground. This problem has been noticed in the case of glass disc insulators [13]. However, our research showed that the mechanical properties of the tested samples of the GRE cores did not deteriorate under the experiment's adopted conditions. Both the long-term subjection to a temperature of about 50 °C and the collective action of temperature and DC voltage of 20 kV did not cause any observable and undesirable effects in the microstructure of the material. A similar statement has been made in earlier studies [9]. The analysis of the obtained research results is presented in the conclusion.

5. Conclusions

The mechanical characteristics of the 3-point bending test for all tested samples (reference and aged) showed a high similarity. Small differences occurred individually for particular samples, but there were no differences in the characteristics of samples that would be typical for any of the sample groups—reference A, or aged B and C.

Notwithstanding the relatively small differences in sample characteristics, all tested samples showed a high repeatability of the maximum stress value. The average maximum stress values for the three sample groups (A, B and C) were almost identical.

The microstructure of the tested GRE material of the HVAC composite insulator carrying rod should be assessed as entirely appropriate, compact and homogeneous.

Tightly arranged glass fibers constituted 2/3 of the material by volume. Areas with just resin were not numerous and their size did not exceed several fibers. This is difficult to avoid in the production process. The ECR glass fibers used were uniform; their diameter distribution was quite narrow and did not raise any objections.

The obtained micro-hardness values were high and proved the quality of the tested material of the HVAC composite insulator carrying rod. A small dispersion of results (±6.5%) should be emphasized. This confirmed the overall good evaluation of homogeneity and cohesiveness of the tested material. It can be concluded that the material hardness of the tested samples from all three series (A, B and C) remained at a similar high level. Therefore, there was no clear impact of ageing neither as a result of temperature nor as the combined action of temperature and voltage on the hardness of the GRE material.

On the basis of the mechanical and microscopic tests presented above (performed on reference samples and aged GRE samples of the HVAC composite insulator carrying rod) it was found that there were no registrable degradation effects. Long-term (6000 h) interaction of temperature 50 °C and the combined action of temperature and DC voltage 20 kV, did not cause any decrease of GRE material mechanical parameters or change in its microstructure. Thus, the test results presented in [9] were fully confirmed.

Author Contributions: Conceptualization: K.W. and P.R.; methodology: K.W., P.R., Z.R., and P.P.; validation: K.W., P.R., Z.R., and P.P.; investigation: K.W. and P.R.; data curation: K.W. and P.R.; writing—original draft preparation: K.W., P.R., Z.R., P.P.; writing—review and editing: K.W., P.R., Z.R., and P.P. All authors have read and agreed to the published version of this manuscript.

Funding: This research received no external funding.

Conflicts of Interest: The authors declare no conflict of interest.

References

1. George, J.; Lodi, Z. Design and selection criteria for HVDC overhead transmission lines insulators. In Proceedings of the 2009 CIGRE Canada Conference on Power Systems, Toronto, ON, Canada, 4–6 October 2009.
2. Kumosa, M.; Armentrout, D.; Burks, B.; Hoffman, J.; Kumosa, L.; Middleton, J.; Predecki, P. Polymer matrix composites in high voltage transmission line application. In Proceedings of the 18th International Conference on Composites Materials (ICCM), Jeju Island, Korea, 21–26 August 2011.
3. Osmanbasic, E. High-Voltage DC Power Transmission: Should HVDC Replace AC in Power Systems? Available online: https://www.allaboutcircuits.com/technical-articles/high-voltage-dc-power-transmission-hvdc-replace-ac-power-systems/ (accessed on 20 October 2020).
4. Raju, M.; Subramaniam, N.P. Comparative study on Disc Insulators Deployed in EHV AC and HVDC Transmission Lines. In Proceedings of the IEEE International Conference on Circuit, Power and Computing Technologies, ICCPCT 2016, Tamilnadu, India, 18–19 March 2016.
5. Gubanski, S. Surface Charge & DC Flashover Performance of Composite Insulators. *INMR*. 13 August 2016. Available online: https://www.inmr.com/surface-charge-flashover-performance-composite-insulators/ (accessed on 20 October 2020).
6. Morshuis, P.; Cavallini, A.; Fabiani, D.; Montanari, G.C.; Azcarraga, C. Stress conditions in HVDC equipment and routes to in service failure. *IEEE Trans. Dielectr. Electr. Insul.* **2015**, *22*, 81–91. [CrossRef]
7. Working Group C4.303 CIGRE. Outdoor insulation in polluted conditions: Guidelines for selection and dimensioning—Part 2: THE DC CASE. In *Technical Brochures CIGRE*; Available online: https://e-cigre.org/publication/518-outdoor-insulation-in-polluted-conditionsguidelines-for-selection-and-dimensioning---part-2-the-dc-case (accessed on 15 December 2020).
8. Yuan, C.; Xie, C.; Li, L.; Xu, X.; Gubanski, S.M.; Zhou, Y.; Li, Q.; He, J. Space charge behavior in silicone rubber from in-service aged HVDC composite insulators. *IEEE Trans. Dielectr. Electr. Insul.* **2019**, *26*, 843–850. [CrossRef]
9. Wieczorek, K.; Jaroszewski, M.; Ranachowski, P.; Ranachowski, Z. Examination of the properties of samples from glass-epoxy core rods for composite insulators subjected to DC high voltage. *Arch. Metall. Mater.* **2018**, *63*, 1281–1286.
10. Kumosa, L.S.; Kumosa, M.S.; Armentrout, D.L. Resistance to brittle fracture of glass reinforced polymer composites used in composite (nonceramic) insulators. *IEEE Trans. Power Deliv.* **2005**, *20*, 2657–2666. [CrossRef]
11. *Glass Textiles—Threads—Markings*; Standard PN-EN ISO2078:2011; Polish Committee for Standarization: Warsaw, Poland, 2011. (In Polish)
12. *Ceramic and Glass Insulating Materials—Part 2*; Standard PN-EN 60672-2:2002; Polish Committee for Standarization: Warsaw, Poland, 2002. (In Polish)
13. Pigini, A. Composite Insulators for DC. *INMR*. 28 June 2018. Available online: https://www.inmr.com/composite-insulators/ (accessed on 20 October 2020).

Publisher's Note: MDPI stays neutral with regard to jurisdictional claims in published maps and institutional affiliations.

Article

Effect of Moisture on the Thermal Conductivity of Cellulose and Aramid Paper Impregnated with Various Dielectric Liquids

Grzegorz Dombek [1],*, Zbigniew Nadolny [1], Piotr Przybylek [1], Radoslaw Lopatkiewicz [2], Agnieszka Marcinkowska [3], Lukasz Druzynski [1], Tomasz Boczar [4] and Andrzej Tomczewski [5]

[1]   Institute of Electric Power Engineering, Poznan University of Technology, 60-965 Poznan, Poland; zbigniew.nadolny@put.poznan.pl (Z.N.); piotr.przybylek@put.poznan.pl (P.P.); lukasz.druzynski@student.put.poznan.pl (L.D.)
[2]   Power Engineering Transformatory Sp. z o.o., 62-004 Czerwonak, Poland; r.lopatkiewicz@petransformatory.pl
[3]   Institute of Chemical Technology and Engineering, Poznan University of Technology, Berdychowo 3, 60-965 Poznan, Poland; agnieszka.marcinkowska@put.poznan.pl
[4]   Institute of Electric Power Engineering and Renewable Energy, Opole University of Technology, Proszkowska 76, 45-758 Opole, Poland; t.boczar@po.edu.pl
[5]   Institute of Electrical Engineering and Electronics, Poznan University of Technology, Piotrowo 3A, 60-965 Poznan, Poland; andrzej.tomczewski@put.poznan.pl
*   Correspondence: grzegorz.dombek@put.poznan.pl; Tel.: +48-61-665-2192

Received: 6 July 2020; Accepted: 26 August 2020; Published: 27 August 2020

Abstract: This paper presents the effect of the impact of moisture in paper insulation used as insulation of transformer windings on its thermal conductivity. Various types of paper (cellulose and aramid) and impregnated (mineral oil, synthetic ester, and natural ester) were tested. The impact of paper and impregnated types on the changes in thermal conductivity of paper insulation caused by an increase in moisture were analyzed. A linear equation, describing the changes in thermal conductivity due to moisture, for various types of paper and impregnated, was developed. The results of measuring the thermal conductivity of paper insulation depending on the temperature are presented. The aim of the study is to develop an experimental database to better understand the heat transport inside transformers to assess aging and optimize their performance.

Keywords: aramid paper; cellulose; dielectric materials; insulation system; mineral oil; moisture; natural ester; synthetic ester; thermal conductivity; transformers

---

1. Introduction

The power system is a collection of devices for the generation, transmission, distribution, storage, and use of electricity. Enabling the delivery of electricity for households, enterprises, and public utilities in a continuous and uninterrupted manner is based primarily on the functional connection and appropriate maintenance of this strategic infrastructure. From the point of view of the transmission and distribution of electricity, power transformers play an important role. Knowledge of the condition of the transformer is necessary to achieve maximum return on investment, as well as to minimize the costs associated with its operation [1,2].

At present, the age of most power transformers working in the power system exceeds 25 years, which may result in the need to revitalize or replace them in the coming years [3,4]. The efficiency of transformers depends mainly on the state of their insulation system [5–7]. The insulation system is a type of "transformer heart". Due to the fact that it consists of solid (cellulose and aramid paper) and

Due to the fact that the main goal was to analyze the impact of moisture on the thermal conductivity of the above-mentioned combinations of papers impregnated with different insulating liquids, samples with different water content in paper (WCP) were prepared. In accordance with [50–52], the water content in paper is defined as a ratio of water weight and dry weight of a paper sample expressed as a percentage.

Measurements of thermal conductivity of the analyzed materials were carried out for temperatures in the range of 25 to 100 °C. The lower temperature range resulted, first, from the capabilities of the measuring system. Secondly, for lower temperatures, thermal aspects do not play a significant role in the operation of transformers. First, the upper temperature range also resulted from the limitations of the measuring system. Secondly, situations where the working temperature of the paper insulation exceeds 100 °C are rare.

2.2. Preparation of Paper Samples with Various Moisture Content

To determine the effect of moisture of fibrous materials impregnated with various insulating liquids on their thermal conductivity, it was necessary to prepare samples of cellulosic and aramid materials with different water content. The sample preparation procedure included the following stages: (I) drying of the samples, (II) moistening the samples, (III) impregnation and conditioning the samples, and (IV) measuring the water content of the samples.

Samples of fibrous materials were dried in a vacuum chamber for eight hours, at a temperature of 90 ± 5 °C, at a pressure of 0.2 to 0.4 mbar. Such drying conditions enabled the water content of the samples to be reduced below 0.2%. Then, to obtain an appropriate level of water content, samples of materials were moistened in a controlled manner by placing them in a climate chamber forcing an appropriate level of relative humidity and temperature, in accordance with the water sorption isotherms in [53,54]. The moistening time of the samples in the chamber was 72 h. In this way, samples with water content in cellulose and aramid paper above 2.0% and 1.5%, respectively, were prepared. The difference in water content between cellulose and aramid samples is due to the different hygroscopicity of both materials.

Due to the inability to set the relative humidity of the air below 10% in the environmental chamber, the remaining samples with a lower moisture level were obtained by placing them in airtight vessels with the appropriate amount of water calculated on the basis of the sample weight, the volume of the vessels, and the assumed moisture level. Samples in sealed vessels were moistened for 168 h. Samples of fibrous materials prepared in this way were impregnated and conditioned in mineral oil, natural ester, or synthetic ester for a period of about 30 days. Before testing, the water content of fibrous material samples was determined using the Karl Fischer Test KFT method in accordance with the International Electrotechnical Commission standard IEC 60814 [53]. Methanol was used to extract water from fibrous samples.

After the thermal conductivity tests, the water content of fibrous materials was measured by control using the KFT method. These studies did not show a significant impact on the conditions of thermal conductivity tests on the change of water content in samples of fibrous materials. The differences in moisture were within the measurement error of the KFT method and did not exceed the 0.2 percentage point.

2.3. Thermal Conductivity Measurements

To measure the thermal conductivity coefficient λ of papers impregnated with insulating liquids, the self-developed measuring system presented in Figure 1 was used. This system has been described in more detail in publications [55,56].

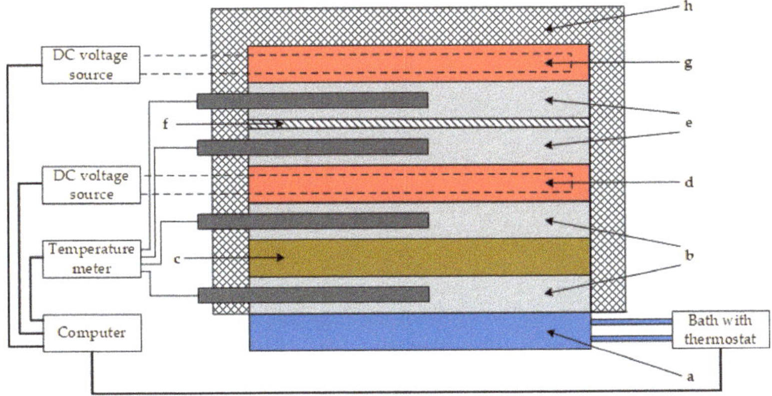

Figure 1. System for measuring the thermal conductivity coefficient λ of solid materials; **a**—cooler, **b,e**—auxiliary plates with measuring probes, **c**—test sample, **d**—main heater, **f**—auxiliary insulation, **g**—auxiliary heater, **h**—main insulation.

The used measuring system is based on the idea of measuring the thermal conductivity coefficient λ using fixed methods. The concept of measurement consists of causing in the test sample, of known thickness d and surface area S, a thermal disturbance ΔT. Based on these quantities, the coefficient of thermal conductivity λ can be determined from the relationship [56]:

$$\lambda = \frac{P}{S} \cdot \frac{d}{\Delta T} \tag{1}$$

where P is the heat of power source (W), S is the surface area of the tested sample (m^2), d is the thickness of the tested sample (m), and ΔT is the temperature drop in the tested sample (°C).

During the measurement, the tested sample (c) was placed between the main heater (d)—supplied with a constant voltage source and the cooler (a)—supplied with water from an external circuit. In the presented measuring system, the tested sample (c) was placed under the main heater (d) in order to eliminate the effect of convection on the temperature drop in the tested sample. The main heater (d) with the power P was designed to generate thermal energy which, penetrating through the sample, causes a decrease in temperature ΔT for a given thickness of sample d. The surface area of the main heater (and also the auxiliary heater) corresponds to the surface area of the tested sample S. The cooler (a) was designed to provide a constant temperature on the bottom surface of the sample. The temperature drop in the tested sample (c) was determined based on the temperature measurement in the auxiliary plates (b), which are made of aluminum and in which the measuring probes (Pt1000 measuring probes connected with the temperature recorder Apek APL 154 (Apek, Warsaw, Poland)) were placed. Auxiliary insulation (f) was placed over the main heater. In turn, an auxiliary heater (g) was placed over the auxiliary insulation to compensate for the heat flow from the main heater (d) in a vertically upward direction. The auxiliary heater (g), like the main heater, is powered by a DC voltage source. The task of the auxiliary heater (g) is to generate such an amount of heat that the indications of the measuring probes placed in the auxiliary plates (e) are the same, which means no heat flow in a perpendicular upward direction. In addition, to isolate heat loss in the measuring system, external insulation was also used (h). Adjusting the settings of both power supplies, as well as recording temperature was done using a special computer program written in the LabView environment (National Instruments, Austin, Texas, TX, USA).

Thermal conductivity error $\Delta \lambda$ was calculated on the basis of the complete differential of Equation (1). The maximal value of the error was smaller than 2% (Appendix A).

3. Experiment Results and Analysis

3.1. Thermal Conductivity Coefficient of Unimpregnated Papers

Table 2 presents the results of measurements of the thermal conductivity of unimpregnated cellulose paper and unimpregnated aramid paper depending on the temperature. The WCP moisture level of both analyzed types of paper was similar and very low—it was 0.52% for cellulose paper and 0.44% for aramid paper, respectively.

Table 2. Thermal conductivity coefficient λ of cellulose and aramid unimpregnated papers at different temperatures T.

Temperature T (°C)	Thermal Conductivity Coefficient λ (W·m^{-1}·K^{-1})	
	Cellulose Paper	Aramid Paper
25	0.076	0.061
40	0.086	0.075
60	0.098	0.088
80	0.110	0.101
100	0.121	0.110

As can be seen from the presented results, the thermal conductivity of unimpregnated cellulose paper was about 9 to 25% higher (depending on the temperature) than the thermal conductivity of unimpregnated aramid paper. This difference decreases with increasing temperature. Both aramid and cellulose paper are polymers, the first of which is a synthetic polymer and the second a natural one. Polymers are characterized by a low value of thermal conductivity coefficient [57]. The thermal conductivity of this group of materials depends on many factors such as structure, molecular weight, density, and degree of crystallinity. As the polymer structure is ordered, its thermal conductivity increases. Both unimpregnated cellulose paper and unimpregnated aramid paper in their structure contain defective structures among other voids, amorphous areas, and entanglements, which impede the spread of heat [58]. They cause a large dispersion of phonons, which reduces heat transport. In addition, voids are filled with air, whose thermal conductivity is very small (0.025 W·m^{-1}·K^{-1}) [59], smaller than the thermal conductivity of papers. A higher value of the thermal conductivity coefficient of cellulose paper results from its density. Both analyzed types of paper were characterized by the same thickness—the thickness of a single layer of paper was 0.005 mm. The density of cellulose paper was 915 kg·m^{-3} and of aramid paper was 709 kg·m^{-3} [46]. This means that cellulose paper in its structure has less difficulty in transporting through heat voids, which are filled with air. Due to the fact that the thermal conductivity of paper is a resultant of the thermal conductivity of paper fibers and air trapped between the fibers, the thermal conductivity of unimpregnated cellulose paper is higher than the thermal conductivity of unimpregnated aramid paper.

3.2. Effect of the Moisture on the Thermal Conductivity of Impregnated Cellulose Paper

Table 3 and Figure 2 present the results of measurements of the thermal conductivity of cellulose paper impregnated with various electrical insulating liquids (mineral oil, synthetic ester, and natural ester) depending on the moisture content.

Comparing the results of the measurements given in Tables 2 and 3, it can be said that the treatment of cellulose paper impregnation with insulating liquids resulted in an increase in its thermal conductivity, which is associated with the replacement of air trapped in the pores of the paper with an insulating liquid which has about one order of magnitude greater thermal conductivity. This conductivity at 25 °C is equal to 0.133 W·m^{-1}·K^{-1} for pure mineral oil, 0.158 W·m^{-1}·K^{-1} for pure synthetic ester, 0.182 W·m^{-1}·K^{-1} for pure natural ester [60,61], and only 0.025 W·m^{-1}·K^{-1} for air.

Based on the measurement results shown in Table 3, it can be concluded that as the temperature increases, the coefficient of thermal conductivity of the impregnated cellulose paper increases, regardless of the type of impregnating liquid. As is well known, the thermal conductivity of pure liquids decreases with temperature (except for water and glycerin). For tested pure insulating liquids in the examined temperature range of 25 to 80 °C, a decrease in their thermal conductivity by 0.007 $W \cdot m^{-1} \cdot K^{-1}$ was observed regardless of their type [60]. However, the thermal conductivity of unimpregnated cellulose paper increased with increasing temperature, and this increase, in the studied temperature range of 25 to 80 °C, was 0.034 $W \cdot m^{-1} \cdot K^{-1}$ (Table 2). This was due to the fact that both the thermal conductivity of gases and solids increase with temperature. Thus, it can be concluded that primarily the cellulose fibers were responsible for the heat transfer in impregnated cellulose paper, not the insulating liquid. In addition, the conductivity of cellulose paper impregnated with insulating liquids was higher than the conductivity of pure insulating liquids. The increase in the thermal conductivity of the oil-impregnated paper is similar for all tested liquids. In such complex systems, many factors influence heat conduction. However, a similar effect of the tested liquids on the interactions in the papers was observed [57,58]. Since the paper seems to be primarily responsible for heat conduction a similar effect is observed.

Table 3. Thermal conductivity coefficient λ of cellulose paper impregnated by various dielectric liquids depending on the water content of paper WCP, at different temperatures T.

Type of Insulation System	Water Content of Paper WCP (%)	Temperature T (°C)				
		25	40	60	80	100
		Thermal Conductivity Coefficient λ ($W \cdot m^{-1} \cdot K^{-1}$)				
Cellulose paper impregnated by mineral oil	0.50	0.152	0.164	0.176	0.191	0.204
	0.87	0.153	0.162	0.176	0.188	0.203
	1.03	0.150	0.162	0.176	0.189	0.206
	1.30	0.152	0.164	0.178	0.194	0.207
	1.68	0.155	0.167	0.184	0.200	0.213
	1.92	0.159	0.169	0.185	0.202	0.215
	2.24	0.158	0.168	0.182	0.198	0.213
	2.84	0.163	0.174	0.191	0.205	0.217
	4.00	0.163	0.173	0.190	0.203	0.217
	4.80	0.166	0.176	0.188	0.204	0.217
Cellulose paper impregnated by synthetic ester	0.18	0.174	0.185	0.198	0.211	0.220
	1.00	0.176	0.187	0.199	0.210	0.221
	2.38	0.178	0.190	0.202	0.212	0.223
	3.12	0.179	0.191	0.202	0.212	0.222
	4.11	0.180	0.194	0.206	0.217	0.225
	4.32	0.184	0.194	0.207	0.216	0.226
	4.85	0.185	0.199	0.208	0.219	0.227
	5.53	0.186	0.198	0.211	0.220	0.228
Cellulose paper impregnated by natural ester	0.31	0.204	0.212	0.221	0.230	0.237
	0.93	0.205	0.212	0.220	0.231	0.239
	2.28	0.206	0.214	0.224	0.236	0.242
	3.15	0.208	0.215	0.225	0.236	0.241
	4.36	0.210	0.217	0.227	0.237	0.242
	4.71	0.211	0.217	0.228	0.238	0.244
	5.27	0.212	0.219	0.229	0.241	0.248
	5.74	0.213	0.221	0.231	0.243	0.249
	6.63	0.215	0.225	0.234	0.246	0.252

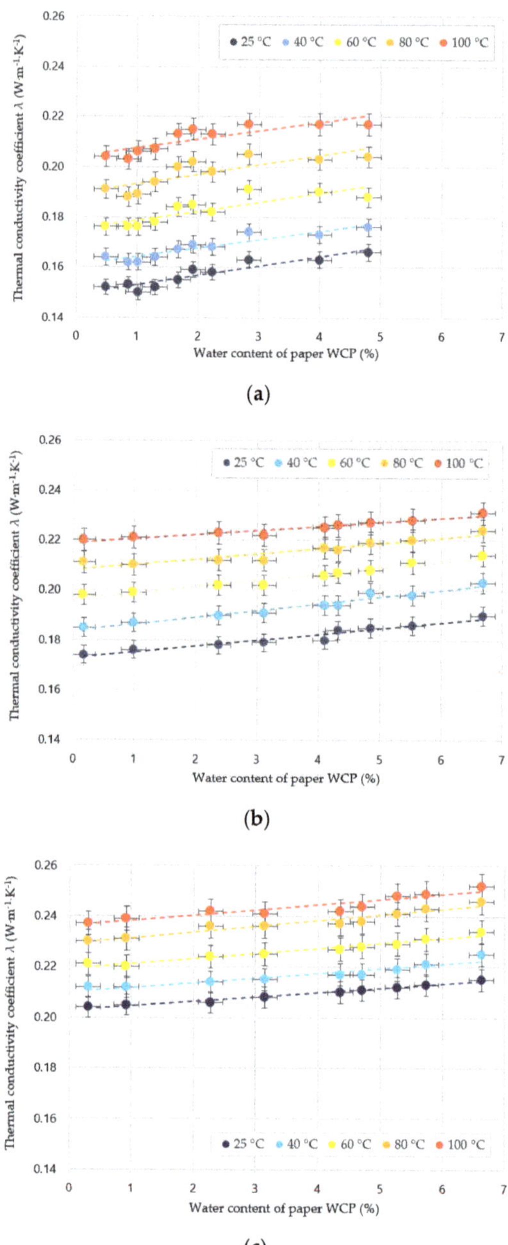

Figure 2. Thermal conductivity coefficient λ of cellulose paper impregnated by various dielectric liquids depending on the water content of paper WCP and temperature T: (**a**) cellulose paper impregnated by mineral oil; (**b**) cellulose paper impregnated by synthetic ester; (**c**) cellulose paper impregnated by natural ester.

Analyzing the measurement results presented in Table 3, it can be seen that the thermal conductivity of cellulose paper impregnated with various insulating liquids increased with increasing moisture content. Water probably penetrated into the pores of the paper. The increase in thermal conductivity of impregnated paper, caused by an increase in moisture, was associated with about four times greater thermal conductivity of water (about 0.60 $W \cdot m^{-1} \cdot K^{-1}$) [60] compared to the thermal conductivity of the analyzed pure insulating liquids (average 0.15 $W \cdot m^{-1} \cdot K^{-1}$) [61]. The increase in the thermal conductivity of the impregnated cellulose paper, accompanied by the increase in moisture, was practically independent of the type of insulating liquid. The average value of the increase in thermal conductivity fluctuated in the range of 5 to 7% for all types of analyzed liquids. However, the increase in paper thermal conductivity, caused by an increase in moisture, depended on the measurement temperature. As the temperature increased, the increases in thermal conductivity of the impregnated cellulose paper were becoming smaller.

In summary, moisture in cellulose insulation increased the thermal conductivity of this insulation. Thus, moisture, in addition to many of the disadvantages described at the beginning of this article, also has positive features. Greater thermal conductivity of cellulose insulation will result in more efficient heat dissipation from the transformer windings to the cooling liquid. This in turn can lower the hot spot temperature.

3.3. Effect of the Moisture on the Thermal Conductivity of Impregnated Aramid Paper

Table 4 and Figure 3 present the results of measurements of the thermal conductivity of aramid paper impregnated with various insulating liquids (mineral oil, synthetic ester, and natural ester) depending on the moisture content.

Based on the results of the measurements, it can be seen that the thermal conductivity of aramid paper, impregnated with insulating liquids, similar to cellulose paper, was higher than the thermal conductivity of unimpregnated aramid paper (containing only air in pores) [60,61]. However, this increase was smaller than for cellulose paper and at a temperature below 60 °C, it did not exceed the thermal conductivity of insulating liquids. It is possible, therefore, that in this case conduction at the liquid-aramid paper interface is less effective.

As in the case of cellulose paper, it can also be seen that the thermal conductivity of all analyzed samples of aramid paper increased with increasing temperature. This means that the heat transfer in impregnated aramid paper, as in the case of cellulose paper, is carried out primarily through the paper fibers, not the insulating liquid.

Table 4. Thermal conductivity coefficient λ of aramid paper impregnated by various dielectric liquids depending on the water content of paper WCP, at different temperatures T.

Type of Insulation System	Water Content of Paper WCP (%)	Temperature T (°C)				
		25	40	60	80	100
		Thermal Conductivity Coefficient λ ($W \cdot m^{-1} \cdot K^{-1}$)				
Aramid paper impregnated by mineral oil	0.07	0.128	0.140	0.149	0.164	0.178
	1.59	0.131	0.142	0.153	0.168	0.181
	1.96	0.132	0.144	0.155	0.169	0.183
	3.84	0.135	0.146	0.157	0.171	0.184
Aramid paper impregnated by synthetic ester	0.50	0.144	0.155	0.168	0.181	0.192
	2.12	0.148	0.160	0.171	0.184	0.196
	2.44	0.149	0.160	0.173	0.184	0.197
	3.94	0.151	0.161	0.173	0.185	0.197
Aramid paper impregnated by natural ester	0.43	0.157	0.169	0.179	0.190	0.200
	1.82	0.161	0.172	0.181	0.193	0.205
	2.09	0.162	0.173	0.183	0.194	0.207
	3.52	0.164	0.174	0.183	0.193	0.208

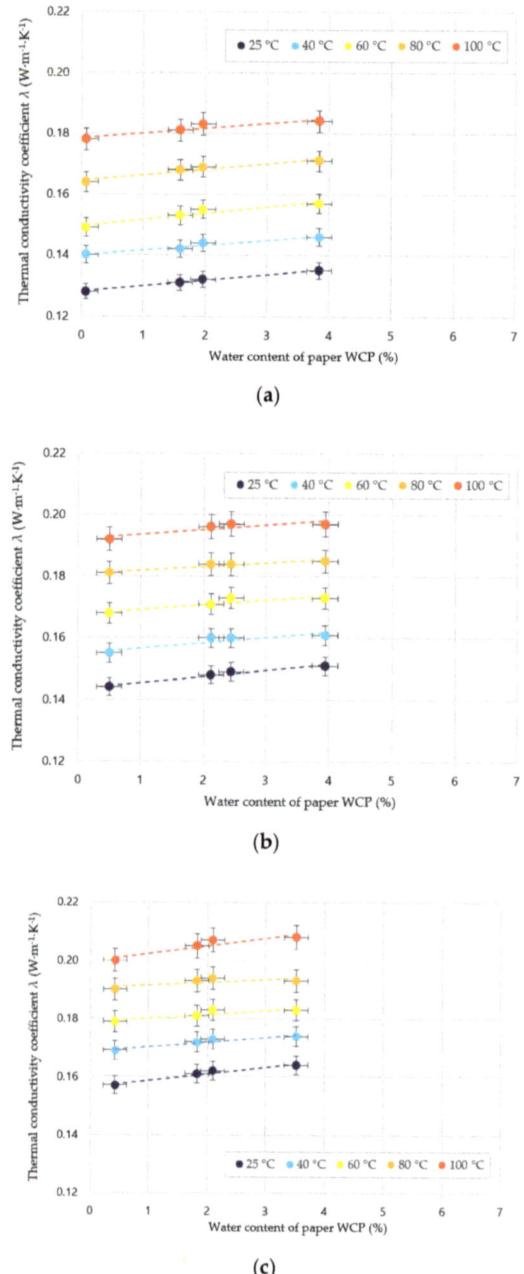

Figure 3. Thermal conductivity coefficient λ of aramid paper impregnated by various dielectric liquids depending on the water content of paper WCP and temperature T: (**a**) aramid paper impregnated by mineral oil; (**b**) aramid paper impregnated by synthetic ester; (**c**) aramid paper impregnated by natural ester.

Analyzing the results of the measurements in Table 4, it can be seen that the thermal conductivity of aramid paper impregnated with various insulating liquids, like in the case of cellulose paper, increased with increasing moisture. Water, which has four times greater thermal conductivity than the thermal conductivity of pure insulating liquids, is responsible for this result [61]. It interacts with the paper, probably binds to it, and penetrates into its pores.

The increase in thermal conductivity of impregnated aramid paper, accompanying the increase in moisture, was practically independent of the type of insulating liquid, and its average value was in the range of 3 to 5% for all types of analyzed liquids. It can be associated with the similar influence of all investigated oils (EN, ES, OM) on aramid paper. However, the increase in thermal conductivity of aramid paper, caused by an increase in moisture, similar to cellulose paper, depended on temperature. For higher temperature values, the increases in thermal conductivity of aramid paper were getting smaller.

In summary, the moisture content of aramid insulation slightly increases its thermal conductivity. Thus, as in the case of cellulose paper, a slightly higher thermal conductivity of aramid insulation will result in more efficient heat dissipation from the transformer windings, which will contribute to a lower value of the hot spot.

3.4. Comparison of Thermal Conductivity of Impregnated Cellulose and Aramid Paper in the Context of Their Moisture

Based on the results of the measurements presented in Tables 3 and 4, it can be stated that the increase in the moisture content of the paper insulation caused an increase in its thermal conductivity, both for cellulose and aramid paper.

The increase in thermal conductivity, caused by moisture, in the case of cellulose paper was 5 to 7%, and in the case of aramid paper, this increase was slightly smaller, equal to 3 to 5%. The reason for this was certainly the upper limit to which the samples could be moistened, which was 5 to 7% WCP for cellulose paper, and only 4% WCP for aramid paper. On this basis, it can be said that moisture in the paper causes a similar increase in thermal conductivity, regardless of the type of paper.

The increase in thermal conductivity caused by moisture was getting smaller as the temperature increased. This regularity was observed for practically all types of impregnating liquid and for both types of analyzed paper (cellulose and aramid).

Figure 4 presents the coefficient of thermal conductivity of paper, depending on moisture, for various types of paper, and impregnating liquid. The values of thermal conductivity are presented for 80 °C, as the most typical temperature for paper insulation of transformer windings.

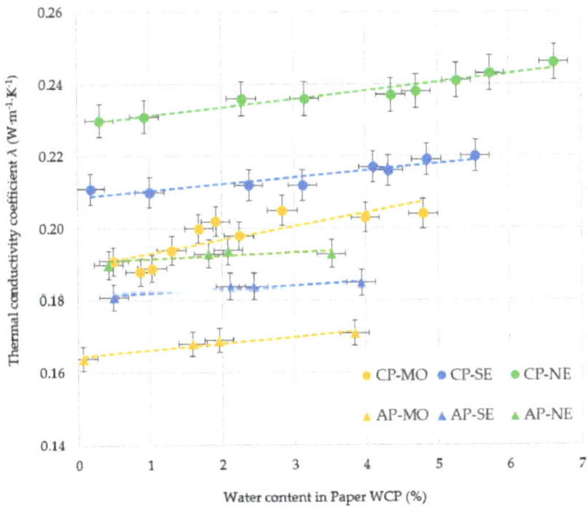

Figure 4. Thermal conductivity of paper insulation depending on moisture WCP measured at 80 °C, for various types of paper and insulating liquid.

As can be seen in Figure 4 the thermal conductivity of paper insulation increases linearly with increasing moisture content. On this basis, a linear equation has been proposed, describing the thermal conductivity of paper insulation depending on moisture:

$$\lambda = a + b \cdot (WCP) \qquad (2)$$

where a is the thermal conductivity of paper insulation for zero moisture (W·m^{-1}·K^{-1}), b is the coefficient determining the impact of moisture on the thermal conductivity of a given material (W·m^{-1}·K^{-1}·%$^{-1}$), and WCP is a percentage of water content in paper insulation (%). The parameters *a* and *b* of the linear equations were obtained for the six combinations of analyzed materials (Table 5).

Table 5. The parameters *a* and *b* of the linear equations for the six combinations of analyzed materials; Δa and Δb mean absolute standard error of parameters *a* and *b*, respectively.

Material	a	Δa	b	Δb
	(W·m^{-1}·K^{-1})		(W·m^{-1}·K^{-1}·%$^{-1}$)	
CP-MO	0.189	0.002	0.0038	0.0009
CP-SE	0.209	0.001	0.0019	0.0003
CP-NE	0.229	0.001	0.0023	0.0002
AP-MO	0.135	0.001	0.0018	0.0004
AP-SE	0.181	0.001	0.0012	0.0003
AP-NE	0.191	0.002	0.0010	0.0007

The *b* factor, determining the effect of moisture on the thermal conductivity of a given material, has very different values depending on the type of paper. For cellulose paper, these values (0.0021 ÷ 0.0038 (W·m^{-1}·K^{-1}·%$^{-1}$)) are much higher than for aramid paper (0.0010 ÷ 0.0018 (W·m^{-1}·K^{-1}·%$^{-1}$)). This means that the same moisture content of paper insulation causes a greater (about two times) increase in the thermal conductivity of cellulose paper than aramid paper. On this basis, it can be concluded that cellulose paper is more sensitive to moisture than aramid paper from the point of view of its thermal conductivity.

The situation is similar when analyzing the *b* factor depending on the type of insulating liquid. For mineral oil, the *b* factor is about twice as high as for both types of analyzed esters. This means that the same moisture content causes a double increase in thermal conductivity of paper impregnated with mineral oil compared to the thermal conductivity of paper impregnated with synthetic or natural ester. On this basis, it can be said that mineral oil is more sensitive to moisture than the analyzed esters from the point of view of thermal conductivity of paper impregnated with these liquids.

In conclusion, it can be said that the most susceptible paper insulation to changes in its thermal conductivity, caused by moisture, was cellulose paper impregnated with mineral oil, for which the coefficient b was 0.0038 (W·m^{-1}·K^{-1}·%$^{-1}$). In turn, the least susceptible paper insulation to changes in its thermal conductivity was aramid paper impregnated with synthetic or mineral ester, for which the coefficient b is four times smaller, equal to 0.0010 ÷ 0.0012 (W·m^{-1}·K^{-1}·%$^{-1}$).

The obtained equations are useful for determining the temperature field of the transformer at the design stage, and especially during its operation when the moisture content of paper insulation increases over the years. In recent years, advanced diagnostic techniques have been developed very dynamically that enable the determination of moisture content in paper insulation. These methods are based on the phenomenon of dielectric spectroscopy (Recovery Voltage Measurement RVM, Frequency Domain Spectroscopy FDS) [62,63].

4. Conclusions

Moisture in the paper insulation increased its thermal conductivity, regardless of the type of paper and insulating liquid that the paper has been impregnated with. In the case of cellulose paper, this increase is 5 to 7%, and for aramid paper, the increase in thermal conductivity fluctuates within

3 to 5%. The increase in thermal conductivity of paper insulation, caused by moisture, is associated with a much higher water conductivity (about 0.60 ($W\cdot m^{-1}\cdot K^{-1}$)) compared to the conductivity of both unimpregnated paper (0.06 ÷ 0.12 ($W\cdot m^{-1}\cdot K^{-1}$)), as well as used pure insulating liquids (about 0.15 ($W\cdot m^{-1}\cdot K^{-1}$)).

The increase in thermal conductivity caused by moisture became smaller as the temperature increased. This relationship was observed for practically all types of impregnating liquids and for both types of analyzed papers.

Based on the obtained results, it was found that the same moisture content of paper insulation caused a greater (about two times) increase in the thermal conductivity of cellulose paper than aramid paper. Thus, cellulose paper was more sensitive to moisture than aramid paper in terms of its thermal conductivity. The same moisture content caused a two-fold increase in the thermal conductivity of paper impregnated with mineral oil compared to the thermal conductivity of paper impregnated with synthetic or natural ester. This means that mineral oil is more sensitive to moisture than the analyzed esters from the point of view of thermal conductivity of paper impregnated with these liquids.

Obtained equations, describing the effect of moisture on the thermal conductivity of paper insulation, can be useful for determining the temperature field of the transformer both at the design stage and during its operation.

Author Contributions: Conceptualization, Z.N. and G.D.; methodology, G.D., Z.N. and R.L; software, R.L.; validation, G.D., P.P. and Z.N.; formal analysis, G.D., Z.N. and A.M.; investigation, G.D., Z.N., R.L. and P.P.; resources, G.D. and Z.N.; data curation, G.D.; writing—original draft preparation, G.D., Z.N., P.P., R.L., A.M., L.D., T.B. and A.T.; writing—review and editing, G.D. and Z.N.; visualization, G.D. and Z.N.; supervision, G.D. and Z.N.; project administration, G.D. and Z.N. All authors have read and agreed to the published version of the manuscript.

Funding: This research received no external funding.

Acknowledgments: This research was funded by the Ministry of Science and Higher Education, grant number 0711/SBAD/4411 and 0912/SBAD/2004.

Conflicts of Interest: The authors declare no conflict of interest.

Appendix A. Thermal Conductivity Error Estimation

Thermal conductivity error was calculated on the basis of the complete differential of equation, which helps to calculate the conductivity:

$$\lambda = f(p, q, \Delta T) = \frac{p \cdot d}{\Delta T} \tag{A1}$$

where λ is the thermal conductivity coefficient ($W\cdot m^{-1}\cdot K^{-1}$), p is the surficial thermal load ($W\cdot m^{-2}$), d is the thickness of sample (m), and ΔT is the temperature difference on the sample surfaces (K).

By expanding the λ function into Taylor series around the measurement values and neglecting words higher than the first row and replacing infinitely small increments of variables with independent finite increments, we will get:

$$\Delta\lambda - \left|\frac{\partial\lambda}{\partial p}\right|\cdot\Delta p + \left|\frac{\partial\lambda}{\partial d}\right|\cdot\Delta d + \left|\frac{\partial\lambda}{\partial\Delta T}\right|\cdot\Delta(\Delta T) \tag{A2}$$

and

$$\Delta\lambda = \left|\frac{d}{\Delta T}\right|\cdot\Delta p + \left|\frac{p}{\Delta T}\right|\cdot\Delta d + \left|-\frac{p\cdot d}{(\Delta T)^2}\right|\cdot\Delta(\Delta T) \tag{A3}$$

It should be noted, that:

$$p = \frac{U \cdot I}{S} \tag{A4}$$

where U is the voltage of heater (V), I is the current of heater (A), and S is the surface of the sample (m^2).

It means, that:

$$\Delta p = \left|\frac{\partial p}{\partial U}\right|\cdot\Delta U + \left|\frac{\partial p}{\partial I}\right|\cdot\Delta I + \left|\frac{\partial p}{\partial S}\right|\cdot\Delta S \tag{A5}$$

and

$$\Delta p = \left|\frac{I}{S}\right|\cdot\Delta U + \left|\frac{U}{S}\right|\cdot\Delta I + \left|-\frac{I\cdot U}{S^2}\right|\cdot\Delta S \tag{A6}$$

It should be noted, that:

$$S = x\cdot y \tag{A7}$$

where x is the length of the sample (m) and y is the width of the sample (m).

It means, that:

$$\Delta S = \left|\frac{\partial S}{\partial x}\right|\cdot\Delta x + \left|\frac{\partial S}{\partial y}\right|\cdot\Delta y \tag{A8}$$

and

$$\Delta S = \left|y\right|\cdot\Delta x + \left|x\right|\cdot\Delta y \tag{A9}$$

Summarizing:

$$\Delta\lambda = \left|\frac{d}{\Delta T}\right|\cdot\left(\left|\frac{I}{S}\right|\cdot\Delta U + \left|\frac{U}{S}\right|\cdot\Delta I + \left|-\frac{I\cdot U}{S^2}\right|\cdot\left(\left|y\right|\cdot\Delta x + \left|x\right|\cdot\Delta y\right)\right) + \left|\frac{p}{\Delta T}\right|\cdot\Delta d + \\ + \left|-\frac{p\cdot d}{(\Delta T)^2}\right|\cdot\Delta(\Delta T) \tag{A10}$$

Parameters d, ΔT, I, U, S, p, x, and y are measurement results or equipment setting values. All Δd, $\Delta(\Delta T)$, ΔU, ΔI, Δx, and Δy mean accuracy of used equipment and measurers, which are known. It is possible to calculate thermal conductivity error $\Delta\lambda$ on the basis of equations (A10). The value of maximal relative error was smaller than 2%.

References

1. Garcia, B.; Villarroel, E.D.; Garcia, D.F. A multiphysics model to study moisture dynamics in transformer. *IEEE Trans. Power Deliv.* **2019**, *34*, 1365–1373. [CrossRef]
2. Sikorski, W.; Szymczak, C.; Siodla, K.; Polak, F. Hilbert curve fractal antenna for detection and on-line monitoring of partial discharges in power transformers. *Eksploatacja i Niezawodność Maint. Reliab.* **2018**, *20*, 343–351. [CrossRef]
3. Gielniak, J.; Graczkowski, A.; Moranda, H.; Przybylek, P.; Walczak, K.; Nadolny, Z.; Moscicka-Grzesiak, H.; Gubanski, S.M.; Feser, K. Moisture in cellulose insulation of power transformers: Statistics. *IEEE Trans. Dielectr. Electr. Insul.* **2013**, *20*, 982–987. [CrossRef]
4. Toudja, T.; Chetibi, F.; Beldjilali, A.; Moulai, H.; Beroual, A. Electrical and physicochemical properties of mineral and vegetable oils mixtures. In Proceedings of the IEEE 18th International Conference on Dielectrics Liquids (ICDL), Bled, Slovenia, 29 June–3 July 2014; pp. 1–4.
5. Jeong, J.-I.; An, J.-S.; Huh, C.-S. Accelerated aging effects of mineral and vegetable transformer oils on medium voltage power transformers. *IEEE Trans. Dielectr. Electr. Insul.* **2012**, *19*, 156–161. [CrossRef]
6. Emsley, M.; Stevens, G.C. Kinetic and mechanisms of the low temperature degradation of cellulose. *Cellulose* **1994**, *1*, 26–56. [CrossRef]
7. Liao, R.; Liang, S.; Sun, C.; Yang, L.; Sun, H. A comparative study of thermal aging of the transformer insulation paper impregnated in natural ester and in mineral oil. *Eur. Trans. Electr. Power* **2010**, *20*, 518–533. [CrossRef]
8. Bandara, K.; Ekanayake, C.; Saha, T.K.; Annamalai, P.K. Understanding the aging aspects of natural ester based insulation liquid in power transformer. *IEEE Trans. Dielectr. Electr. Insul.* **2016**, *23*, 246–257. [CrossRef]
9. Hao, J.; Liao, R.; Chen, G.; Ma, Z.; Yang, L. Quantitative analysis ageing status of natural ester-paper insulation and mineral oil-paper insulation by polarization/depolarization current. *IEEE Trans. Dielectr. Electr. Insul.* **2012**, *19*, 188–199. [CrossRef]

10. Dombek, G.; Nadolny, Z. Thermal properties of mixtures of mineral oil and natural ester in terms of their application in the transformer. In Proceedings of the International Conference Energy, Environment and Material Systems (EEMS), Polanica-Zdroj, Poland, 13–15 September 2017; pp. 01040-1–01040-6.
11. Seytashmehr, A.; Fofana, I.; Eichler, C.; Akbari, A.; Borsi, H.; Gockenbach, E. Dielectric spectroscopic measurements on transformer oil-paper insulation under controlled laboratory conditions. *IEEE Trans. Dielctr. Electr. Insul.* **2008**, *10*, 903–917. [CrossRef]
12. Dombek, G.; Nadolny, Z.; Marcinkowska, A. Effects of nanoparticles materials on heat transfer in electro-insulating liquids. *Appl. Sci.* **2018**, *8*, 2538. [CrossRef]
13. IEC 60076-2:2011. *Power Transformers—Part 2: Temperature Rise for Liquid-Immersed Transformers*; International Electrotechnical Commission (IEC): New York, NY, USA, 2011.
14. Godina, R.; Rodrigues, E.M.G.; Matias, J.C.O.; Catalao, J.P.S. Effect of loads and other key factors on oil-transformer ageing: Sustainability benefits and challenges. *Energies* **2015**, *8*, 12147–12186. [CrossRef]
15. Wang, C.; Wu, J.; Wang, J.; Zhao, W. Reliability analysis and overload capability assessment of oil-immersed power transformers. *Energies* **2016**, *9*, 43. [CrossRef]
16. Lopatkiewicz, R.; Nadolny, Z.; Przybylek, P.; Sikorski, W. The influence of chosen parameters on thermal conductivity of windings insulation describing temperature distribution in transformer. *Przeglad Elektrotechniczny* **2012**, *88*, 126–129. (In Polish)
17. Dombek, G.; Nadolny, Z. Electro-insulating nanofluids based on synthetic ester and TiO_2 or C_{60} nanoparticles in power transformer. *Energies* **2018**, *11*, 1953. [CrossRef]
18. Xia, G.; Wu, G.; Gao, B.; Yin, H.; Yang, F. A new method for evaluating moisture content and aging degree of transformer oil-paper insulation based on frequency domain spectroscopy. *Energies* **2017**, *10*, 1195. [CrossRef]
19. Betie, A.; Meghnefi, F.; Fofana, I.; Yeo, Z. Modelling the insulation paper drying process from thermogravimetric analyses. *Energies* **2018**, *11*, 517. [CrossRef]
20. N'chos, J.S.; Fofana, I.; Hadjadj, Y.; Beroual, A. Review of physicochemical-based diagnostic techniques for assessing insulation condition in aged transformers. *Energies* **2016**, *9*, 367. [CrossRef]
21. Bhalla, D.; Bansal, R.K.; Gupta, H.O.; Hari, O.M. Preventing power transformer failures through electrical incipient fault analysis. *Int. J. Perform. Eng.* **2013**, *9*, 23–31. [CrossRef]
22. Sparling, B.; Aubin, J. Assessing water content in insulating paper of power transformers. *Electr. Energy T&D Mag.* **2007**, *11*, 30–34.
23. Cui, Y.; Ma, H.; Saha, T.; Ekanayake, C.; Wu, G. Multi-physics modelling approach for investigation of moisture dynamics in power transformers. *IET Gener. Transm. Dis.* **2016**, *10*, 1993–2001. [CrossRef]
24. Smolka, J.; Nowak, A.J. Experimental validation of the coupled fluid flow, heat transfer and electromagnetic numerical model of the medium-power dry-type electrical transformer. *Int. J. Therm. Sci.* **2008**, *47*, 1393–1410. [CrossRef]
25. Yatsevsky, V.A. Hydrodynamics and heat transfer in cooling channels of oil-filled power transformers with multicoil windings. *Appl. Therm. Eng.* **2014**, *63*, 347–353. [CrossRef]
26. Mithun, M.; Thankachan, B. A novel, cost effective capacitive sensor for estimating dissolved moisture in transformer oil. In Proceedings of the 10th International Conference on Sensing Technology (ICTS), Nanjing, China, 11–13 November 2016; pp. 1–6.
27. Oommen, T.V.; Prevost, T.A. Cellulose insulation in oil-filled power transformers: Part II maintaining insulation integrity and life. *IEEE Electr. Insul. Mag.* **2006**, *22*, 5–14. [CrossRef]
28. Su, Q.; James, R.E. *Condition Assessment of High Voltage Insulation in Power System Equipment*, 1st ed.; The Institution of Engineering and Technology: London, UK, 2008; pp. 121–158.
29. Wickman, S.H. Transformer bushings and oil leaks. *Transform. Mag. Spec. Ed. Bushing* **2017**, *4*, 148–154.
30. Sokolov, V.; Aubin, J.; Davydov, V.; Gasser, H.P.; Griffin, P.; Koch, M.; Lundgaard, L.; Roizman, O.; Scala, M.; Tenbohlen, S.; et al. *Moisture Equilibrium and Moisture Migration within Transformer Insulation Systems*; Cigré Technical Brochure 349; International Council on Large Electric Systems (CIGRE): Paris, France, 2008.
31. Li, Y.; Zhou, K.; Zhu, G.; Li, M.; Li, S.; Zhang, J. Study on the influence of temperature, moisture and electric field on the electrical conductivity of oil-impregnated pressboard. *Energies* **2019**, *12*, 3136. [CrossRef]
32. Saha, T.K.; Purkait, P. Understanding the impacts of moisture and thermal aging on transformer's insulation by dielectric response and molecular weight measurements. *IEEE Trans. Dielectr. Electr. Insul.* **2008**, *15*, 568–582. [CrossRef]

33. Huang, L.; Wang, Y.; Li, X.; Wei, C.; Lu, Y. Effect of moisture on breakdown strength of oil-paper insulation under different voltage types. In Proceedings of the IEEE International Conference on High Voltage Engineering and Application (ICHVE), Athens, Greece, 10–13 September 2018; pp. 1–4.

34. Wang, Y.; Xiao, K.; Chen, B.; Li, Y. Study of the impact of initial moisture content in oil impregnated insulation paper on thermal aging rate of condenser bushings. *Energies* **2015**, *8*, 14298–14310. [CrossRef]

35. Sikorski, W.; Walczak, K.; Przybylek, P. Moisture migration in an oil-paper insulation system in relations to online partial discharge monitoring of power transformers. *Energies* **2016**, *9*, 1082. [CrossRef]

36. *Experiences in Service with New Insulating Liquids*; Cigré Technical Brochure 436; International Council on Large Electric Systems (CIGRE): Paris, France, 2010.

37. Christina, A.J.; Salam, M.A.; Rahman, Q.M.; Wen, F.; Ang, S.P.; Voon, W. Causes of transformer failures and diagnostic methods—A review. *Renew. Sustain. Energy Rev.* **2018**, *82*, 1442–1456. [CrossRef]

38. Shroff, D.H.; Stannett, A.W. A review of paper aging in power transformers. *IEEE Proc. C Gener. Transm. Distrib.* **1985**, *132*, 312–319. [CrossRef]

39. Lundgaard, L.E.; Hansen, W.; Linhjell, D.; Painter, T.J. Aging of oil-impregnated paper in power transformers. *IEEE Trans. Power Deliv.* **2004**, *19*, 230–239. [CrossRef]

40. Mosinski, F. The influence of water and oxygene on load and lifetime of power transformers. In Proceedings of the Transformer During Operation Conference, Kolobrzeg, Poland, 20–22 April 2005; pp. 117–119. (In Polish)

41. Daghrah, M.; Wang, Z.; Liu, Q.; Hiker, A.; Gyore, A. Experimental study of the influence of different liquids on the transformer cooling performance. *IEEE Trans. Power Deliv.* **2019**, *34*, 588–595. [CrossRef]

42. Velandy, J.; Garg, A.; Narasimhan, C.S. Thermal performance of ester oil transformers with different placement of cooling fan. In Proceedings of the 9th Power India International Conference (PIICON), Sonepat, India, 29 February–1 March 2020; pp. 1–7.

43. Velandy, J.; Garg, A.; Narasimhan, C.S. Continuos thermal overloading capabilities of ester oil transformer in oil directed cooling conditions. In Proceedings of the 9th Power India International Conference (PIICON), Sonepat, India, 29 February–1 March 2020; pp. 1–7.

44. Cabeza-Prieto, A.; Camino-Olea, M.S.; Rodriguez-Esteban, M.A.; Llorente-Alvarez, A.; Saez-Perez, M. Moisture influence on the thermal operation of the late 19th century brick façade, in a historic building in the city of Zamora. *Energies* **2020**, *13*, 1307. [CrossRef]

45. Nikitsin, V.I.; Alsabry, A.; Kofanov, V.A.; Backiel-Brzozowska, B.; Truszkiewicz, P. A model of moist polymer foam and a scheme for the calculation of its thermal conductivity. *Energies* **2020**, *13*, 520. [CrossRef]

46. DuPontTM Nomex® 926. Technical Data Sheet. Available online: https://www.dupont.com/fabrics-fibers-and-nonwovens/nomex-electrical-insulation.html (accessed on 3 May 2020).

47. Nynas Nytro Draco Technical Data Sheet. Available online: http://www.smaryoleje.pl/nynas.html (accessed on 4 May 2020).

48. Midel 7131 Synthetic Ester Transformer Fluid. FIRE Safe and Biodegradable. Available online: https://www.midel.com/app/uploads/2018/05/MIDEL-7131-Product-Brochure.pdf (accessed on 4 May 2020).

49. Cargill Dielectric Fluids. EnvirotempTM FR3TM Fluid. Available online: https://www.cargill.com/doc/1432160189547/fr3-dielectric-fluid-data-sheet.pdf (accessed on 4 May 2020).

50. Martin, D.; Perkasa, C.; Lelekakis, N. Measuring paper water content of transformers: A new approach using cellulose isotherms in nonequilibrium conditions. *IEEE Trans. Power Deliv.* **2013**, *28*, 1433–1439. [CrossRef]

51. Atanasova-Höhlein, I.; Agren, P.; Beauchemin, C.; Cucek, B.; Darian, L.; Davidov, V.; Dreier, L.; Gradnik, T.; Grisaru, M.; Koncan-Gradnik, M.; et al. *Moisture Measurement and Assessment in Transformer Insulation—Evaluation of Chemical Methods and Moisture Capacitive Sensors*; International Council on Large Electric Systems (CIGRE): Paris, France, 2018.

52. *IEC 60814. Insulating Liquids—Oil-Impregnated Paper and Pressboard—Determination of Water by Automatic Coulometric Karl Fischer Titration*; International Electrotechnical Commission (IEC): New York, NY, USA, 1997.

53. Przybylek, P. Water saturation limit of insulating liquids and hygroscopicity of cellulose in aspect of moisture determination in oil-paper insulation. *IEEE Trans. Dielectr. Electr. Insul.* **2016**, *23*, 1886–1893. [CrossRef]

54. Przybylek, P. A Comparison of bubble evolution temperature in aramid and cellulose paper. In Proceedings of the IEEE International Conference on Solid Dielectrics (ICSD), Bologna, Italy, 30 June–4 July 2013; pp. 983–986.

55. Lopatkiewicz, R.; Nadolny, Z.; Przybylek, P. The influence of water content on thermal conductivity of paper used as a transformer windings insulation. In Proceedings of the IEEE International Conference on the Properties and Applirations of Dielectric Materials (ICPADM), Bangalore, India, 24–28 July 2012; pp. 1–4.

56. Lopatkiewicz, R.; Nadolny, Z.; Przybylek, P. Influecne of water content in paper on its thermal conductivity. *Przeglad Elektrotechniczny.* **2010**, *86*, 55–58. (In Polish)

57. Chen, H.; Ginzburg, V.V.; Yang, J.; Yang, Y.; Liu, W.; Huang, Y.; Du, L.; Chen, B. Thermal conductivity of polymer-based composites: Fundamentals and applications. *Prog. Polym. Sci.* **2016**, *59*, 41–85. [CrossRef]

58. Chae, H.G.; Kumar, S. Minaking strong fibres. *Science* **2008**, *319*, 908–909. [CrossRef]

59. Vargaftik, N.B.; Filippov, L.P.; Tarzimanov, A.A.; Totski, E.E. *Handbook of Thermal Conductivity of Liquids an Gases*, 1st ed.; CRC Press: Cleveland, OH, USA, 1994; pp. 48–75.

60. Dombek, G.; Nadolny, Z. Liquid kind, temperature, moisture, and ageing as an operating parameters conditioning reliability of transformer cooling system. *Eksploatacja i Niezawodność—Maintenance and Reliability* **2016**, *18*, 413–417. [CrossRef]

61. Dombek, G.; Nadolny, Z. Influence of paper type and liquid insulation on heat transfer in transformers. *IEEE Trans. Dielectr. Electr. Insul.* **2018**, *25*, 1863–1870. [CrossRef]

62. Kuang, Y.C.; Chen, G.; Jarman, P. Recovery voltage measurement on oil-paper insulation with simple geometry and controlled environment. In Proceedings of the IEEE International Conference on Solid Dielectrics (ICSD), Toulouse, France, 5–9 July 2004; pp. 1–4.

63. Wang, S.; Wei, J.; Yang, S.; Dong, M.; Zhang, G. Temperature and thermal aging effects on the frequency domain spectroscopy measurement of oil-paper insulation. In Proceedings of the IEEE 9th International Conference on the Properties and Applications of Dielectric Materials (ICPADM), Harbin, China, 19–23 July 2009; pp. 329–332.

Article

Analysis of Polarization and Depolarization Currents of Samples of NOMEX®910 Cellulose–Aramid Insulation Impregnated with Mineral Oil

Stefan Wolny *  **and Adam Krotowski**

Faculty of Electrical Engineering, Automatic Control and Computer Science, Opole University of Technology, Proszkowska 76 B2, 45-758 Opole, Poland; d565@doktorant.po.edu.pl
* Correspondence: s.wolny@po.edu.pl

Received: 3 November 2020; Accepted: 18 November 2020; Published: 20 November 2020

Abstract: The article presents results of laboratory tests performed on samples of NOMEX®910 cellulose–aramid insulation impregnated with Nynas Nytro 10× inhibited insulating mineral oil using the polarization and depolarization current analysis method (PDC Method). In the course of the tests, the insulation samples were subjected to a process of accelerated thermal degradation of cellulose macromolecules, as well as weight-controlled dampening, thereby simulating the ageing processes occurring when using the insulation in power transformers. The effects of temperature in the ranges typical of normal transformer operation were also taken into account. On the basis of the obtained data, the activation energy was then fixed together with dominant time constants of cellulose–aramid insulation relaxation processes with respect to the temperature and degree of moisture, as well as thermal degradation of cellulose macromolecules. It was found that the greatest and predictable changes in the activation energy value were caused by the temperature and the degree of moisture in the samples. A similar conclusion applies to the dominant time constant of the relaxation process of cellulose fibers. Degree of thermal degradation samples was of marginal importance for the described parameters. The final outcome of the test results and analyses presented in the article are regression functions for the activation energy and the dominant time constants depending on the earlier listed parameters of the experiment, which may be used in the future diagnostics of the degree of technical wear of cellulose–aramid insulation performed using the PDC method.

Keywords: dielectric polarization; relaxation methods; activation energy; cellulose–aramid paper; moisture insulation; ageing effect; power transformer insulation testing

---

1. Introduction

Power transformers are undoubtedly one of key elements of the electric energy distribution system. Their unfailing operation determines not only continuous energy deliveries for recipients, but also stable operation of the whole power system. The effects of a possible breakdown of a power transformer are always multidimensional, starting from purely economic ones, to logistics, to threats to property and life, as well as possible environmental contamination. For this reason, transformer units which are regarded as key (e.g., power plants' step-up transformers or distribution transformers working in critical power system nodes), are equipped with automatic security systems, monitoring systems for a number of parameters determining the technical condition of the unit, and are subjected to diagnostic procedures recommended by respective standards or operating instructions. Bearing in mind the complex design of power transformers as well as their almost unique structure, ensuring unfailing continuous operation for a few dozen years is a difficult problem, which requires extensive scientific and expert knowledge.

Particularly for new power transformers, the technical condition of the electric insulation saturated with dielectric fluids is regarded as the basic hazard to safe operation, where the key factors are the level of moisture as well as the degree of ageing, defined as the degree of the materials' thermal degradation [1–6]. Despite decades of designing and manufacturing power transformers, cellulose electric paper saturated with insulating mineral oil continues to be used as the basic electric insulation system. However, various kinds of synthetic and natural esters have become more and more recognized as substitutes for mineral oil [4–8]. One of the key factors accelerating the process of departing from mineral oils as the impregnating fluid, apart from improved electric, physical and chemical parameters of esters as compared to mineral oil, is their considerably better biodegradability. Likewise, we are constantly looking for a material that would effectively replace the commonly applied cellulose. At present, the highest hopes are related to synthetic papers manufactured on the basis of aramid fibers. This material was invented and patented by DuPont™ in the 1960s. DuPont™ produce a broad assortment of products to be used as electric insulation material in electric machines (also in power transformers) under the commercial name NOMEX® [6–9]. As compared to the classic cellulose paper, aramid paper is characterized by considerably better electrical properties (e.g., higher dielectric strength, higher volume and surface resistivity), physical properties (e.g., considerably higher breaking strength) and chemical properties (better resistance to ageing as a result of higher thermal resistance over continuous operation). Possible disadvantages of the aramid material include a higher price, lower ability to absorb dielectric fluids (worse impregnation as compared to cellulose), significantly higher rigidity than cellulose paper of similar thickness, which substantially complicates the process of manufacturing the insulation system for a transformer with complex geometry. In addition, intensification of the phenomenon of generating electrostatic charges as a result of stream electrification in transformers with induced circulation of dielectric fluid has also been observed [10,11]. Too high a presence of electrostatic charges in the solid insulation of a working transformer may lead to the development of partial discharges, accelerated ageing processes and, consequently, to complete discharge.

In order to use the unquestionable advantages of aramid papers in the insulation of power transformers with induced circulation of dielectric fluid, DuPont™ proposed a hybrid material, in the form of cellulose paper coated on both sides with a thin layer of synthetic aramid paper. This has resulted in a higher temperature class as compared to clean cellulose paper, an improved material dielectric fluid absorption capacity, and the impregnation capacity as compared to clean aramid paper, among other benefits. The article presents tests conducted on samples of insulation manufactured from this kind of material, which DuPont™ sells under the trade name NOMEX®910 [12]. A more detailed description of cellulose–aramid paper is presented in Section 2 of this article.

Because NOMEX®910 electro-technical paper contains a considerable amount of cellulose, its use in the insulation of a power transformer involves practically the same hazards as in a classic oil-saturated cellulose–aramid insulation. In brief, over the course of use, cellulose fibers, as a result of the presence of oxygen molecules, water, and elevated temperature, are subjected to the processes of oxidization, hydrolysis and pyrolysis, resulting in the bonds of cellulose macromolecules being torn apart (reduced degree of polymerization) and deterioration, first of all, of mechanical parameters (e.g., breaking strength). In addition, the described cellulose ageing processes generate a continuous increase in the degree of the material's moisture due to the fact that water is one of the products of decay of the bonds of cellulose macromolecules. The increased moisture of cellulose paper significantly reduces electrical parameters (e.g., dielectric strength or volume resistivity), becoming a significant threat to further safe operation of the transformer's insulation. In order to monitor the so-called degree of insulation's technical wear in transformers, a number of diagnostic methods are applied, which estimate the degree of moisture as well as the ageing of cellulose paper impregnated with dielectric fluids. Therefore, a question arises: can the diagnostic methods applied so far, developed for impregnated cellulose papers, also be applied for new cellulose–aramid papers? The article presents the results of tests using the polarization diagnostic method analyzing the profiles of polarization and depolarization currents over time. This method is broadly used in diagnostic practice, and is referred to using the abbreviation PDC [7,13–15].

2. Materials Used for the Tests

As earlier described in the article's abstract, for experimental tests the authors used solid insulation made of NOMEX®910 cellulose–aramid electro-technical paper, manufactured by DuPont™. The basis of this material is a high quality thermally improved cellulose pulp. A sheet of electro-technical paper formed from it is then coated on both sides with a thin layer of a binder from a high temperature meta-aramid polymer (NOMEX®). This procedure leads to an increase in the temperature class of the new material, which, for operation in mineral oil, increases to 130 °C (for operation in the environment of esters, it is 140 °C) [12]. In the case of electro-technical paper made of cellulose only, for Kraft type papers, the temperature class is 110–120 °C for operation in mineral oil. Figure 1 presents the cross-section of NOMEX®910 material structure and a real photograph of a small sample.

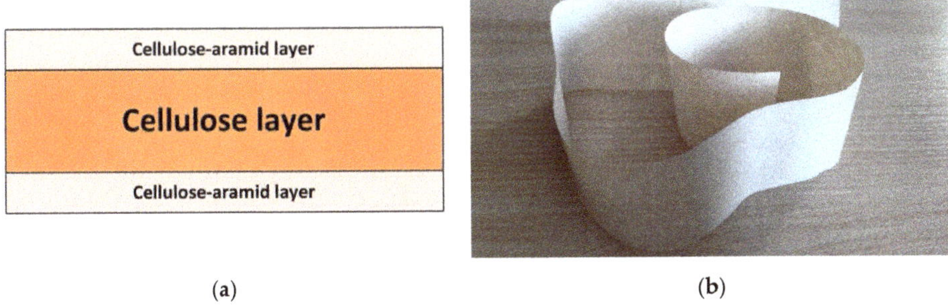

(a)   (b)

Figure 1. NOMEX®910 material structure: (**a**) cross-section; (**b**) photograph of a small sample.

An approx. 10 °C increase in the temperature class of cellulose–aramid papers as compared to Kraft type cellulose papers is obviously not their single advantage. The basic objective of the designers was to obtain an increased mechanical strength of the new material, more precisely, tearing, breaking and tensile strength. It was assumed that, while in the case of the new cellulose–aramid material the listed parameters values will be similar to classical cellulose paper, along with the maintenance-related ageing process, the new material will definitely keep the values required by the standards for a longer time. This results from the fact that aramid fibers demonstrate a considerably higher resistance to the effect of temperature (the temperature class of electro-technical papers made of aramid only for operation in the air is 220 °C, e.g., NOMEX®410 [9]), at the same time, they provide significant strengthening for thermally weakened cellulose. As a result, during operation of the transformer and the inevitably increasing degradation of cellulose (chains of cellulose macromolecules being torn apart as a result of the oxidization, hydrolysis and pyrolysis phenomena), the aramid component will continue to be intact, ensuring the paper's strengthened mechanical structure, and, consequently, will extend the technical life of the insulation and the whole transformer. The material's good saturability with dielectric fluids was observed, as well as an increased dielectric strength during electric tests in the environment of mineral oil, which is particularly important [16]. At present, the substantially higher price as compared to Kraft type electro-technical papers should be considered the only disadvantage of NOMEX®910 material. However, because the material appeared on the market just several years ago, there are no statistically significant data describing many aspects related to its maintenance in power transformers. The most important aspects include: migration water processes between oil–aramid–cellulose and the time of setting the hydrodynamic balance depending on the temperature of the whole insulation; heat dissipation processes from the inside of the insulation (particularly when the surface of cellulose–aramid paper is contaminated with insulating oil ageing products); and, finally, the effectiveness of the methods applied so far to diagnose the level of moisture and ageing of the impregnated insulation. In the latter case, this article attempts to answer this very significant aspect with reference to the PDC method. Doubts in this respect can still be a valid factor that discourages potential transformer manufacturers from applying the insulation concerned to a broader extent. Table 1 presents basic technical parameters of 0.08 mm thick NOMEX®910 paper, which has been used in the tests.

Table 1. Typical mechanical and electrical properties of NOMEX®910 paper (thickness 0.08 mm). Data from official DuPont™ online data sheet available online [16].

Property	Units	Value	Test Method
Basis weight	g/m^2	80	ASTM D646
Burst strength	N/cm^2	27	ASTM D828
Tensile strength, MD [1]	N/cm	70	ASTM D828
Tensile strength, XD [2]	N/cm	17	ASTM D828
Elongation, MD	%	2.2	ASTM D828
Elongation, XD	%	6.9	ASTM D828
Tear strength, MD	N	0.45	TAPPI 414
Tear strength, XD	N	0.7	TAPPI 414
AC rapid rise breakdown in mineral oil	kV/mm	87	ASTM D149
Dielectric constant 60 Hz, 23 °C, mineral oil	-	3.2	ASTM D150
Dissipation factor 60 Hz, 23 °C, mineral oil	%	0.9	ASTM D150

[1] Machine direction, [2] Cross machine direction.

Nynas inhibited mineral insulating oil with commercial symbol Nytro 10× was chosen as the impregnating fluid. Mineral oils of this Swedish company are very often applied in power transformers, and installed in European electric energy distribution systems. Table 2 states basic technical parameters of this dielectric fluid.

Table 2. Typical physical, chemical, and electrical properties of Nynas Nytro 10× mineral oil. Data are from the official Nynas data sheet, available online [17].

Property	Units	Value	Test Method
Density, 20 °C	kg/dm^3	0.876	ISO 12185
Viscosity, 40 °C	mm^2/s	7.6	ISO 3104
Viscosity, −30 °C	mm^2/s	730	ISO 3104
Flash point	°C	144	ISO 2719
Pour point	°C	−60	ISO 3016
Neutralization value	mg KOH/g	<0.01	IEC 296
Water content	mg/kg	<20	IEC 814
Interfacial tension	mN/m	45	ISO 6295
Dissipation factor, 90 °C	-	<0.001	IEC 247
Breakdown voltage -before treatment	kV	40–60	IEC 156
-after treatment	kV	>70	IEC 296

3. Sample Preparation Method

The insulation samples were made of 80 µm thick Nomex®910 transformer cellulose–aramid paper. It is the most frequently used type of cellulose–aramid insulation applied in contemporary new power transformers. The insulation paper was cut into 1300 mm × 100 mm strips. Then, before impregnation, the samples were subjected to a process of accelerated ageing by placing them in a sterilizer and heating at four different temperatures (130 °C, 150 °C, 170 °C and 190 °C) with the access of air for a definite time. In this way, 5 degrees of sample ageing were obtained: 0–1 h of ageing (fresh paper); 2–25 h of ageing at 130 °C; 3–25 h of ageing at 150 °C; 4–25 h of ageing at 170 °C; and 5–25 h of ageing at 190 °C. After the stage of accelerated ageing, the samples were placed in vacuum and heated at 120 °C for 2 h to be dried before impregnation.

Directly after the accelerated thermal ageing processes and drying in vacuum, the samples were subjected to a process of weight-controlled dampening, which consisted of the paper absorbing moisture directly from the surrounding air for a definite time. The degree of sample moisture was determined as the percentage growth in paper weight to a predefined amount. This has resulted in 4 degrees of sample moisture for all 5 degrees of ageing: 1—initial residual moisture (sample impregnation directly after the vacuum drying process, the adopted degree of sample moisture was close to 0%); 2—1.5% sample moisture; 3—2% sample moisture; and 4—2.5% sample moisture. Directly after the weight-controlled dampening process, the samples were subjected to the process of impregnation with the dielectric fluid. Nynas inhibited mineral insulating oil with commercial symbol Nytro 10× was used for this purpose. Before the impregnation process, insulating oil was initially degassed and dried in a vacuum at a temperature of 60 °C for 2 h, reducing the content of water dissolved in oil to approx. 10 ppm. Figure 2 shows the entire sample preparation process as a diagram.

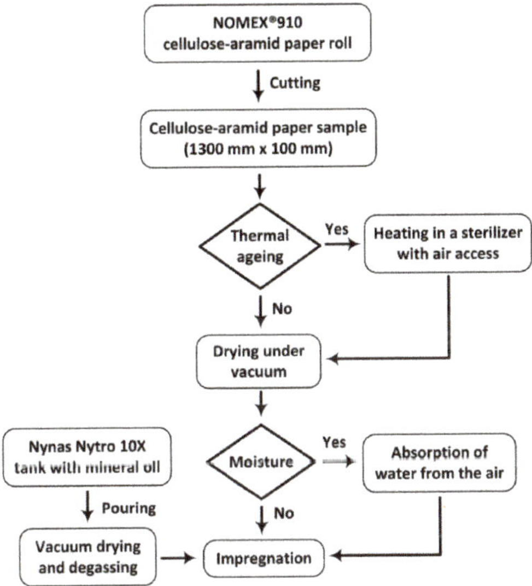

Figure 2. Sample preparation diagram.

The earlier described parameters of the sample ageing and moisture processes are certainly not accidental. Accelerated thermal ageing of cellulose paper in the environment of atmospheric air causes a significant loss in the degree of polymerization of cellulose macromolecule chains. On the basis of the authors' past long-term studies and literature data [18,19] it can be stated that the selected ageing time as well as temperature range resulted in a gradual, proportional loss in the degree of cellulose polymerization from a value equal to approx. 1000 (non-aged samples) to approx. 200 (ageing in 190 °C for 25 h). The described scope of changes in the degree of cellulose polymerization corresponds to the whole technical life of cellulose insulation in power transformers [20]. Likewise, it can be assumed that the degree of moisture of cellulose–oil insulation in the range between approx. 0% and 2.5% in practice corresponds to moisture recorded in the statistical majority of power transformers working in electric energy distribution systems of many states [21,22]. Due to some technical problems and the long process of sample moistening, the decision was made to resign from values exceeding 2.5% of water in paper, knowing that values reaching as much as approx. 5% are recorded in strongly worn out transformers. It was decided to consider higher degrees of sample moisture in future tests.

After the impregnation process, the samples were wound on the low potential electrode, which was a brass roll of 160 mm length and 40 mm diameter. In this way, 10 layers of insulation were obtained. The high potential electrode was made of a thin 80 mm aluminum foil. The foil was wound on the roll with the sample. Figure 3 shows a cross section of the electrode system and Figure 4 shows the system ready for testing. The tests were carried out in a hermetic chamber equipped with a system for temperature adjustment and stabilization. The tests were performed in the temperature range from 20 °C to 60 °C, which is consistent with the values typical of PDC diagnostics [13–15].

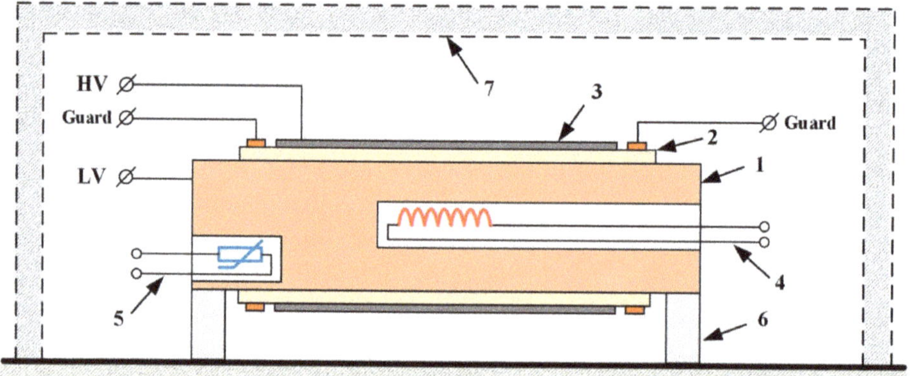

Figure 3. View of the measuring electrode system and the insulation sample studied: 1—a roll made of brass (LV—low potential electrode), 2—aramid–oil insulation sample, 3—metal foil (HV—high potential electrode), 4—heater, 5—temperature sensor, 6—insulator, 7—hermetic vessel plus thermal insulation.

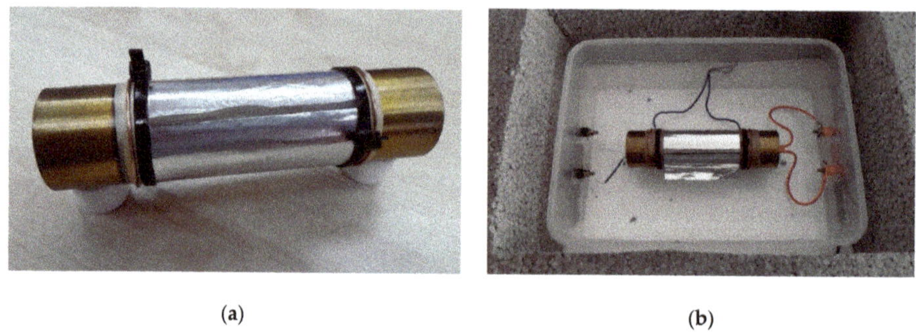

(**a**)   (**b**)

Figure 4. Photograph of the insulation sample prepared for testing: (**a**) zoom on sample; (**b**) after placing the sample in the measuring system.

4. PDC Method

Figure 5 presents a diagram of the connections used in diagnostics of the condition of the paper–oil insulation using the PDC method and time characteristics of currents and voltages which were recorded while performing the tests. The PDC method consists of applying to the examined object (in the industrial diagnostics, these are clamps of a power transformer) a source of DC voltage and measuring and recording I_P polarization current for a certain period of time t_P. After time t_P, the voltage source is disconnected, and the examined object's measuring clamps are shortened. Then, the measurement and registration of I_D depolarization current begins, also for a certain period of time t_D. I_P current decreases over the time of impact of the voltage source until the course is fixed at a certain level I_K resulting from a finite resistivity value of the insulation being examined. On the other hand, I_D current has the opposite sign to I_P current and also decreases over time. Finally, the course of I_D current decreases until zero is achieved, when the examined system discharges completely.

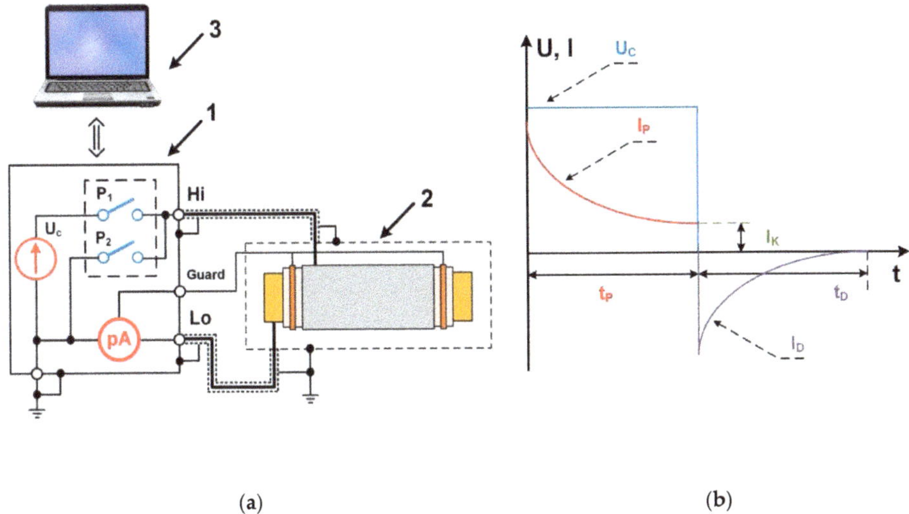

(a)   (b)

Figure 5. Diagnostics of the state of paper–oil insulation samples using the polarization and depolarization current analysis (PDC) method: (**a**) connections diagram; (**b**) time characteristics of currents and voltages. 1—measuring-switching system, 2—object being examined, 3—computer.

From the point of view of diagnosing the condition of oil insulation in power transformers, it is extremely important to select the right value of U_C charging voltage and I_P and I_D current recording times. On the one hand, too low a value of U_C voltage will cause substantial measuring difficulties (in the PDC method typical recorded current values are nA fractions), and thereby growth in interference (external and resulting from the presence of electrostatic loads in the insulation being examined). On the other hand, too high a U_C value will cause the "masking" effect of the conductivity element of I_P current (problems in precise registration of the absorption element of I_P current) and can cause non-linear relaxation phenomena, significantly hindering the correct interpretation of the received measurement results. For this reason, it is recommended for the U_C voltage value in the diagnostics of the condition of the paper–oil insulation in power transformers not to exceed 1 kV. The relaxation mechanism time constant values existing in such insulation require the polarization and depolarization time not to be shorter than min. 1000 s.

When relay P_1 is switched on (Figure 5a), the U_C charging voltage is put on the object being examined and I_P polarization current flowing in the measuring circuit is registered (Figure 5b). After time t_P, the control system switches P_1 relay off and P_2 on. Then, I_D depolarization current is recorded in the circuit. The measurement is ended after time t_D, when P_2 relay is opened. The PDC method requires very careful shielding of the measurement cables, the examined object, and the measurements to be made with reference to the joint earth potential.

In the first approx. 100 s of the polarization and depolarization current measurement, the recorded values are mainly determined by the properties of the insulating oil, i.e., first of all its conductivity, provided that in this case important factors are the following: degree of moisture, contamination, acid value or the effect of temperature. Increasing oil conductivity causes almost proportional increases in the initial polarization current values, therefore it is possible to fix oil conductivity from I_P values registered directly after applying the U_C voltage. The condition of cellulose–aramid insulation is determined for considerably longer times, sometimes even exceeding 1000 s [13–15]. The degree of cellulose moisture determines the current leakage value, which increases along with the increasing moisture. A much faster depolarization current decay is also observed.

With sample polarization current time characteristics, it was decided to determine activation energy E_A at which the low-frequency cellulose–aramid insulation relaxation process is subject to change. The purpose of the calculations was the assumption that the activation energy E_A value is determined by the degree of moisture as well as the ageing of the samples, defined as the degree of cellulose macromolecule thermal degradation. The Low-Frequency Dispersion Jonscher equation was used as the sought sample polarization current regression function [23] in the form of:

$$I_P(t) \propto A_1 \cdot t^{-N_1} + A_2 \cdot t^{-N_2}, \tag{1}$$

where A_1, A_2, N_1, N_2—function parameters.

The activation energy E_A can be calculated utilizing linear approximation of the Arrhenius temperature graph [24], applying the following dependencies:

$$ln(t_A) = f\left(\frac{1000}{T}\right), \tag{2}$$

$$E_A = 1000 \cdot a \cdot k, \tag{3}$$

where t_A—characteristic time (after which the relaxation process change occurs), T—sample temperature (in Kelvin degrees), E_A—activation energy (eV), a—directional coefficient of linear regression function, and k—Boltzmann constant.

Characteristic time was calculated using the formula:

$$t_A = \sqrt[-N_1+N_2]{\frac{A_2}{A_1}}, \tag{4}$$

Using sample depolarization current time characteristics, it was decided to analyze the dominant time constants separately for the relaxation processes of aramid and cellulose fibers, depending on the degree of moisture as well as ageing of the cellulose–aramid insulation being examined. To this end, a Debye equation with two relaxation times was used as the sought depolarization current regression function:

$$I_D(t) \propto B_1 \cdot e^{-\frac{t}{\tau_1}} + B_2 \cdot e^{-\frac{t}{\tau_2}}, \tag{5}$$

where B_1, B_2—function parameters, τ_1, τ_2—dominant time constants of the relaxation processes, for aramid fibers and cellulose fibers, respectively.

5. Experimental Results

An MIC-15k1 high resistance meter from Sonel® was used for the measurements of the polarization and depolarization currents. Thanks to the embedded battery of accumulators, the meter provided a solid and stable U_C charging voltage value, which amounted to 500 V for all the experiments. The registration of the polarization and depolarization currents in time was realized with the dielectric discharge factor measurement function (DD factor—Dielectric Discharge). The meter's software allowed any adjustment of the polarization and depolarization times, while data communication was over a wireless Bluetooth connection. The meter's sampling frequency was approximately 2 Hz, which, in the case of measurements of low-frequency currents, was a fully sufficient value.

5.1. Effect of Temperature

Figure 6 presents exemplary time characteristics of the polarization and depolarization currents for a non-aged insulation sample with an average degree of moisture (residual moisture plus 1.5% paper weight increase as a result of water absorption from the environment, before the impregnation process).

We can observe on the characteristics how the sample temperature affects the recorded current values, in the range from 20 °C to 60 °C, at steps every 10 °C. The scope of temperature changes of the samples adopted in the experiments is typical of diagnosing the condition of oil insulation in power transformers performed using polarization diagnostic methods (including the PDC method) [13–15,18–22]. It is important to remember that polarization diagnostic methods require the transformer to be detached from the network (off-line state) and, consequently, the insulation temperature must be lower than typical, which is assumed at approx. 60 °C for the transformer's normal operation mode. If the transformer is disconnected for a long time from the network, the insulation temperature may be reduced even further to an ambient temperature, e.g., 20 °C.

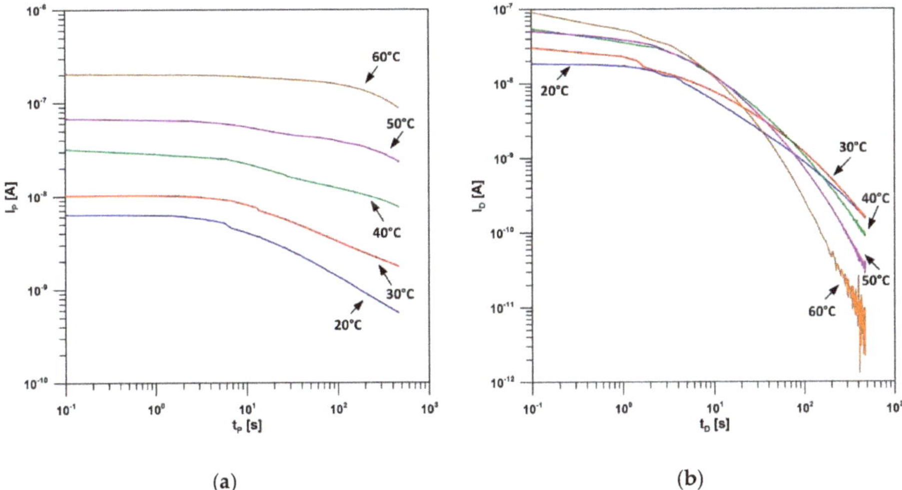

(a)   (b)

Figure 6. Effect of temperature on the characteristics of the PDC method for a selected cellulose–aramid insulation sample (non-aged with 1.5% moisture) mineral oil-impregnated: (**a**) polarization current; (**b**) depolarization current.

The characteristics presented in Figure 6a prove that, along with an increase in insulation sample temperature, the depolarization current also increased across the whole time range. The reason for this phenomenon is the declining resistivity of mineral oil and cellulose itself. The observed change was almost proportional in relation to temperature, which is typical of insulation made of mineral oil-impregnated cellulose only [14,15]. In the case of the depolarization current (Figure 6b), temperature growth generated, just as before, an initial increase in the current, but a much faster decay was observed for times exceeding approx. 20 s. The reason is that the phenomenon related to the destructive effect of temperature on the polarization process of dielectric dipoles, making the following depolarization faster, because temperature facilitates the achievement of the initial state of dipoles disorder. As before, it can be stated that this is typical of mineral oil-impregnated cellulose insulation, which also proves the correctness of the completed measurements. It turned out that a small layer of aramid fibers does not significantly disturb the characteristics of the registered polarization and depolarization currents, with reference to the previously known effect of temperature in the PDC method for insulation made of mineral oil-impregnated cellulose paper only. The above conclusion was confirmed for all the examined samples, i.e., regardless of the degree of their moisture or ageing.

Figure 7a presents Debye regression functions according to Formula (5), which were used in order to fix time constants of two relaxation mechanisms based on the depolarization current of the cellulose–aramid insulation sample being examined. Certainly, the data presented in Figure 7a are only exemplary, while the regression functions being described were used for all the insulation samples, i.e., for each degree of moisture and ageing and for all temperatures. The sample of the insulation being examined was a thin layer of aramid fibers and mineral oil-impregnated cellulose, therefore it was assumed that the time characteristics of the depolarization current would contain at least three dominant time constants of the relaxation processes of the previously mentioned materials, i.e., aramid, cellulose and oil. Because the impregnation of the samples was made using fresh inhibited mineral oil, the time constant of the relaxation process of this poorly polar fluid was very small (below 1 s). Therefore, bearing in mind the primary nature of the tests being conducted (cellulose–aramid material), it was decided to ignore it. Recognizing that water collected in the insulation is stored mostly in cellulose (considerably higher absorbability compared to aramid fibers), a longer time constant τ_2 was needed to describe this process only. Then, a shorter time constant τ_1 was assigned to the relaxation process of aramid fibers. Figure 7b presents the effect of temperature on the value of the previously described time constants for a selected cellulose–aramid insulation sample, initially thermally aged in 150 °C with the smallest moisture.

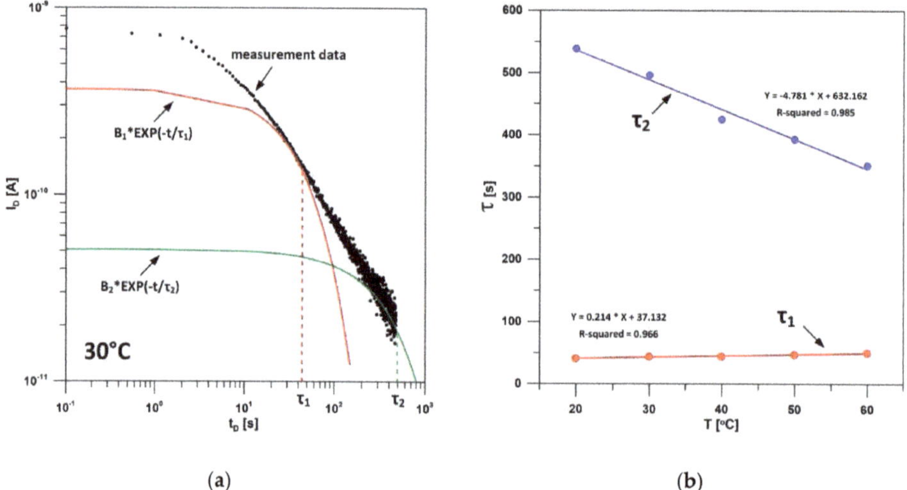

(a)   (b)

Figure 7. Analysis of time characteristics of the depolarization current by means of Debye regression function for a selected cellulose–aramid insulation sample (aged at 150 °C with the smallest moisture), mineral oil-impregnated: (**a**) depolarization current in the measurement temperature 30 °C; (**b**) temperature dependence of the dominant time constants for two relaxation processes.

Analyzing the characteristics from Figure 7b, two opposing tendencies can be noticed. Temperature growth in the sample being examined resulted in an insignificant increase in time constant τ_1 value and a significant decrease in time constant τ_2 value. In the first case, the change in time constant τ_1 can be explained by the fact that the growing measurement temperature increased at the same time as the relative permeability of aramid fibers, and slightly reduced the material's volume resistivity [9]. Assuming that the time constant value in the electric equivalent circuit defines the product of the current capacity and resistance of the material, permeability growth probably determined a slight change in time constant τ_1 value. In the case of time constant τ_2, which was correlated with the relaxation of cellulose fibers and the water collected in the material, temperature growth stimulated the process of water migration to the impregnating fluid, i.e., mineral oil [25]. Therefore, the amount of water in the insulation decreased (highly polar liquid with high permeability), and at the same time volume resistivity also decreased. The result of these phenomena was a considerable decrease in time constant τ_2 value. The described observations were confirmed for all the examined insulation samples, regardless of the degree of ageing and moisture.

5.2. Effect of the Degree of Moisture

Figure 8 presents exemplary characteristics of the polarization current measured for cellulose–aramid insulation samples with initial thermal ageing at 150 °C and with various moisture degrees. For comparison purposes, Figure 8 presents the results of tests for two different temperatures. The increased moisture of the samples resulted in a significant growth in the polarization current values at the initial stage of the measurement (until approx. 10 s) and a growth in the current conductivity element, which could be observed for considerably longer times (close to 1000 s). It is a typical phenomenon of the effect of increased moisture for the classic cellulose–oil insulation, widely described in many publications [13–15]. Therefore, it can be assumed that practically all the water is stored in the Nomex®910 paper cellulose layer. The measurement temperature growth stimulates the process of water migration from the cellulose layer to mineral oil, thereby reducing the differences in the polarization current characteristics being described (Figure 8b). The term "initial moisture" means the degree of cellulose–aramid insulation sample moisture, which was left in the material after the vacuum drying process, still before the oil impregnation process. As the drying method utilized for cellulose materials in industry, was applied, it may be assumed that the initial moisture of Nomex®910 paper did not exceed 0.5%.

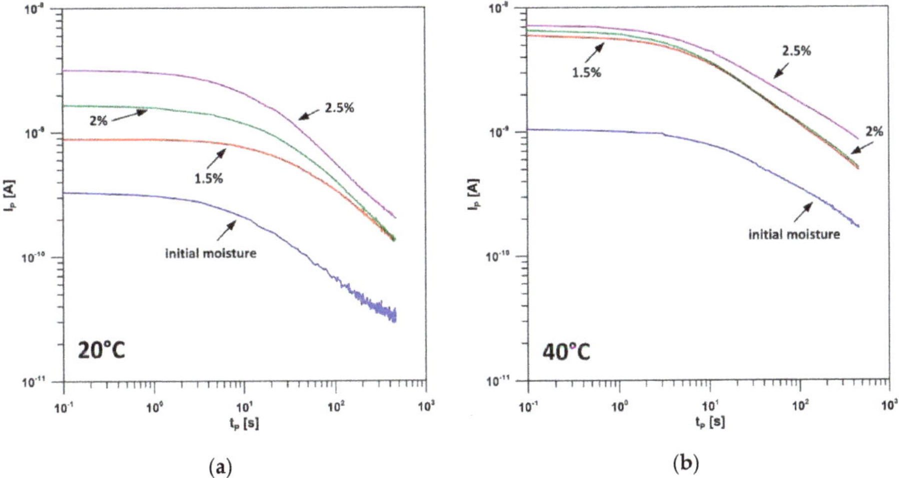

Figure 8. Characteristics of the polarization current of cellulose–aramid insulation sample with initial thermal ageing at 150 °C and with various moisture: (**a**) polarization currents in the measurement temperature 20 °C; (**b**) polarization currents in the measurement temperature 40 °C.

In a similar manner, it is also possible to interpret the depolarization current characteristics, which are presented in Figure 9 for the same selected insulation samples. The increased moisture in the samples also generated an increased value of the depolarization current in the initial part of the analysis (until approx. 10 s); sometime later, a faster decay process of the depolarization current was observed for these samples.

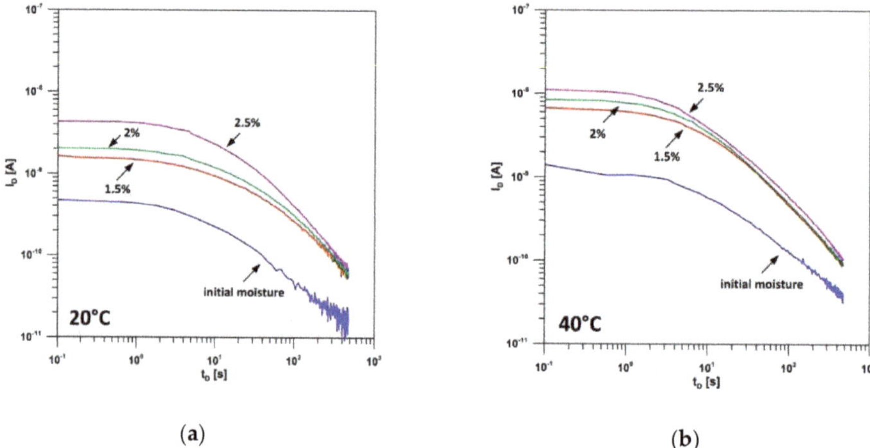

Figure 9. Characteristics of the depolarization current of cellulose–aramid insulation samples with initial thermal ageing at 150 °C and with various moisture: (**a**) depolarization currents in the measurement temperature 20 °C; (**b**) depolarization currents in the measurement temperature 40 °C.

The measurement temperature growth (Figure 9b) also reduced the differences between the depolarization currents of samples with various moisture degrees within approx. 10 s, and this is due to the water migration process from cellulose to oil intensifying with increased temperature. The phenomenon described was observed for all the examined insulation samples, regardless of the degree of ageing (thermal degradation of cellulose fibers). The characteristics presented on Figures 8 and 9 are only general in nature.

Using the sample depolarization current characteristics depending on the measurement temperature, it was decided to analyze, using Equations (1)–(4), the dependence of the activation energy E_A on the degree of moisture of the cellulose–aramid insulation. Figure 10 presents the manner of fixing characteristic time t_A (Figure 10a) and the activation energy E_A value (Figure 10b) with the use of the Arrhenius graph [24] for a selected non-aged sample with the smallest degree of moisture.

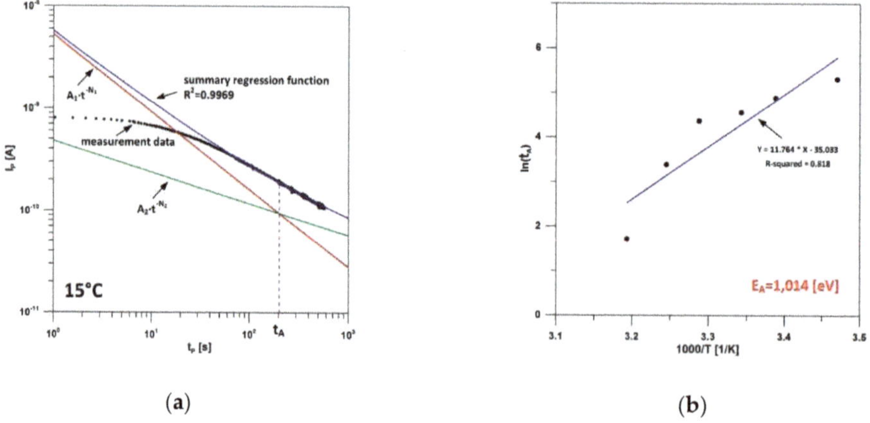

Figure 10. Method of fixing the activation energy E_A for a selected non-aged cellulose–aramid insulation sample with the smallest degree of moisture: (**a**) way of fixing t_A characteristic time from the polarization current; (**b**) Arrhenius graph.

The result of calculation of the activation energy E_A depending on moisture for a selected non-aged cellulose–aramid insulation sample is presented in Figure 11a. A significant increase in the activation energy value along with the growing degree of insulation moisture should be noted. A similar phenomenon was observed for the classic mineral oil-impregnated cellulose insulation; however, the calculated E_A values are slightly higher here [26]. Therefore, it can be assumed that the introduction of additional layers of aramid fibers in Nomex®910 paper raises the activation energy value at which the low-frequency insulation relaxation process is subject to change.

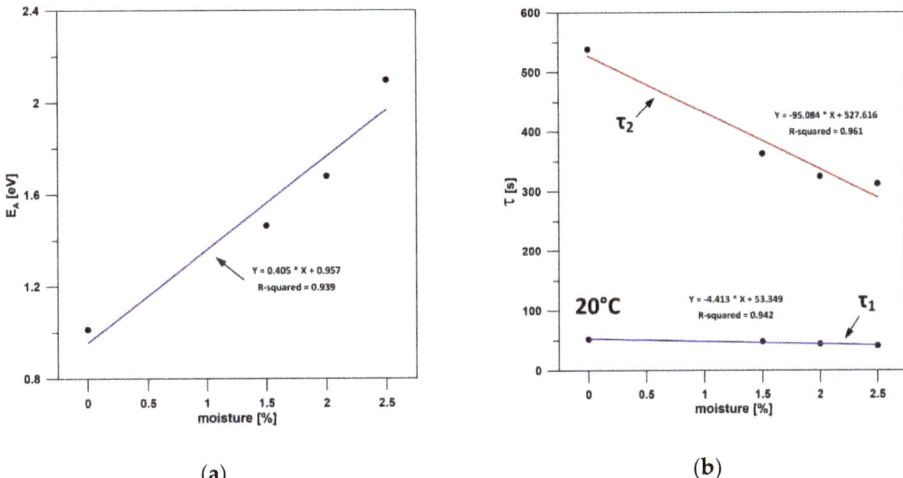

Figure 11. Dependence of the activation energy E_A (**a**) and dominant time constants τ_1 and τ_2 (**b**) on the degree of moisture for a selected non-aged cellulose–aramid insulation sample.

Figure 11b presents the effect of the degree of sample moisture on the value of two dominant time constants of the relaxation processes, determined on the basis of the depolarization current characteristics according to Equation (5). Increased moisture resulted in a slight decrease in time constant τ_1, which was correlated with the relaxation process of aramid fibers, and, on the other hand, a considerably more intensive decrease in time constant τ_2, which was correlated with the relaxation of cellulose fibers. The presented difference is probably related to a significantly higher water absorbability by cellulose than aramid fibers.

5.3. Effect of the Degree of Ageing

Figure 12 presents exemplary characteristics of the polarization currents (Figure 12a) and the depolarization currents (Figure 12b) measured for unmoistened samples with various ageing degrees. It was decided to analyze the effect of the degree of ageing in the temperature range of initial degradation of cellulose fibers (before the impregnation process) from 130 °C to 190 °C, with steps every 20 °C.

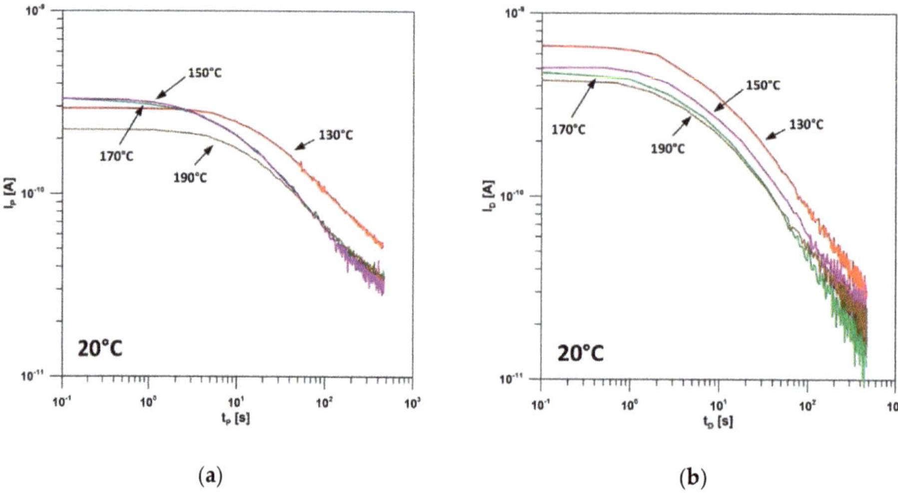

Figure 12. Characteristics of the polarization currents (**a**) and the depolarization currents (**b**) measured for unmoistened cellulose–aramid insulation samples for different degrees of initial ageing in the measurement temperature 20 °C.

The greatest and the predictable effect of the degree of sample ageing was observed for the depolarization current within approx. 10 s (Figure 12b). Therefore, it can be concluded that a growth in the degree of thermal degradation of cellulose fibers significantly reduces the depolarization current value in this time range. For longer times, mostly due to small values of the registered depolarization currents at the level of 20–40 pA, the characteristics overlap, which makes their correct analysis difficult. In the case of the polarization currents (Figure 12a), within the observation time range up to approx. 10 s of the measurement, only the application of the initial degradation temperature of cellulose fibers equal to 190 °C caused a significant reduction in the current value as compared to the other samples. As thermal degradation of cellulose, namely a decrease in the degree of polymerization of its macromolecules, results mainly in a significant loss of mechanical properties (e.g., breaking strength), while the material's volume resistivity remains at a similar level, the polarization current characteristics in longer time ranges (approx. 1000 s) will stabilize close to the value equal to the leakage current. Similar conclusions were achieved for the other measurement temperatures, i.e., up to 60 °C inclusive. Unfortunately, the introduction of additional sample moisture in the range from 1.5% to 2.5% rendered the effect of the degree of ageing practically unfeasible to be observed on the polarization and depolarization current characteristics. Water molecules, because of their strongly polar nature, effectively conceal any minor effect of the degree of ageing, with any possible changes in the characteristics being within the boundaries of the meter's measuring error. This phenomenon was already observed in earlier publications from a co-author of this article [7,19].

Figure 13 presents the effect of the degree of sample ageing on the activation energy E_A value (Figure 13a) and values of two dominant time constants of the relaxation processes (Figure 13b), which were determined on the basis of the depolarization current characteristics according to Equation (5).

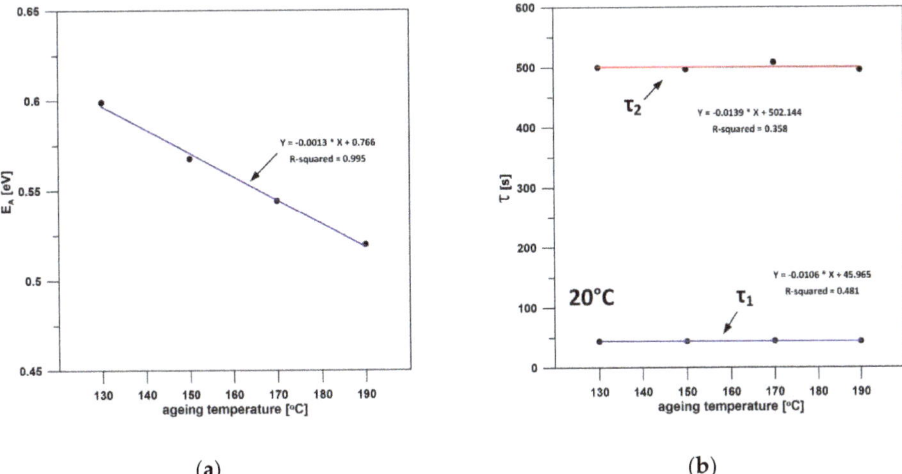

Figure 13. Dependence of the E_A activation energy (**a**) and dominant time constants τ_1 and τ_2 (**b**) on the degree of ageing for unmoistened cellulose–aramid insulation samples.

The most important change that was observed on the characteristics from Figure 13 is that the growing degree of thermal degradation of cellulose fibers resulted in a minor, but constant, decrease in the activation energy value, after which the low-frequency relaxation process of the insulation being examined was subject to change. A probable reason is that the activation of cellulose macromolecules with a smaller degree of polymerization as a result of ageing is less energy-intensive in the polarization process. In the case of the dominant time constants from Figure 13b, it was observed that the growing degree of sample ageing does not cause any significant changes. The explanation of this fact for time constant τ_1, which is correlated with the relaxation processes of aramid fibers, is quite obvious. The initial sample ageing temperature is simply too low to cause any significant changes in the structure of this material's fibers [9]. In the case of time constant τ_2, correlated mainly with the cellulose fiber relaxation process, the change is also negligible due to the similar process of decay of the depolarization current (Figure 12b) and the earlier described measuring difficulties.

6. Conclusions

The results of the tests presented in the article confirm that Nomex®910 cellulose–aramid electro-technical paper produced by DuPont™ is a material that can successfully be used and safely operated in the electro-insulation systems of power utilities for a long time. The basic goal of the company's engineers, i.e., strengthened cellulose material structure by two-sided covering of the paper surface with a thin layer of aramid, has certainly been accomplished. This article authors' opinion is particularly strong because of the fact that, after heating Nomex®910 paper in the temperature of 190 °C for 25 h with air access, the material changed to a darker color, however the structure of the surfaces did not change to a significant extent as compared to the samples heated at lower temperatures. The same process applied to classic Kraft type cellulose paper resulted in the material cracking when being handled after it was taken out of the furnace. Certainly, this statement applies to roll papers, with a thickness comparable to the cellulose–aramid paper used for the tests (0.08 mm).

At present however, there is still the problem of practical adaptation of the diagnostic methods applied so far for the classic cellulose–oil insulation, used, for example, in power transformers, as compared to hybrid semi-synthetic insulation, which Nomex®910 cellulose–aramid paper from DuPont™ is. It seems obvious that that the future of electro-insulation systems of power transformers will be synthetic materials. In the case of liquid materials, in many new and already operating transformer units, oil of mineral origin is being replaced by biodegradable esters. Likewise, aramid material is being introduced to solid electric insulation systems at the stage of transformer design and production. Unfortunately, because the new materials are more expensive than the classic materials (cellulose and mineral oil), the number of transformers with semi-synthetic insulation operating at present is still small, limiting expert knowledge connected with the maintenance of these units.

The test results presented in the article have proved that the PDC diagnostic method (polarization and depolarization method) can be successfully used to estimate the degree of moisture of cellulose–aramid insulation impregnated with insulating mineral oil. The profiles of polarization and depolarization currents depending on moisture of the samples are predictable and similar to the characteristics measured for the classic cellulose–oil insulation. The introduction of a thin layer of aramid in Nomex®910 paper does not induce any significant changes in the characteristics of the hydrodynamic balanced paper–oil, which is certainly, from the point of view of adaptation of the PDC method, a very promising feature. In the analysis of the depolarization current using the regression method with Debye double function, described by Formula (5), in the case of a longer time constant of τ_2 relaxation processes, its practically linear dependence on the degree of sample moisture was observed (Figure 11b). This gives hope for the use of this information in future practical diagnostics of the degree of moisture in transformer insulation, although the article authors are aware that in the case of such complex systems, the linear dependence can change. The characteristics of the activation energy E_A depending on moisture (Figure 11a) and depending on the degree of ageing (Figure 13a) are only cognitive in nature. In practical diagnostics, measurements of polarization and depolarization currents for several temperatures of the transformer's insulation are usually impossible. The process of cooling down the transformer's insulation system in offline mode is very time-consuming, which would require multiple diagnoses using the PDC method to be performed over several days, while, for the needs of diagnostics, transformers are switched off for as short a time as possible, subject to the company's economic calculations. At present, like for the classic cellulose–oil insulation, the estimation of the degree of ageing for samples made of Nomex®910 paper impregnated with insulating mineral oil is a great challenge for the PDC method. The cellulose ageing processes are always accompanied by increased moisture, because water is one of the products of decay of its macromolecules. Water is a strongly polar liquid, and therefore has a "masking" effect for significantly smaller changes in the characteristics of, for example, the depolarization currents that are caused by ageing changes. In this respect, there is a need to continue further research.

Author Contributions: Conceptualization, S.W.; methodology, S.W.; formal analysis, S.W. and A.K.; investigation, S.W. and A.K.; writing—original draft preparation, S.W.; visualization, A.K.; supervision, S.W. All authors have read and agreed to the published version of the manuscript.

Funding: This research received no external funding.

Acknowledgments: The NOMEX®910 for testing was provided by DuPont™ Poland Sp. z o.o., ul. Postepu 17b, 02-676 Warszawa, Poland.

Conflicts of Interest: The authors declare no conflict of interest.

References

1. Liao, R.; Hao, J.; Chen, G.; Ma, Z.; Yang, L. A Comparative Study of physicochemical, dielectric and thermal properties of pressboard insulation impregnated with natural ester and mineral oil. *IEEE Trans. Dielectr. Electr. Insul.* **2011**, *18*, 1626–1637. [CrossRef]
2. Fofana, I.; Borsi, H.; Gockenbach, E.; Farzaneh, M. Aging of transformer insulating materials under selective conditions. *Electr. Energy Syst.* **2007**, *17*, 450–470. [CrossRef]
3. Aslam, M.; Basit, A.; ul Haq, I.; Saher, S.; Khan, A.D.; Khattak, A.N. Improved Insulation Durability to Improve Transformer Aging. *Int. J. Emerg. Electr. Power Syst.* **2020**, *21*, 20190173. [CrossRef]
4. Munajad, A.; Subroto and Suwarno, C. Study on the Effects of Thermal Aging on Insulating Paper for High Voltage Transformer Composite with Natural Ester from Palm Oil Using Fourier Transform Infrared Spectroscopy (FTIR) and Energy Dispersive X-ray Spectroscopy (EDS). *Energies* **2017**, *10*, 1857. [CrossRef]

5. Przybylek, P.; Moranda, H.; Moscicka-Grzesiak, H.; Cybulski, M. Laboratory Model Studies on the Drying Efficiency of Transformer Cellulose Insulation Using Synthetic Ester. *Energies* **2020**, *13*, 3467. [CrossRef]

6. Dombek, G.; Nadolny, Z.; Przybylek, P.; Lopatkiewicz, R.; Marcinkowska, A.; Druzynski, L.; Boczar, T.; Tomczewski, A. Effect of Moisture on the Thermal Conductivity of Cellulose and Aramid Paper Impregnated with Various Dielectric Liquids. *Energies* **2020**, *13*, 4433. [CrossRef]

7. Wolny, S.; Lepich, M. Influence of ageing and moisture degree of aramid-oil insulation on depolarization current. In Proceedings of the IEEE International Conference on Dielectrics, Montpellier, France, 3–7 July 2016; pp. 1163–1166.

8. Wolny, S. Analysis of High-frequency Dispersion Characteristics of Capacitance and Loss Factor of Aramid Paper Impregnated with Various Dielectric Liquids. *Energies* **2019**, *12*, 1063. [CrossRef]

9. DuPont™ Nomex® 400 Series. Technical Data Sheet. Available online: https://www.dupont.com/products/nomex-400-series.html (accessed on 3 November 2020).

10. Zdanowski, M. Streaming Electrification of Mineral Insulating Oil and Synthetic Ester MIDEL 7131. *IEEE Trans. Dielectr. Electr. Insul.* **2014**, *21*, 1127–1132. [CrossRef]

11. Zdanowski, M. Streaming Electrification Phenomenon of Electrical Insulating Oils for Power Transformers. *Energies* **2020**, *13*, 3225. [CrossRef]

12. DuPont™ Nomex® 900 Series. Technical Data Sheet. Available online: https://www.dupont.com/products/nomex-900-series.html (accessed on 3 November 2020).

13. Hao, J.; Liao, R.; Chen, G.; Ma, Z.; Yang, L. Quantitative analysis ageing status of natural ester-paper insulation and mineral oil-paper insulation by polarization/depolarization current. *IEEE Trans. Dielectr. Electr. Insul.* **2012**, *19*, 188–199.

14. Wolny, S.; Kedzia, J. The assessment of the influence of temperature of selected parameters of the approximation method of depolarization current analysis of paper–oil insulation. *J. Non-Cryst. Solids* **2010**, *356*, 809–814. [CrossRef]

15. Saha, T.K.; Purkait, P. Investigation of Polarization and Depolarization Current Measurements for the Assessment of Oil-paper Insulation of Aged Transformers. *IEEE Trans. Dielectr. Electr. Insul.* **2004**, *11*, 144–154. [CrossRef]

16. DuPont™ Nomex®T910. Technical Data Sheet. Available online: http://protectiontechnologies.dupont.com/Nomex-910-transformer-insulation (accessed on 3 November 2020).

17. Nynas Nytro 10X. Technical Data Sheet. Available online: https://www.reinhardoil.dk/Nytro-10-X.pdf (accessed on 3 November 2020).

18. Wolny, S.; Adamowicz, A.; Lepich, M. Influence of Temperature and Moisture Level in Paper-Oil Insulation on the Parameters of the Cole-Cole Model. *IEEE Trans. Power Deliv.* **2014**, *29*, 246–250. [CrossRef]

19. Wolny, S. Aging degree evaluation for paper-oil insulation, carried out using the recovery voltage method. *IEEE Trans. Dielectr. Electr. Insul.* **2015**, *22*, 2455–2462. [CrossRef]

20. Gorgan, B.; Notingher, P.V.; Wetzer, J.M.; Verhaart, H.F.A.; Wouters, P.A.A.F.; van Schijndel, A.; Tanasescu, G. Calculation of the remaining lifetime of power transformers paper insulation. In Proceedings of the 2012 13th International Conference on Optimization of Electrical and Electronic Equipment (OPTIM), Brasov, Romania, 22–24 May 2012; pp. 293–300.

21. Jadav, R.B.; Ekanayake, C.; Saha, T.K. Understanding the Impact of Moisture and Ageing of Transformer Insulation on Frequency Domain Spectroscopy. *IEEE Trans. Dielectr. Electr. Insul.* **2014**, *21*, 369–379. [CrossRef]

22. Zukowski, P.; Rogalski, P.; Koltunowicz, T.N.; Kierczynski, K.; Subocz, J.; Zenker, M. Cellulose Ester Insulation of Power Transformers: Researching the Influence of Moisture on the Phase Shift Angle and Admittance. *Energies* **2020**, *13*, 5511. [CrossRef]

23. Jonscher, A. The Universal Dielectric Response and Its Physical Significance. *IEEE Trans. Electr. Insul.* **1992**, *27*, 407–423. [CrossRef]

24. Zukowski, P.; Koltunowicz, T.N.; Kierczyński, K.; Rogalski, P.; Subocz, J.; Szrot, M.; Gutten, M.; Sebok, M.; Jurcik, J. Permittivity of a composite of cellulose, mineral oil, and water nanoparticles: Theoretical assumptions. *Cellulose* **2016**, *23*, 175–183. [CrossRef]

25. Oommen, T.V. Moisture equilibrium in paper-oil insulation systems. In Proceedings of the 1983 EIC 6th Electrical/Electronical Insulation Conference, Chicago, IL, USA, 3–6 October 1983; pp. 162–166.
26. Zenker, M.; Mrozik, A. Dielectric response of aramid paper impregnated with synthetic ester. *Electr. Rev.* **2018**, *10*, 164–167. (In Polish)

Publisher's Note: MDPI stays neutral with regard to jurisdictional claims in published maps and institutional affiliations.

 *energies*

Article

The Influence of the Window Width on FRA Assessment with Numerical Indices

Szymon Banaszak *, Eugeniusz Kornatowski and Wojciech Szoka

Faculty of Electrical Engineering, West Pomeranian University of Technology, 70-310 Szczecin, Poland;
Eugeniusz.kornatowski@zut.edu.pl (E.K.); wojciech.szoka@zut.edu.pl (W.S.)
* Correspondence: szymon.banaszak@zut.edu.pl

Abstract: Frequency response analysis is a method used in transformer diagnostics for the detection of mechanical faults or short-circuits in windings. The interpretation of test results is often performed with the application of numerical indices. However, usually these indices are used for the whole frequency range of the recorded data, returning a single number. Such an approach is inaccurate and may lead to mistakes in the interpretation. An alternative quality assessment is based on the estimation of the local values of the quality index with the moving window method. In this paper, the authors analyse the influence of the width of the input data window for four numerical indices. The analysis is based on the data measured on the transformer with deformations introduced into the winding and also for a 10 MVA transformer measured under industrial conditions. For the first unit the analysis is performed for various window widths and for various extents of the deformation, while in the case of the second the real differences between the frequency response curves are being analysed. On the basis of the results it was found that the choice of the data window width significantly influences the quality of the analysis results and the rules for elements number selection differ for various numerical indices.

Keywords: transformer winding; deformation; frequency response analysis (FRA); numerical index; window width

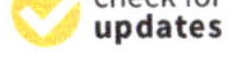 check for updates

Citation: Banaszak, S.; Kornatowski, E.; Szoka, W. The Influence of the Window Width on FRA Assessment with Numerical Indices. *Energies* **2021**, *14*, 362. https://doi.org/10.3390/en14020362

Received: 3 December 2020
Accepted: 8 January 2021
Published: 11 January 2021

Publisher's Note: MDPI stays neutral with regard to jurisdictional claims in published maps and institutional affiliations.

1. Introduction

In electric power systems transformers are very important elements. Their technical condition has a direct influence on the reliability of a power system. A power transformer's design and construction are quite complex and must meet many requirements related to the electrical, mechanical, thermal or environmental properties. In recent decades, the average age of a transformer in operation continues to increase, which means a growing demand for the development of diagnostic methods that can determine the actual technical condition of the transformer. Diagnostics plays a major role in the technical and economic aspects of power distribution companies' asset management [1]. Modern approaches in this field introduce health indices, which allow the managing staff to plan the operation and repairs [2]. Some such indices are used in complex systems introduced in power companies, taking into account the importance of the transformer in the power system [3]. The key issue in many methods is the interpretation of the obtained results, which includes the application of various signal processing methods. One such method is a moving window approach, used in linear digital filtration [4]. This method can be successfully used also in other numerical indices, including frequency response analysis.

The heart of a transformer is its active part (a core with all windings) with its designed mechanical strength. The design of the active part of transformers should be resistant to many mechanical forces, especially those caused by short-circuit currents. The strength of the structure is ensured by the appropriate connection of elements and the clamping system of the windings. However, with the passage of time, the mechanical integrity of the windings deteriorates due to the aging of the insulation and the cumulative effects of the previous network or mechanical events (e.g., transport). In some cases, production errors [5] and the resulting insufficient resistance to factors not significantly different from the nominal conditions are revealed. When a short-circuit current of considerable magnitude occurs in the winding, as a consequence, significant electrodynamic forces also appear, which are proportional to the square of this current [6]. These forces deform the original shape of the winding and can also damage various components. One of the consequences of this process is the occurrence of electrical discharges in the reduced insulation gap. In other cases, initially a slight deformation of the windings does not necessarily lead to an immediate failure of the transformer. Such insulation can continue its operation, however, reduced gaps (e.g., similar turns of adjacent coils) and the disturbance of the turn insulation due to its aging lead to a reduction in electrical strength, which may end in a catastrophic failure [7].

One of the diagnostic methods used for testing of power transformers is frequency response analysis (FRA), which is widely used for the analysis of the mechanical condition of a transformer's active part, especially its windings. It is possible to detect faults, like radial deformations, axial displacements or short-circuits in windings. The FRA measurement results are usually presented in the form of Bode plots, where the amplitude is calculated as a scalar ratio of the signal measured at the output of the circuit to the signal given to the input, and presented in the form of attenuation (in dB). The phase shift of the frequency response results from the difference between these signals and is presented in degrees [8]. FRA is a comparative method, where two curves in a frequency domain are compared and the observed differences can be an indicator of a failure, for example a local deformation or a short-circuit in a winding. For this reason, there are no simple criteria for the determination of the winding's condition, the comparison is done in a wide frequency spectrum, which consists of many local series and parallel resonances, capacitive or inductive slopes of the frequency response (FR) curve. It should be mentioned that the proper connection scheme also influences the FR shape [9].

The assessment of the measurement data is a hard task, usually performed by experienced personnel. This assessment is aided by the application of several numerical indices, which usually return a single value that describes the condition of a winding. These indices and their applicability to FRA assessment are compared and analysed in many publications. The biggest set of indices—over 30—is gathered, described and compared in [10]. In [11] some tests of numerical indices with real FRA data coming from controlled deformations are performed, while in [12] these indices are tested with six case studies. In [13] the sensitivity tests of FRA results are performed and [14] proposes the criteria for interpretation of basic indices. From the point of view of the industrial user of the FRA method, it is hard to use over 20 different numerical indices, which can provide contrary conclusions, because they are sensitive to various differences between the compared frequency response curves. In addition, these differences are related to unknown faults, because—depending on the construction of the transformer, its power rating, winding connection, etc.—there is no universal direct correlation between a fault's type, size or location and its influence on the FR curve changes.

The authors of this paper performed the comparison of 14 of the most popular indices in [15], which was based on data coming from controlled deformations introduced into transformer windings and also from some industrial measurements. This allowed grouping indices into four categories and the selection of a single index from each category, which covers the behaviour of all indices from the given group. Thus, by using only four indices it is possible to perform a comprehensive assessment of the FRA data. This approach is called the grouped indices method—GIM. However, the result is still a single number returned by each index for the analysed frequency range. This limitation can be avoided by calculation of given index value not for the whole analysed range, but for a defined data window. In such a way it is possible to obtain a curve in a frequency domain for each numerical index. The question is what should be the width of such a window (number of analysed elements)?

Another problem in using numerical indices is the selection of the input data. The authors use only a medium frequency (MF) range, in which local deformations in the windings would create visible differences. The borders of this range depend on the geometrical size of the transformer and its construction; generally, the bigger the transformer, the lower in the frequency domain its response lies. The low frequency range is strongly influenced by an iron core, so—for example—all short-circuits in windings can be easily detected, as they generate a large difference between the compared curves in this range. A high frequency range shows the influence of wave phenomena in the windings and the connection setup of the diagnostic equipment and is usually omitted in the analysis. Details of the selection of the borders for an MF range are described in [15] by authors of this paper and wider in doctoral thesis [16]; this starts with the inflection point on the capacitive slope of the first deep resonance (visible for an end-to-end open test circuit and being the result of interaction of the bulk capacitance of the winding with magnetizing inductance of the core) and ends with the beginning of the wave phenomena influence on the FR curve (visible as the character change of the curve)—the example is presented in Figure 1, which presents the FRA data measured on the transformer used for tests described in Section 3 of this paper (details of the measurement configuration are also given there).

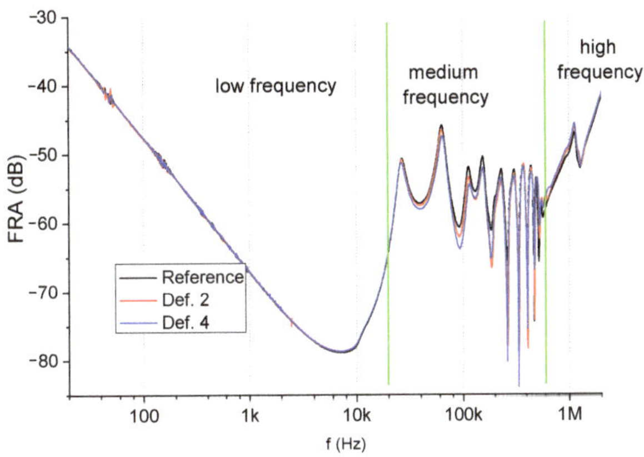

Figure 1. FRA data measured in "end-to-end open" test circuit on a transformer used for the laboratory tests (Section 3 of this paper).

Although the input data are limited to the MF range, is still too wide to allow GIM indices to performing an accurate analysis with just a single (global) value. The borders of this range should be used rather as the beginning and end of the assessment performed in narrower windows. The efficient comparative analysis necessary for FRA results assessment needs determination of the local differences between the curves: reference and analysed ones. In such a case, a "moving window" technique may be introduced. This is well known in digital signal processing and digital filtration algorithms, widely described in the literature (for example, see [17]). Similarly, the problems of digital signal quality assessment can also be found in many papers [18].

In this paper, the authors introduce the "moving window" technique to FRA data analysis with numerical indices. The research results presented in the article include:

- presentation of the assumptions of the FRA method,
- discussion of the properties of numerical quality criteria (numerical indices) used in the GIM method,
- assumptions of the "moving window" method applied to the analysis of numerical data obtained from FRA measurements,
- analysis of the probabilistic properties of four chosen quality criteria, taking into account the influence of the window size on the variance of the values of these quality

criteria, and consequently—the impact of the window size on the readability of the FRA analysis results, and

- results of experimental research using a moving window and quality criteria used in the GIM method.

The data used for the analysis comes from controlled deformations performed on the active part of the distribution transformer and from industrial measurements coming from 10 MVA transformer. The aim of the research was to check the influence of a number of the window elements on the quality and applicability of a frequency response comparative analysis with chosen quality criteria.

2. Frequency Response Analysis and Assessment of Measured Data

In FRA results assessment usually only the amplitude is taken under investigation. It is commonly designated as FRA, so it can be written as follows:

$$FRA\ (dB) = 20\log(U_{out}/U_{in}), \tag{1}$$

where U_{in} is the voltage on the input and U_{out} is the voltage on the output.

Differences between the curves can be observed as frequency shifts of resonances or changes in curve damping [19].

The FRA measurements methodology is standardized [20]. The standard provides details of performing the FR measurements to obtain repeatable results, allowing the comparison of test results measured by various diagnostic companies or by equipment coming from various producers. All measurement data used in this paper was obtained according to this standard and using the end-to-end open test setup, presented in Figure 2.

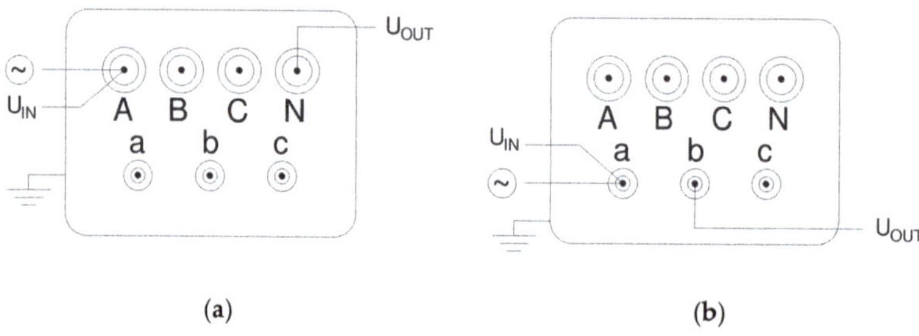

Figure 2. FRA "end-to-end open" measurement configurations on typical transformer: (**a**) Y-connected winding; (**b**) D-connected winding.

The FRA data was measured with a FRAnalyzer test device from Omicron (Austria), using standard settings, usually chosen by personnel for performing industrial measurements. This means that the total number of points, measured in the logarithmic scale of frequency (from 20 Hz to 2 MHz) was 1000. This number can be modified by advanced users, which allows for focusing on a given frequency range. These points are divided into frequency ranges as presented in Table 1.

Table 1. Standard distribution of test points in frequency ranges for the Omicron FRAnalyzer test device.

From Frequency	To Frequency	Points	Sweep Mode
20 Hz	<100 Hz	50	logarithmic
100 Hz	<1 kHz	210	logarithmic
1 kHz	<10 kHz	210	logarithmic
10 kHz	<100 kHz	210	logarithmic
100 kHz	<1 MHz	210	logarithmic
1 MHz	2 MHz	110	logarithmic

FRA diagnostics using local assessment of quality indices is the actual scientific problem. The research in this field can be found in papers from top publishers [21,22] or from international conferences, for example [23]. An assessment based only on a single value returned by any numerical index may be misleading and result in a wrong interpretation of the measured FR data.

The application of the moving window technique generates a series of quality indices QI(f) (where f—frequency) for subsequent positions of the algorithm window. The values of QI(f) are calculated by a chosen quality criterion, for example MSE (mean squared error), CC (correlation coefficient), ASLE (absolute sum of logarithmic error) or SD (standard deviation) as proposed in the GIM method [10]. The formulas representing these indices are given in Table 2.

Table 2. Numerical indices used in the grouped indices method (GIM) [15].

Name	Formula	(Number)
Correlation coefficient	$CC = \dfrac{\sum_{i=1}^{N} Y_{0i} Y_{1i}}{\sqrt{\sum_{i=1}^{N} [Y_{0i}]^2 \sum_{i=1}^{N} [Y_{1i}]^2}}$	(2)
Standard deviation	$SD = \sqrt{\dfrac{\sum_{i=1}^{N} (Y_{0i} - Y_{1i})^2}{N}}$	(3)
Mean squared error	$MSE = \dfrac{\sum_{i=1}^{N} (Y_{0i} - Y_{1i})^2}{N}$	(4)
Absolute sum of logarithmic error	$ASLE = \dfrac{\sum_{i=1}^{N} |20 log_{10} Y_{1i} - 20 log_{10} Y_{0i}|}{N}$	(5)

The input values are two sequences of FRA measurement data: The reference $FRA_{Y0}(f)$ = {Y0: $y0_i$ | i = 1, 2, … , I}, and the diagnosed one, $FRA_{Y1}(f)$ = {Y1: $y1_i$ | i = 1, 2, … , I}, where I is the index of the subsequent element of the set and also index (number) of FRA test frequency. Regardless of used quality index F (Table 2), the analysis result—in general—will be a set QI(f) = {QI: qi_i | i = N < i < I − N}. Calculation of the elements of this set is performed by the general formula:

$$QI_i = F(y0_{i-N<i<i+N}, y1_{i-N<i<i+N}) \tag{2}$$

with the assumption that the algorithm window contains K = 2 N + 1 elements and N ≤ i ≤ I − N.

The process of QI(f) calculation is similar to an algorithm of Finite Impulse Response filter (FIR) with two inputs [4]. The input data are sets Y0 and Y1. The input data are used for the successive calculation of QI, according to the rule F, using the window containing K = 2 N + 1 of consecutive elements of data vectors (values Y0 and Y1)—see Figure 3a. The window is being moved along the input vectors by one element and in the calculation process there is created the vector of output data QI, containing 2·N elements less than each of the input vectors. The process of selection of elements from input vectors, which are used for the calculation of QIi value is shown in Figure 3b.

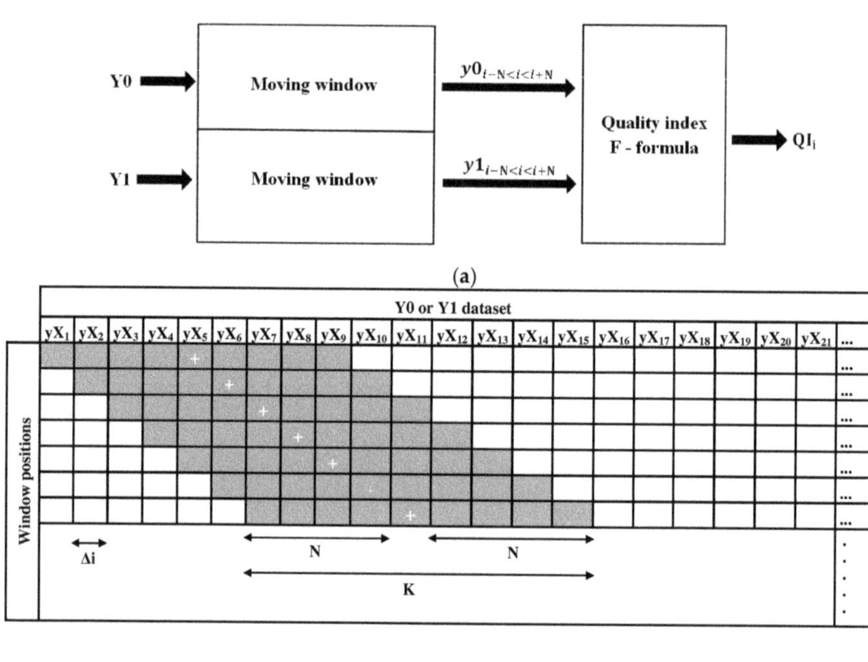

Figure 3. (**a**) A flow chart of calculation of local values of the quality index, (**b**) the moving window algorithm: successive positions of the window in the process of elements acquisition from data vectors Y0 and Y1. The symbol "+" marks the central (current) window element, while total window width is K = 9 (N = 4).

The application of the moving window technique needs the determination of two parameters: a step of the algorithm and a number of window elements. If changes of index *i* are done with the step $\Delta i = 1$, the frequency resolution of calculated index QI(*f*) is maximal, equal to the number of measured points. If $\Delta i > 1$, the frequency resolution is lower.

The choice of the window size is not an easy task. It can be predicted that large values of K will lead to a smoothening of the results. On the contrary, the low values of this parameter would lead to a very precise QI representation of the differences between FRA_{Y0} and FRA_{Y1}. In such a case, obtained result QI(*f*) may be hard to interpret, similarly to the raw input data (measured curves of FRA(*f*)). The question arises: what should be the value of K?

If all probabilistic properties of the FRA dataset are taken under consideration, as well as the definitions of the chosen quality indices CC, SD, MSE and ASLE, the dependency of the indices variance changes as a function of N parameter (defining the number of window elements K = 2 N + 1) are given in Figure 4. The vertical axis represents—normalized to maximal values—the variance of the chosen quality indices. It can be seen that in the case of SD, MSE and ASLE, the application of a small size window would lead to a precise representation by these indices of the changes between the reference and diagnosed FRA datasets. For windows having large number of elements (e.g., with K > 11 (N > 5) elements), the curves representing these indices value changes would be smoothened, which in FRA diagnostics—in some cases—may be advantageous.

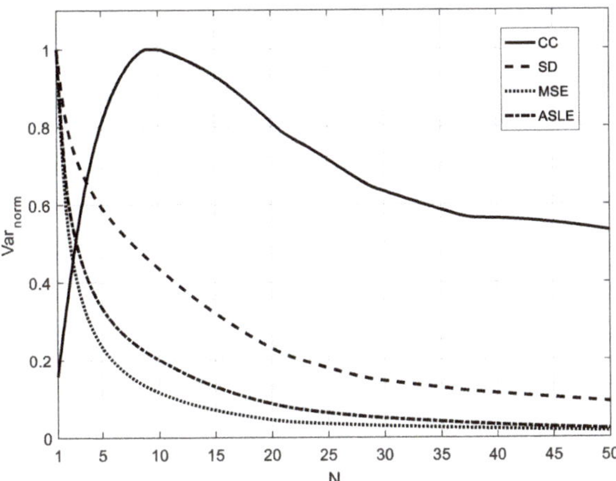

Figure 4. The dependency of normalized variance of quality indices values from N (the number of window elements K = 2 N + 1).

The variance values for each QI quality index (CC, SD, MSE or ASLE) were calculated using the definition of unbiased estimation of variance:

$$Var(\text{QI}) = \frac{1}{L-1} \sum_{1}^{L} (\text{QI}_i - \overline{\text{QI}})^2 \tag{3}$$

where:

QI_i are the values of CC, SD, MSE or ASLE calculated for all positions of the window having K = 2 N + 1 elements, "shifted" along the input data vector according to formula (6), which contains L elements, $1 \leq i \leq L$,

$\overline{\text{QI}}$ is the mean value of CC, SD, MSE or ASLE, calculated for given window size K = 2 N + 1.

Finally, for such calculated vectors of variance (separately for individual quality indicators), graphs shown in Figure 4 were prepared, normalizing the results against the maximum QI value for each of the quality criteria.

A different character of window size influence on variance may be observed for the index CC. A maximum can be clearly seen on its graph in Figure 4. This means that for the N equal to approximately 10 the variance of CC index values is maximal, which results in the most exact representation of its changes in the frequency function. The curves shown on Figure 4 were obtained by averaging the results of the normalized variances calculations of four quality criteria for eight transformers (which have various power ratings) and for the standard measurement point distribution in frequency ranges, as shown in Table 1. If the frequency spectrum will be divided in a different way, the maximum value of CC index normalized variance is for N different than 10. In the case of the remaining indices curves would be similar to those presented on Figure 4 and will still be valid for the rule: "small" window—big variance, "large" window—small variance.

The conclusion from this comparison is that the optimal value of N cannot be defined. It depends on the expected final effect: a very exact representation of the changes or a smoothened analysis result. In the first case for three criteria: SD, MSE and ASLE, the value of N shall be lower than 5 (the increase of the variance in Figure 4). Where the analysis results need smoothening, it is the contrary. A significant increase in the value of N (much more than 5) will not influence the quality of the results: the smoothening level will change slowly, while the amount of calculations will change proportionally to the number of elements in the algorithm window.

In the case of the CC index window width estimation, the character of its variance changes needs to be considered in the function of N. A choice of N = 10 would allow for detailed representation of its values changes in the frequency domain, so its "sensitivity" would be maximal (for a standard distribution of test point in frequency ranges given in Table 1).

3. The Influence of the Window Size on GIM Indices

The four indices, chosen for the GIM method, were used for assessment of the data coming from the deformational experiment, performed on a 6/0.4 kV, 800 kVA, Dyn transformer. Its active part was removed from the tank and various deformations were introduced into windings. One of them was chosen for testing the abovementioned indices with different sizes of analysis window. The results measured in controlled deformations are especially useful for described numerical indices analysis, because they are connected to a known deformation in the winding. This deformation was based on compression of the top disks, by removing spacers between them—it is presented in Figure 5. It was applied to disks 1-2 (Def. 1), 1-3 (Def. 2), 1-4 (Def. 3) and 1-5 (Def. 4), so there were four levels of this deformation, compared to the reference measurement.

Figure 5. Example of deformation introduced into windings of a transformer: compression (reduced inter-disc distances in three gaps)—Def. 3.

The analysis of the four GIM indices was conducted in the medium frequency range, which for this transformer was set from 20 kHz to 600 kHz (according to [15,16]). Various values of N were analysed. In the following section, three of these are presented in the graphs: N = 1, 10 and 20. The presentation of more results would make the graphs unclear.

Figure 6 presents the FR data measured on this unit and used for the analysis. The graph shows only the MF range, used for further analysis (it is a part of curves from Figure 1). At the lower and higher frequencies no significant changes between the curves are present. To make the graph clear, only two deformations are presented (Def. 2 and Def. 4) with the reference curve (healthy winding). The differences between the curves are visible as damping shifts of the whole frequency ranges (for example from 70 kHz to 200 kHz) or at the resonant points (270 kHz, 340 kHz, 470 kHz).

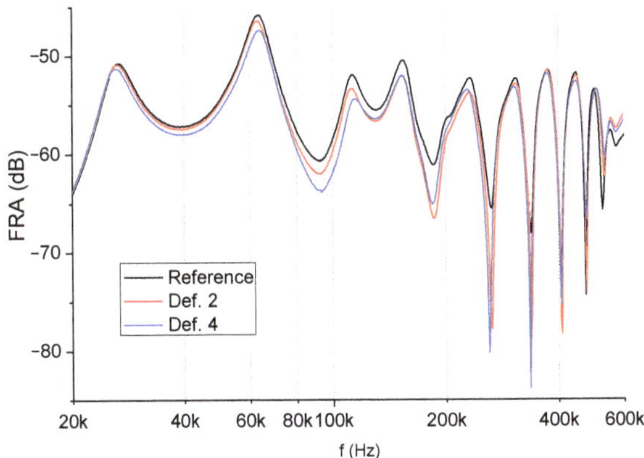

Figure 6. FR curves measured on the healthy winding of 6/0.4 kV, 800 kVA transformer and with axial deformations.

Figures 7–10 present the values of the four numerical indices for various values of N (K = 2 N + 1). Each case also contains the average value, which is a single value calculated from the given index for the whole analysed range. In other words, it represents the typical industrial approach to analysis with numerical indices and is presented on the graphs with dashed line. The curves used for comparison with the indices are the reference one and Def. 4—the biggest deformation.

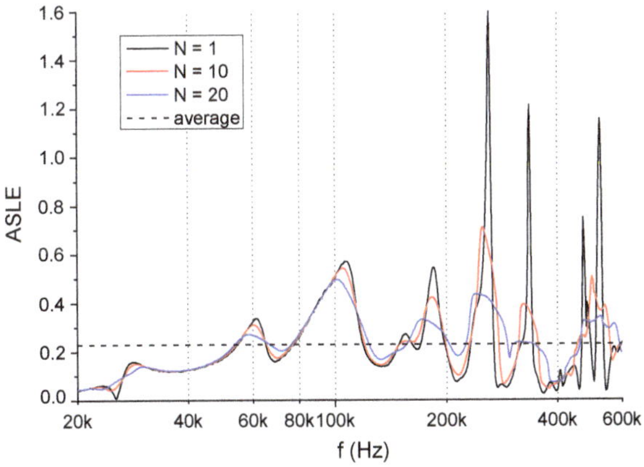

Figure 7. The results of assessment with ASLE for various values of N. The dashed line represents the average value (global).

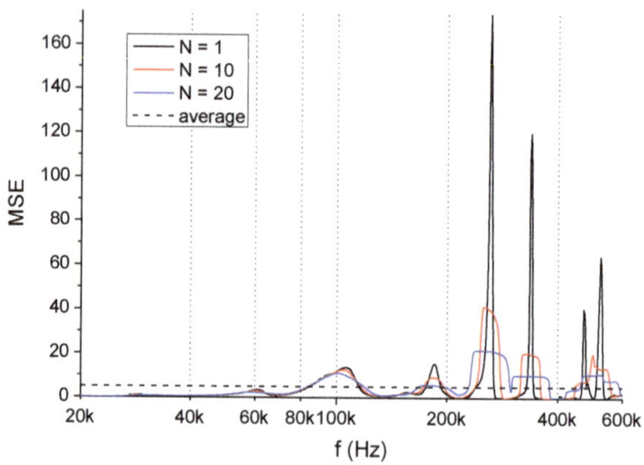

Figure 8. The results of assessment with MSE for various values of N. The dashed line represents the average value (global).

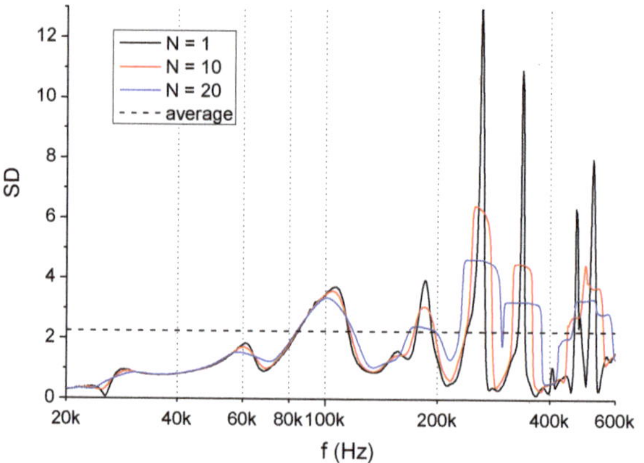

Figure 9. The results of assessment with SD for various values of N. The dashed line represents the average value (global).

It is expected that with the increase of the value of N, the graph will get flattened. The cause of such behaviour is illustrated in Figure 4: for the large values of N, the variance of the indices values decreases, so the smoothening level gets higher.

The first example (Figure 7) is calculated with the ASLE index. The differences between N = 1 and N = 20 are clearly visible, especially in the frequency range where a narrow and steep value change appears—for example at approx. 270 kHz or 340 kHz. The smoothening is very strong, if compared to less steep areas (for example at approx. 100 kHz).

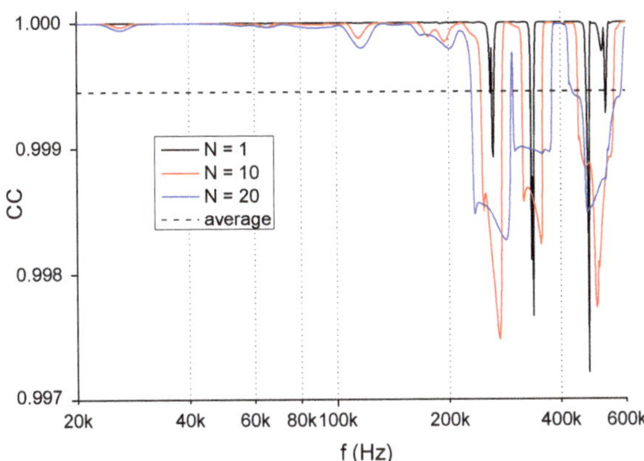

Figure 10. The results of assessment with CC for various values of N. The dashed line represents the average value (global).

The next graph (Figure 8) presents the calculation conducted with the MSE index. In this case, the graph flattening is also very strong in the case of narrow and steep areas, such as at approx. 270 kHz. The difference between the average value (dashed line) and the maximum value for N = 1 is huge.

Figure 9 presents the results of assessment with the SD index. The conclusions are similar to the two previous cases, however, the smoothening is not so radical with the increase of N elements. For example, at approx. 270 kHz the change of the window size from 3 (N = 1) to 21 (N = 10) results in the drop of the SD value by approximately half.

The last example is the values of the CC index, shown in Figure 10. For this numerical index, the influence of the number of N is not so obvious.

Depending on the frequency range, the number of N influences to a different extent. For example, the resonance of the FR data at 270 kHz gives the biggest change to the CC result if the value of N = 10, while the two other resonances (340 kHz, 470 kHz) give the biggest change in CC for N = 1. This is related to the width of the resonance (see Figure 6), as the first of these (270 kHz) does not have slopes as steep as the other two.

To summarize the results from Figures 7–10, it can be stated that the window size depends on ones needs, whether it is necessary to smoothen or sharpen the output curve. However, the best value for the average approach is N = 10—especially in the case of CC index. For this quality index the value N = 10 guarantees the optimal precision of local values changes representation (Figure 2), which advantageously influences the legibility of CC (f) curve and the accuracy of diagnostic conclusions. In order to compare indices ASLE, SD, MSE and CC this value is used for testing all the indices for various extents of the deformation, which is presented in Figures 11–14. The curves presented on these graphs are the results of the assessment of the FR data with the four indices. Each curve represents the values calculated from comparison of the data measured in the healthy state with the data measured after introducing the deformation. For example, the curve marked as Def. 1 is the result of analysis with a given index, for N = 10, of the reference line and the line for deformation 1. Figures contain the results for Def. 1, Def. 3 and Def. 4 (minimal deformation, maximal deformation and one in-between) to make them easier to analyse. In addition, for each case the global value of the given index is given, in other words the average value, for the window width equal to the total number of points in the analysed MF range.

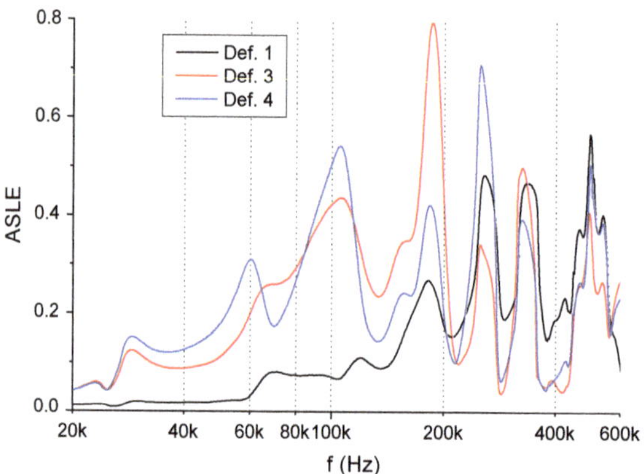

Figure 11. The results of assessment with ASLE for various deformations.

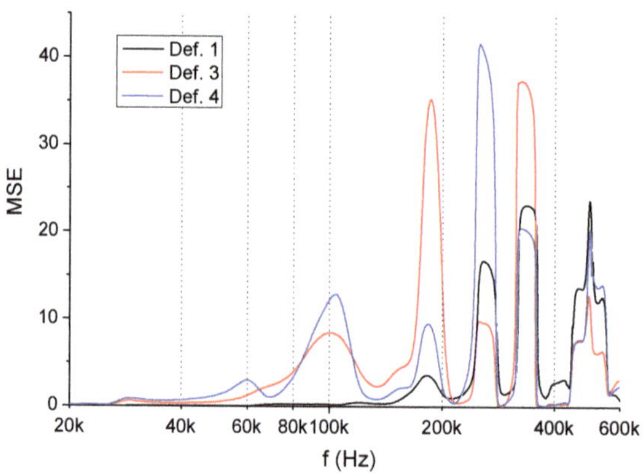

Figure 12. The results of assessment with MSE for various deformations.

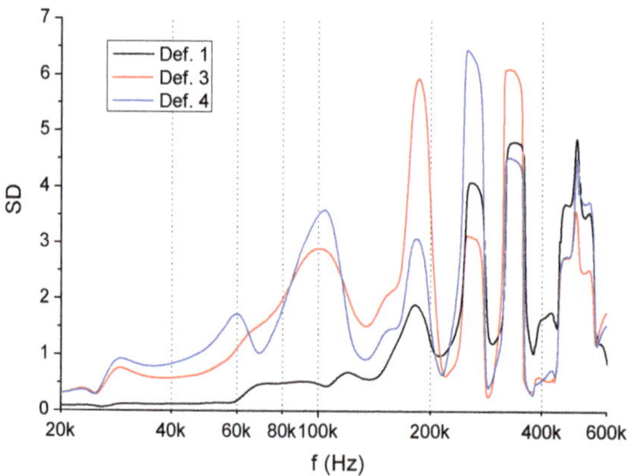

Figure 13. The results of assessment with SD for various deformations.

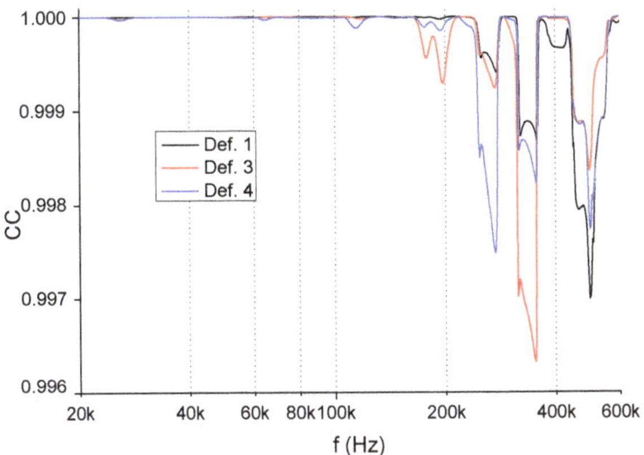

Figure 14. The results of assessment with CC for various deformations.

The first example is shown in Figure 11, presenting the results obtained with the ASLE index. Depending on the frequency range (input FR data), there are visible various behaviours of curves calculated for the three deformation scales. However, if these shapes are compared to the global values (for the whole range), which are Def. 1—0.140, Def. 3—0.225 and Def. 4—0.230, it can be seen that latter two do not reflect the complexity of the changes visible on the graph, where—for example—Def. 3 has its maximum at 180 kHz, while for Def. 4 it is at 240 kHz. Their global values are almost the same, so an analysis based on them would not show any differences between these two cases. In other words application of the moving window technique is proven to be effective in the analysis of the FRA data.

A similar effect can be observed in Figure 12, where the MSE index was used for the assessment. The global values are in this case: Def. 1—2.92, Def. 3—5.11, Def. 4—5.04, so, again, the latter two are almost similar, while the curves show the maximums for different frequencies. The same conclusions may be drawn for the third index, SD (see Figure 13), where the global values are, respectively: 1.71, 2.26 and 2.25.

The last case, the CC index, also has similar behavior. Def. 1 has its maximum at approx. 540 kHz, Def. 3 at 340 kHz, while Def. 4 is at 270 kHz. In this case, the global values are: 0.99961, 0.99944 and 0.99945, so again Def. 3 and Def. 4 are very similar—see Figure 14.

From the above examples, it can be seen that for N = 10 all four indices act in the same way: their local extremums depend on the actual deformation and its influence on the FR curve. The comparison to global values clearly shows that analysis done for the narrower window, not for the whole analysed range, gives more information. It can be seen that for deformations introduced into the winding, the global value of the quality index may be insufficient to detect the fault. The quality indices for Def. 3 and Def. 4 are almost similar, while the measurements were taken for different geometries of windings.

4. The Application to Industrial FRA Results

The research described in the Section 3 was repeated for data coming from the industrial measurements. The tested transformer was 15/6 kV, 10 MVA unit, which has clearly visible differences between two measurements, namely reference and fault curves. The medium frequency range, which in this case is from 20 kHz to 600 kHz, can be observed in Figure 15. The exact cause of differences seen on this Figure is unknown for the authors, because the owner of the transformer did not share decisions taken on transformer further operation and possible results of the internal inspection. However, these results are good for the moving window technique analysis, because differences between both curves cover all possible scenarios, which can be encountered in the transformers' frequency response analysis: a shift in the frequency domain (60–120 kHz) related to capacitance changes, a change of a shape with a new resonance (approx. at 200 kHz), which can be affected by many factors that constitute that parallel resonance (e.g., capacitance and inductance interaction in that frequency) and a damping change (approx. at 360 kHz) being an effect of changed parameters forming given resonance, e.g., capacitance and turn-to-turn magnetic couplings, if there was a deformation in the winding or local resistance change in the case of partial (via small resistance) local short-circuit. With such data the comparison of changing window widths showed all possible behaviours of the output data.

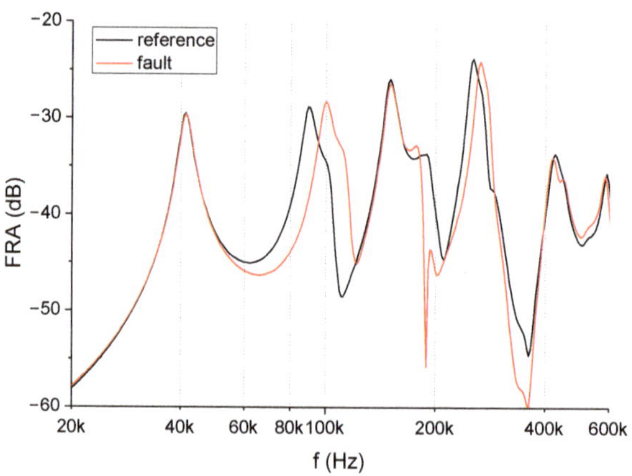

Figure 15. FR curves of industrial transformer 15/6 kV, 10 MVA in the medium frequency range.

The first tested numerical index was ASLE. Its results are shown in Figure 16. For value N = 1 the output results are very sharp, strongly pointing all present differences between curves. By increasing the N value to 10 and 20 the curve is smoothened and follows the frequencies of changes visible in FR graph from Figure 15. For diagnostic purposes the better result is obtained for value N = 10, because for changes at 200 kHz and 360 kHz is shown two separate maximums. Additionally, if compared to the average value, used in the standard approach to numerical indices application, there are clearly visible frequency ranges in which curves vary from this value.

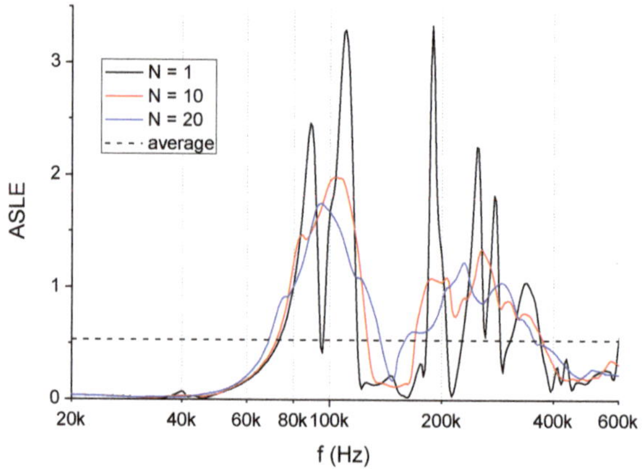

Figure 16. The results of assessment of data from Figure 15 with ASLE for various values of N. The dashed line represents the average value (global).

Similar conclusions may be drawn for data presented in Figure 17 and representing the assessment with MSE index. For N = 1 the output graphs reaches very high amplitudes and is strongly sensitive on any differences between two input curves. Again, for N = 10 some local maximums may be observed in frequency ranges correlating to FR curves measured on the transformer and comparison to the average value clearly indicated "suspicious" frequency ranges.

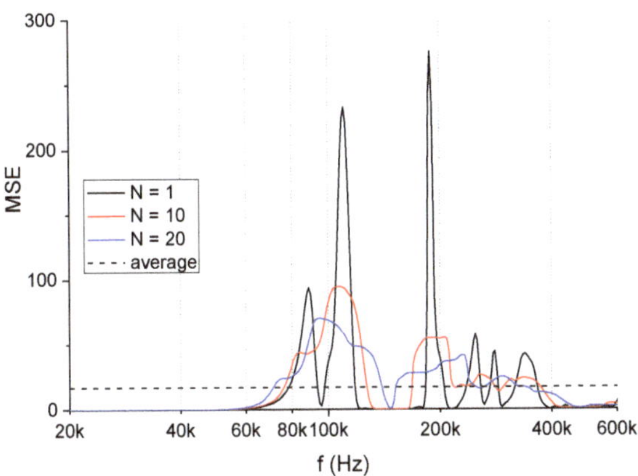

Figure 17. The results of assessment of data from Figure 15 with MSE for various values of N. The dashed line represents the average value (global).

The third numerical index is SD, with results for various window sizes presented on Figure 18. Also this case shown the best efficiency in the detection of differences between input curves for N = 10, both indicating frequency ranges connected to faults and giving the good contrast if compared to the average value (global).

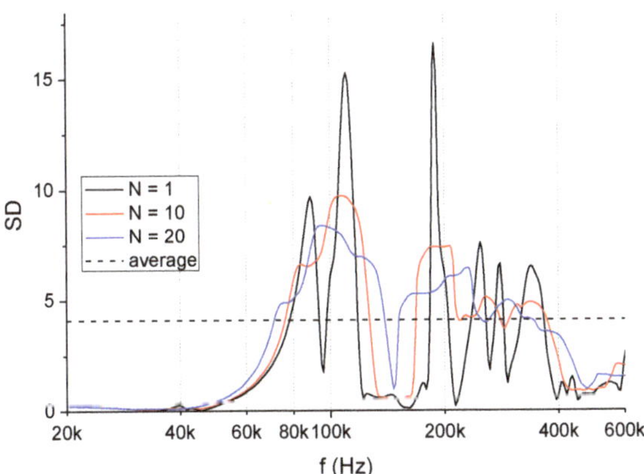

Figure 18. The results of assessment of data from Figure 15 with SD for various values of N. The dashed line represents the average value (global).

The last case is CC numerical index, analysed in Figure 19 and showing different behaviour. The value N = 1 gives the smaller sensitivity to differences between input curves, pointing out mainly the change of the shape (new resonance). However, the increase of N to 10 or 20 returns more differences in the curve shape. Additionally, in this case N = 10 gives better results, because the index is more sensitive to local changes. For example there are visible local minimums at approx. 200 kHz and 250 kHz. All such local "suspicious" frequency ranges differ from the average value of this index.

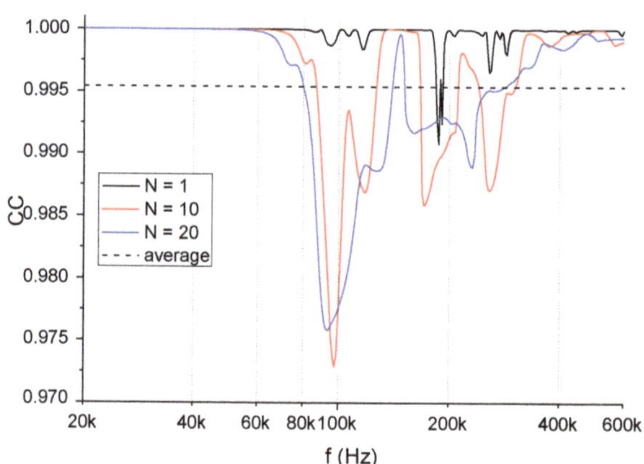

Figure 19. The results of assessment of data from Figure 15 with SD for various values of N. The dashed line represents the average value (global).

The analysis of results presented on Figures 16–19 leads to the conclusion that the best quality of detection of differences between compared curves is achieved for the value N = 10, for which all categories of changes are indicated in the graphs: resonance shifts, damping changes and curve shape change.

5. Summary

The paper presents the research on the data window width for the analysis of frequency response results of transformer windings performed with numerical indices. Four quality indices were chosen to test the results measured on the active part of the transformer with deformations introduced into the winding. The results were analysed for various window widths (K = 2 N + 1).

In the first stage, the variance of the four indices was tested, depending on the N value, which allowed for testing their sensitivity to this parameter. In the case of the correlation coefficient (CC) index, the results were not unequivocal.

The analysis of data for various extents of deformation introduced into laboratory tested transformer showed that the approach based on moving the window is more accurate and allows for detection of the winding geometry changes, which are undetected by a standard approach with a single (global) value of a given index. This is clearly presented in the case of deformations 3 and 4. The choice of the N value depends on the needs of the analysis, however, the proposed N = 10 seems to be the most universal and useful for most practical analyses, however, some users of the frequency response analysis (FRA) method may need other effects (smoothening vs. sharpening). The similar conclusions may be drawn from observation of the industrial measurements performed on the second transformer. The value N = 10 allowed for the best effect of differences detection between compared frequency response curves (datasets). For such value the output curve of each numerical index clearly showed the local difference of the value, strongly differing from the average value, for any type of possible changes to the output data: a frequency shift, damping change or change in the shape (new resonance).

For practical application of obtained results authors propose to compare output data of each index in frequency range for N = 10 to average value (local maximum to average value ratio). By setting the fault detection criteria, for example on level 30%, it will be possible to identify the frequency ranges of input frequency response data having "suspicious" values differences. It can be applied to the automatic detection systems. The results of experimental tests presented in Section 3 relate to an exemplary transformer with a power rating of 800 kVA. It should be clearly emphasized that the proposed "optimal" window sizes apply to transformers with such power and design features and—in particular—with the adopted frequency resolution of the frequency response analysis method.

The value of fault detection criteria should be further analysed, because it depends for example on geometrical size of the unit (power rating) or its construction. This topic is the subject of further research of the authors. In this paper, the authors wanted to draw the attention of other researchers dealing with diagnostics using the frequency response analysis method to the problem of window size selection when using objective (numerical) quality indicators. The readability of the obtained results for local frequency ranges depends on the number of window elements and there is no intuitive rule: the precision of mapping local numerical changes in quality indicators is inversely proportional to the number of window elements. This is evidenced by the presented analysis of the variance of changes in the correlation coefficient value, for which the maximum variance occurs for a given number of window elements.

Author Contributions: Conceptualization, S.B. and E.K.; Formal analysis, E.K.; Investigation, S.B., E.K. and W.S.; Methodology, S.B., E.K. and W.S.; Project administration, S.B.; Supervision, S.B. and E.K.; Validation, S.B. and E.K.; Writing—original draft, S.B., E.K. and W.S. All authors have read and agreed to the published version of the manuscript.

Funding: This research received no external funding.

Institutional Review Board Statement: Not applicable.

Informed Consent Statement: Not applicable.

Conflicts of Interest: The authors declare no conflict of interest.

References

1. Tenbohlen, S.; Coenen, S.; Djamali, M.; Müller, A.; Samimi, M.H.; Siegel, M. Diagnostic Measurements for Power Transformers. *Energies* **2016**, *9*, 347. [CrossRef]
2. Scatiggio, F.; Pompili, M.; Calacara, L. Transformers Fleet Management Through the use of an Advanced Health Index. In Proceedings of the 2018 IEEE Electrical Insulation Conference (EIC), San Antonio, TX, USA, 17–20 June 2018; pp. 395–397.
3. Bohatyrewicz, P.; Płowucha, J.; Subocz, J. Condition Assessment of Power Transformers Based on Health Index Value. *Appl. Sci.* **2019**, *9*, 4877. [CrossRef]
4. Antoniou, A. *Digital Signal Processing*; Mcgraw-Hill: New York, NY, USA, 2016.
5. Sharafi, D. Manufacturing defect in a group of western power transformers. In Proceedings of the Workshop on Diagnostic Measurements on Power Transformers, Klaus, Austria, 4 April 2008.
6. Waters, M. *The Short-Circuit Strength of Power Transformers*; Macdonald&Co. Ltd.: London, UK, 1966.
7. Bjerkan, E. High Frequency Modeling of Power Transformers. Stresses and Diagnostics. Ph.D. Thesis, Norwegian University of Science and Technology, Trondheim, Norway, 2005.
8. Alsuhaibani, S.; Khan, Y.; Beroual, A.; Malik, N.H. A Review of Frequency Response Analysis Methods for Power Transformer Diagnostics. *Energies* **2016**, *9*, 879. [CrossRef]
9. Samimi, M.H.; Tenbohlen, S.; Akmal, A.A.S.; Mohseni, H. Effect of Different Connection Schemes, Terminating Resistors and Measurement Impedances on the Sensitivity of the FRA Method. *IEEE Trans. Power Deliv.* **2017**, *32*, 1713–1720. [CrossRef]
10. Samimi, M.H.; Tenbohlen, S. FRA interpretation using numerical indices: State-of-the-art. *IJEPES* **2017**, *89*, 115–125. [CrossRef]
11. Rahimpour, E.; Jabbari, M.; Tenbohlen, S. Mathematical comparison methods to assess transfer functions of transformers to detect different types of mechanical faults. *IEEE Trans. Power Deliv.* **2010**, *25*, 2544–2555. [CrossRef]
12. Badgujar, K.P.; Maoyafikuddin, M.; Kulkarni, S.V. Alternative statistical techniques for aiding SFRA diagnostics in transformers. *IET Gener. Transm. Distrib.* **2012**, *6*, 189–198. [CrossRef]
13. Bak-Jensen, J.; Bak-Jensen, B.; Mikkelsen, S.D. Detection of faults and ageing phenomena in transformers by transfer functions. *IEEE Trans. Power Deliv.* **1995**, *10*, 308–314. [CrossRef]
14. Jong-Wook, K.; ByungKoo, P.; Seung Cheol, J.; Sang Woo, K.; PooGyeon, P. Fault diagnosis of a power transformer using an improved frequency-response analysis. *IEEE Trans. Power Deliv.* **2005**, *20*, 169–178.
15. Banaszak, S.; Szoka, W. Transformer frequency response analysis with the grouped indices method in end-to-end and capacitive inter-winding measurement configurations. *IEEE Trans. Power Deliv.* **2019**, *35*, 571–579. [CrossRef]
16. Velásquez Contreras, J.L. Intelligent Monitoring and Diagnosis of Power Transformers in the Context of an Asset Management Model. Ph.D. Thesis, Polytechnic University of Catalonia UPC, Barcelona, Spain, 2012.

17. Rao, K.D.; Swamy, M.N.S. *Digital Signal Processing. Theory and Practice*; Springer Nature Singapore Pte Ltd.: Berlin/Heidelberg, Germany, 2018.
18. Shnayderman, A.; Gusev, A.; Eskicioglu, A. An SVD-Based Gray-Scale Image Quality Measure for Local and Global Assessment. *IEEE Trans. Image Proc.* **2006**, *15*, 422–429. [CrossRef] [PubMed]
19. Sofian, D.M.; Wang, Z.D.; Li, J. Interpretation of Transformer FRA Responses—Part II: Influence of Transformer Structure. *IEEE Trans. Power Deliv.* **2010**, *25*, 2582–2589. [CrossRef]
20. International Electrotechnical Commission. *IEC 60076-18: Power Transformers—Part 18: Measurement of Frequency Response*; IEC Standard: Geneve, Swietzerland, 2012.
21. Tarimoradi, H.; Gharehpetian, G.B. Novel Calculation Method of Indices to Improve Classification of Transformer Winding Fault Type, Location, and Extent. *IEEE Trans. Ind. Inform.* **2017**, *13*, 1531–1540. [CrossRef]
22. Kornatowski, E.; Banaszak, S. Frequency Response Quality Index for Assessing the Mechanical Condition of Transformer Windings. *Energies* **2020**, *13*, 29. [CrossRef]
23. Miyazaki, S.; Mizutani, Y.; Taguchi, A.; Murakami, J.; Tsuji, N.; Takashima, M.; Kato, O. Proposal of Objective Criterion in Diagnosis of Abnormalities of Power-Transformer Winding by Frequency Response Analysis. In Proceedings of the 2016 International Conference on Condition Monitoring and Diagnosis (CMD), Xi'an, China, 25–28 September 2016; pp. 74–77.

Article

Using the Method of Harmonic Distortion Analysis in Partial Discharge Assessment in Mineral Oil in a Non-Uniform Electric Field

Alper Aydogan [1], Fatih Atalar [1], Aysel Ersoy Yilmaz [1] and Pawel Rozga [2,*]

[1] Department of Electrical and Electronic Engineering, Istanbul University-Cerrahpaşa, 34320 Avcilar, Istanbul, Turkey; aydoganalper@gmail.com (A.A.); fatih.atalar@istanbul.edu.tr (F.A.); aersoy@istanbul.edu.tr (A.E.Y.)
[2] Institute of Electrical Power Engineering, Lodz University of Technology, 90-924 Łódź, Poland
* Correspondence: pawel.rozga@p.lodz.pl; Tel.: +48-609-725-622

Received: 21 August 2020; Accepted: 14 September 2020; Published: 15 September 2020

Abstract: In high-voltage equipment, it is vital to detect any failure in advance. To do this, a determination of the partial discharges occurring at different voltage types as well as at different electrode configurations is essential for observing the oil condition. In this study, an experimental setup consisting of a needle–semi-sphere electrode configuration immersed in mineral oil is prepared for laboratory experiment. In such a way, a non-uniform electric field is created and the leakage currents are monitored from the grounded electrode. A total of six different electrode configurations are analyzed during the tests by the use of hemispheres of different diameters as grounded electrodes and copper and steel pointed (medical) needle high-voltage electrodes. In the experiments, the partial discharges occurring at four different voltage levels between 5.4 and 10.8 kV are measured and recorded. The effect of the different electrode configurations and voltage levels on the harmonic distortion are noted and discussed. It is experimentally confirmed that it is possible to measure the leakage current caused by the partial discharges of the corona type in oil at the different metal points, creating high-voltage electrodes and different electric field distributions based on the proposed non-invasive measurement technique. The studies showed that there is a significant rise of even harmonic components in the leakage current during the increase in the partial discharge intensity with the 5th harmonic as dominant.

Keywords: partial discharge; mineral oil; harmonic distortion; non-uniform electric field; discrete Fourier transform

1. Introduction

The insulation quality of power transformers, cables, or circuit breakers, being the main devices used in high-voltage transmission systems, is very important from the point of view of ensuring sustainable and reliable electrical energy for end users [1]. Dielectric liquids, which have been applied in the mentioned devices as a sole insulation or as a part of a combined insulation system, have been extensively used since the 1850s. The main dielectric liquid applied in high-voltage equipment has been mineral oil, however other liquids such as silicone oils, synthetic esters, and natural esters have been used as well [2]. Although dielectric liquids are easy to implement in the insulating systems and provide electrical insulation effectively, they are very sensitive to impurities due to their specific chemical structure. In particular, pollutant materials within the structure such as solid particles [3,4], gas bubbles [5], or water droplets [6,7] greatly reduce the dielectric performance of insulating liquids [8]. There are many methods, such as, for example, the AC breakdown voltage and dielectric dissipation factor measurement, gas chromatography, and the determination of the moisture content in oil, which are used to estimate the condition of a given dielectric liquid [9,10]. In turn, the experiments focused on the obtainment of the electrical strength and dielectric characteristics of liquids, which are the determinants of their use as proper insulators in high-voltage insulating systems [11,12].

When the dielectric liquid is used as insulation partial discharges (PDs), this constitutes a significant problem for the functioning of the insulating system. Partial discharges refer to the events that occur in the insulating system, deteriorating partially the existing insulation [13], and are also defined as partial breakdowns in the cavities in the insulation due to the high electric field [14]. When assessing partial discharges, including the detection, measurement, and location of PD sources, various effects and physical phenomena accompanying partial discharges are identified, such as the presence of current pulses and the associated emission of electromagnetic waves, chemical changes in insulating materials, the emission of acoustic waves, the emission of light radiation, local temperature increases in the discharge channel or increasing the pressure in the discharge channel [15–17]. The fast aperiodic current pulses with a rise time of the order of nanoseconds that accompany the partial discharges are the source of radiated and conducted disturbances. When the electromagnetic waves are considered, they may cover the range of light radiation. The light emission results from de-excitation processes and ion recombination. Depending on the medium in which the discharge takes place (air, SF_6, oil) and atmospheric factors (temperature, pressure), the emission spectrum is in the range of UV and infrared wavelengths. Part of the energy released during partial discharges changes into mechanical energy in the form of the above-mentioned acoustic signals propagating in all directions, subject to the phenomena of attenuation, reflection, refraction, and dispersion. The amplitude of the acoustic emission pulses is inversely proportional to the distance from the discharge source, and the intensity of the acoustic emission is inversely proportional to the square of the distance. A single partial discharge, which can be compared to a micro-explosion, affects the area covered by the phenomenon, causing a change in the physical properties of the dielectric. As the physical properties of insulating material change, they reflect in changes in the dielectric dissipation factor (tg δ). The system in which PDs do not occur is characterized by a relatively constant value of tg δ, while when the PDs appear their intensity causes an increase in the tg δ. The chemical transformations resulting from partial discharges largely depend on the design of the device and the dielectric materials used in it. The most important chemical transformations resulting from partial discharges are gas evolution, the surface erosion of materials, or electric wood formation. Considering mineral oil-based insulation, the processes connected with PDs give reflection in the generation of specific types of gases as a result of the chemical degradation of oil or oil-paper combined insulation due to electrical or thermal stresses. In a more macroscopic scale, the partial discharges have also a significant influence on the energy and power losses resulting from ionization processes associated with PDs or disturbances in the supplying voltage and current (the possible occurrence of harmonic distortion) [15–33].

There are few direct or indirect methods which are commonly used to detect partial discharges in insulating systems with liquid dielectrics and to assess their intensity. The widely known indirect method is gas chromatography (called dissolved gas analysis—DGA), in which the gases generated due to partial discharges (or other electrical and thermal events) may be collected and analyzed. This method is also referred to as a chemical method. The oil condition is assessed from the measured gas concentrations on the basis of comparison with typical values given in standards or on the basis of a developed analysis method, such as, for example, the Duval triangle method [9,18]. During oil analysis from gas chromatography, it is important to point out that also other oil parameters must be evaluated, such as the moisture content, dielectric dissipation factor, AC breakdown voltage, etc. Another non-electrical PD measurement method is the acoustic emission (AE) method [19,20]. It is a non-invasive method allowing for the determination of the place of partial discharge occurrence and also its type (corona discharge, surface discharge, internal discharge in cavities) by registering the acoustic wave generated by the discharge. PDs may be analyzed also using Ultra High Frequency (UHF) detection. This method has been extensively studied in recent years, mainly as a promising method for use in the condition monitoring of power transformers. However, many tests of the UHF method have been performed on experimental systems with mineral oil as the dielectric liquid [21,22].

In this method, the partial discharge data of different frequency components are evaluated in the range of ultra-high frequency radiation, and the identification of PD types may be performed on the basis of the information obtained in this way. Some works report also the use of light emitted by partial discharges for the assessment of discharge intensity. A qualitative assessment may be conducted on the basis discharge images, as in [23], or the spectra of emitted light may be compared to conclude on the discharge behavior [24].

In addition to the above-mentioned indirect, non-invasive methods of PD analysis of the oil-insulated systems, the most reliable method is still the electrical method based on IEC 60270 Standard [7]. By means of a specially designed testing system, a quantitative assessment of PD intensity is possible. The apparent charge, measured typically in picocoulombs (pC), and the phase-resolved partial discharge (PRPD) patterns recorded during measurement may give the valuable data on the insulating system condition determined by the insulating components' quality [9,16,25]. Typically, the process of data collection is determined by increasing the testing voltage gradually and the registration of the PD signals during the whole process or in time intervals. Partial discharge inception voltage (PDIV) is determined, and PD behavior at higher values of voltage than PDIV is observed. In experimental laboratory studies, a comparison of the different factors influencing the PD intensity (electrode configurations, liquid kind) may be carried out in this way.

There are many works which have been devoted to the problem of PD detection and analysis, with many different points of view on this issue presented by the authors [16,26–32]. Chen et al. experimentally modeled PD behavior in oil-impregnated paper-insulated transformers. Sudden increases in PDs were the determinant in the evaluation of partial discharge-based failures. The authors of the mentioned paper found that average value of PDs is more determinant in asymmetric electrode systems [26]. Gulski et al., in a fundamental work [27], analyzed the deterioration behavior of different insulating materials in high-voltage electrical systems. In the study, the surface discharges and corona discharges in transformer oil were investigated as well. They created and monitored PDs and analyzed the PRPD patterns obtained from the data collected. They found that it is possible to determine the type of partial discharge failure as a result of this form of analysis [27,28]. Cavallini et al. examined the characteristic patterns of PDs in a current transformer and resin. They showed that it is possible to detect other sources of noise on the basis of their studies [29,30]. In addition, Contin et al. have studied major mathematical models used to characterize partial discharge waveforms. They analyzed and evaluated the behavior of the high-frequency current transformer sensors used to collect the original partial discharge signal (symmetrical PD pulse signal and asymmetric PD pulse signal) found in mathematical models [31]. Alvarez et al. have obtained results that successfully measure common PD sources (corona, surface discharges, and internal discharges) by defining the source of signals collected from different partial discharge and noise sources in distribution systems and pulse sources that were mathematically evaluated and classified [32]. The statistical-based approach was presented, however, in [16], where different PD analysis techniques (electrical method, acoustic emission, and UHF) were compared. The authors found that every method was characterized by different properties and restrictions, so the different fields of application and specific measurement condition requirements need to be supported. Additionally, they indicated on a need for conducting further research works connected with the development of a method of PD source detection and identification in insulating systems with oil insulation.

Taking this latter into account, in the studies presented in this paper the behavior of partial discharges was investigated by means of the authors' approach based on the leakage current monitoring from the grounded electrode of the experimental system considered. Thus, the goal of the studies was to confirm the possibility of analyzing the PD intensity quantitively on the basis of the collected current signals, which will be then analyzed using discrete Fourier transform. A non-uniform electric field system was created to do that by using a needle–hemisphere electrode configuration immersed in mineral oil. Two different pointed-type high-voltage electrodes (one made of steel and the second made of copper) and three different-in-radius hemisphere electrodes (2, 5, and 6 cm) were applied to note the possible influence of different factors on the registered courses. The current signals at the selected voltage levels were sampled at a frequency level of 6250 Hz.

2. Mathematical Formulation

Harmonic analysis is one of the most frequently used methods to find the failure of high-voltage, medium-voltage, and low-voltage electrical and electronic systems. When the transmission and distribution lines are examined, external factors such as severe atmospheric conditions and some other stresses like electric or thermal problems may cause insulation deterioration. This leads to the formation of non-sinusoidal wave shapes. Harmonic components can be used to estimate the physical and dielectric conditions. To calculate the harmonic components, generally Fourier transform is used. A Fourier transform of a function $f(t)$ is shown in well-known Equation (1):

$$F(\omega) = \int_{-\infty}^{\infty} f(t)e^{-j\omega t}dt \tag{1}$$

where $F(\omega)$ is the function in the frequency domain for $f(t)$ and ω is the angular frequency of the system. The function may be also converted from the frequency domain to the time domain by Equation (2).

$$f(t) = \frac{1}{2\pi} \int_{-\infty}^{\infty} F(\omega)e^{-j\omega t}dt \tag{2}$$

When n samples are taken for each period, periodically sampled functions can be obtained by using Equations (1) and (2). Discrete Fourier transform is shown in Equations (3) and (4), in which normal and inverse transformation are possible, respectively.

$$F(k\Delta\Omega) = \sum f(n\Delta T)e^{-j2\pi kn/N} \tag{3}$$

$$F(n\Delta T) = \sum f(k\Delta\Omega)e^{-j2\pi kn/N} \tag{4}$$

Considering Equation (3), k, $n = 0, 1, \ldots, N - 1$, where n is the sampling number in the time domain and k is the sampling number in the frequency domain, N represents the total number of samples.

$\Delta\Omega = 2\pi/\Delta T$ and $\Delta T = T/N$ are used to calculate the change in time and change in frequency, respectively. Discrete Fourier Transform (DFT) is often used in harmonic measurements, as it is taken as measurement information which can be obtained as the sampling function in the time domain. Since DFT is set off as measurement information in the form of the sampling function in the time domain, it is often used for measuring the harmonic component values.

Fast Fourier Transform (FFT) is used to facilitate discrete Fourier transform, as shown in Equation (5), by using the similarity of the elements in the matrix $[W^{kn}]$.

$$W = e^{-2j\pi/N} \tag{5}$$

Following this result, the number of transactions decreases to the $(N{\cdot}log_2(N))$ level. To analyze the non-sinusoidal current values, Fourier transform is used, as in Equation (6).

$$i(t) = \sum_{n=1}^{\infty} i_n(t) = \sum_{n=1}^{\infty} \sqrt{2}I_n sin(n\omega_1 t + \delta_n) \tag{6}$$

In the above equation, the DC terms are neglected. In the time domain, the nth degree is the instantaneous value of the harmonic current. It is the effective value of harmonic current at the nth degree, ω_1 is the angular frequency of the base frequency, and δ_n is the phase angle of the harmonic current at the nth degree.

The harmonic distortion (*HD*) of any current signal can be calculated in Equation (7).

$$HD_n = \frac{I_n}{I_1} \tag{7}$$

The corresponding current for the *n* order harmonic is expressed as the relative harmonic current.

3. Experimental Setup and Measurement Methodology

An experimental setup for leakage current measurement under a high voltage was established for the laboratory studies. A schematic representation of this setup is shown in Figure 1.

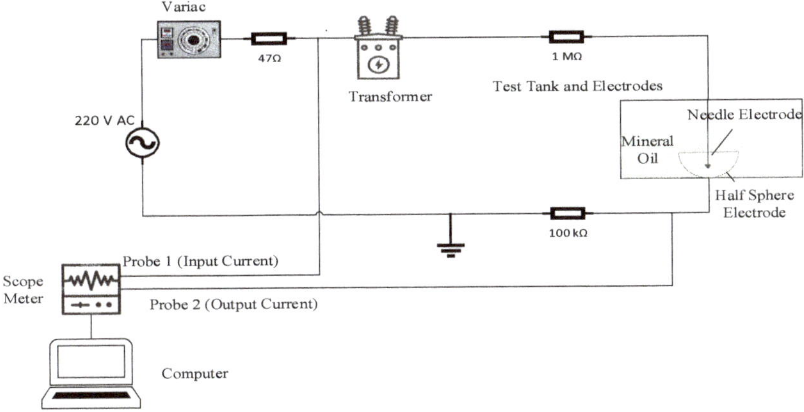

Figure 1. Experimental setup and circuit diagram.

The voltage adjusted by a Variac autotransformer (Varsan brand) with 5 kVA power and a 0–220 V range at the 50 Hz frequency level was applied to the primary winding of the voltage transformer. The voltage transformer VTB 30–200 (Esitas brand) has a conversion ratio of 220 V/40 kV, with a rated power of 1.5 kVA. No deterioration in the signal of the voltage applied was observed during the pre-studies. A 1 MΩ limiting resistance was placed on the high-voltage side of the transformer, as it is typically applied in high-voltage laboratory measurements to protect against high current during breakdown in the testing system [11,12,16,33]. The input current was measured via a 47 Ω resistor at the Variac output, while the output current was measured via a 100 kΩ resistor connected to the ground electrode. The input and output currents corresponding with PD events were measured with a Fluke meter 190–504/EU/S, which has 200 MHz bandwidth oscilloscope with a sampling rate of 2.5 GS/s (Giga Sample/second), with an accuracy of 2.1% of reading + 0.04 × range/div. The current probes used were, however, in both cases (Probe 1 and Probe 2 from Figure 1) the Fluke-branded vps410-ii-v models. Since the approach of the studies was to analyze the signals of leakage current from the grounded electrode, the signal from Probe 2 was further analyzed using FFT. The data of the measured values were taken by FlukeView® SW160 for the MultiFunction Counter interface software. It is important to point out herein that the 47 Ω resistor used does not affect the experimental results, understood as a distortion of the source signal. The primary and secondary voltage of the HV transformer were, however, measured by the Fluke 80 K–40 HV Probe and True RMS Multimeter. The actual photograph of the circuit used is presented in Figure 2.

During each test, a total number of 30,000 data points was obtained during the 4.8 s of measurement. Thus, the sampling frequency was 6250 Hz, and the leakage current data were obtained and read solely by this specific mode of the oscilloscope used. The full sampling size of the oscilloscope was not applied. None of the smoothing windows methods were used, and signals were drawn over the −/+2 Hz adjacent of the frequency for the specified harmonic values. For example, the signals between 148 and 152 Hz for the 3rd harmonic were evaluated. Additionally, frequency leakage was not a concern.

The sensitivity of the experiments was kept stable in terms of unwanted effects. All of the data considered just belonged to leakage currents and harmonic distortions of the PDs. Additionally, there was no deterioration in the applied testing voltage that may affect the experiment. In the main feed of the laboratory, clean sine was provided through filters.

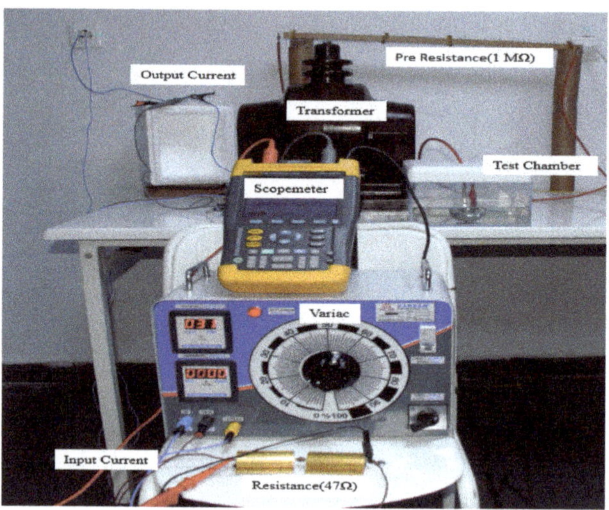

Figure 2. Photograph of the experimental setup.

A classical FFT analysis of the data collected was driven with Matlab software. It was determined how much amplitude the signal has and at which frequency.

The test chamber from Figure 2 is made of plexiglass and has dimensions of $25 \times 25 \times 10$ cm. Mineral oil with a dielectric constant of 2.2 was used throughout the experiments as the dielectric liquid under test. The oil before the beginning of the experiment fulfilled the requirements for the mineral oil specified in the IEC 60296 Standard [34].

In this study, three different hemispheres made of pure copper with the diameters of 6, 5, and 2 cm were used as the grounded electrodes, respectively. Steel (medical) and copper-pointed needles especially prepared for the experiment, both with a tip radius of 0.1 µm, were however used as high-voltage (HV) electrodes. All the electrode configurations are listed in Table 1.

Table 1. Electrode configurations.

Electrode Configuration	Electrode 1 (Half-Sphere Diameter)	Electrode 2 (Tip Point Radius)
A	6 cm copper	0.1 µm copper
B	6 cm copper	0.1 µm steel
C	5 cm copper	0.1 µm copper
D	5 cm copper	0.1 µm steel
E	2 cm copper	0.1 µm copper
F	2 cm copper	0.1 µm steel

For each electrode configuration five electrode ranges (gap distances) were considered from 1 mm to 5 mm with increasing steps of 1 mm. The term "electrode range" is defined as the distance between the end of the needle and the deeper point of the semi-sphere consisting of a grounded electrode. The HV electrodes were changed in every different electrode gap. This was done because the tip point of the needle electrode was worn down due to the high voltage stress. As was reported in many publications [11,16,33,35,36], the degree of damage to the needle tip is believed to be a reflection of the extremely high temperatures prevailing during the discharge processes, especially at breakdown (electrical arc). An electrical arc has a high energy because the electron avalanches increase in a cumulative way, and at the end of this phenomenon a conductive bridge occurs from the tip point of the electrode to the deep mid-point of the hemisphere-grounded electrode. Hence, the temperature of the tip point of the HV electrode gradually increases. While the PDs continue, more and more electrons give way from the tip point of the HV electrode. Excessive increment in both the temperature and numbers of moving electrons causes damage to the point of the HV electrode.

The experiments were conducted under four pre-selected voltage levels, which were 5.4, 7.2, 9, and 10.8 kV, respectively. The selected values resulted from the limitations of the used supplying system, where voltage might be set with 1.8 kV steps. At the 5.4 kV voltage level, the meaningful and comparable data started to be obtained. In turn, for the tests performed at voltages of over 10.8 kV, healthy data could not be obtained with the measuring instruments applied. Therefore, to protect the measuring instruments, the experiment was terminated at the voltage level of 10.8 kV.

The test chamber used in the experiment was specially sketched and designed. All the dimensions were predefined for the purpose of the experiment. In addition, the positions and distances between the electrodes were specially designed in a millimeter range. Hence, hemispherical and needle electrodes could adapt their position in this delicate range. During the experiment, the HV electrodes were aligned exactly to the midpoint of the hemisphere. The above approach enabled the repeatability of the measurement process.

Sample data of 20 periods were selected and analyzed according to each voltage level.

4. Results and Discussion

At first, Table 2 presents the complex results of the HD values for each electrode configuration at characteristic voltage levels obtained separately for each electrode range. However, in the further part of the paper, the distinctive configurations are analyzed separately with some relations to each other.

Table 2. HD values for each electrode configuration at characteristic voltage levels.

Electrode Configuration	Voltage (kV)	d (mm)	HD_2	HD_3	HD_4	HD_5	HD_6	HD_7	HD_8	HD_9
A	5.4	1	0.0299	0.0094	0.0196	0.0736	0.0252	0.0238	0.0199	0.0373
A	5.4	2	0.0424	0.0062	0.0571	0.0911	0.0534	0.0265	0.0427	0.0427
A	5.4	3	0.0444	0.0122	0.0476	0.108	0.0401	0.0125	0.0333	0.0538
A	5.4	4	0.0497	0.0108	0.0429	0.076	0.0527	0.0343	0.0482	0.0267
A	5.4	5	0.0359	0.0192	0.0344	0.0809	0.0267	0.0258	0.021	0.0247
A	7.2	1	0.0448	0.007	0.0446	0.0701	0.0424	0.0192	0.0332	0.0368
A	7.2	2	0.0492	0.0101	0.0458	0.0768	0.0452	0.0202	0.0345	0.0379
A	7.2	3	0.0331	0.0091	0.0364	0.0794	0.0382	0.0341	0.0325	0.0307
A	7.2	4	0.0222	0.0121	0.0394	0.0766	0.0378	0.035	0.0338	0.0325
A	7.2	5	0.0418	0.0112	0.0253	0.079	0.0296	0.0267	0.0219	0.0223
A	9	1	0.0186	0.0068	0.0135	0.0855	0.0168	0.0213	0.0174	0.0348
A	9	2	0.0261	0.0071	0.0218	0.0783	0.0241	0.03	0.018	0.0383
A	9	3	0.0304	0.0091	0.0323	0.0761	0.0377	0.0337	0.0299	0.0359
A	9	4	0.0299	0.0117	0.022	0.0831	0.0258	0.0283	0.0179	0.0343
A	9	5	0.0375	0.0086	0.0294	0.0723	0.0322	0.0265	0.0188	0.0372
A	10.8	1	0.0125	0.0066	0.0115	0.078	0.018	0.0264	0.0165	0.0313
A	10.8	2	0.0245	0.0079	0.0169	0.0864	0.0246	0.0252	0.0126	0.0464
A	10.8	3	0.0254	0.0063	0.0233	0.0691	0.0251	0.0179	0.0223	0.0337
A	10.8	4	0.025	0.009	0.0154	0.085	0.0279	0.0238	0.0273	0.0326
A	10.8	5	0.0288	0.0084	0.0271	0.078	0.032	0.0257	0.029	0.0315
B	5.4	1	0.0622	0.0138	0.0408	0.065	0.053	0.0214	0.0527	0.023
B	5.4	2	0.0842	0.0106	0.052	0.0833	0.0268	0.026	0.0464	0.0195
B	5.4	3	0.0667	0.0112	0.0457	0.089	0.0301	0.0297	0.0354	0.0293
B	5.4	4	0.1028	0.0096	0.0621	0.0668	0.0283	0.0329	0.0377	0.0256
B	5.4	5	0.0611	0.0164	0.0578	0.0965	0.039	0.0357	0.0288	0.0219

Table 2. *Cont.*

Electrode Configuration	Voltage (kV)	d (mm)	HD_2	HD_3	HD_4	HD_5	HD_6	HD_7	HD_8	HD_9
B	7.2	1	0.0352	0.0071	0.0177	0.0662	0.0246	0.0106	0.0274	0.0361
B	7.2	2	0.0524	0.0108	0.025	0.0891	0.0239	0.023	0.033	0.0214
B	7.2	3	0.0492	0.0067	0.0264	0.0659	0.0341	0.0136	0.045	0.0299
B	7.2	4	0.0425	0.0111	0.028	0.0752	0.0229	0.0159	0.0286	0.0333
B	7.2	5	0.0551	0.0091	0.0271	0.0693	0.0443	0.0278	0.0676	0.0348
B	9	1	0.0227	0.0082	0.0176	0.0826	0.0269	0.0112	0.0256	0.0343
B	9	2	0.056	0.0133	0.0336	0.0895	0.0241	0.018	0.031	0.028
B	9	3	0.0403	0.0075	0.0183	0.0696	0.0345	0.0191	0.0427	0.03
B	9	4	0.0356	0.0102	0.016	0.067	0.0237	0.0046	0.0336	0.0312
B	9	5	0.0343	0.0091	0.0209	0.083	0.0192	0.0225	0.0207	0.0296
B	10.8	1	0.0273	0.0093	0.0094	0.0622	0.0175	0.0247	0.0183	0.0284
B	10.8	2	0.0433	0.012	0.013	0.0812	0.0221	0.0135	0.0184	0.0275
B	10.8	3	0.032	0.0062	0.0101	0.0745	0.0163	0.0216	0.0222	0.0343
B	10.8	4	0.0385	0.0118	0.0204	0.0756	0.016	0.0188	0.018	0.0328
B	10.8	5	0.0304	0.0059	0.0182	0.0819	0.0184	0.0114	0.0233	0.03
C	5.4	1	0.0591	0.0137	0.0474	0.0526	0.0508	0.0453	0.0429	0.0334
C	5.4	2	0.0652	0.0134	0.0408	0.0626	0.037	0.0287	0.0302	0.0404
C	5.4	3	0.069	0.0146	0.0509	0.0842	0.0538	0.0164	0.0376	0.0532
C	5.4	4	0.0725	0.0115	0.0593	0.092	0.0756	0.0311	0.067	0.0369
C	5.4	5	0.054	0.0118	0.0403	0.0764	0.0521	0.0347	0.0466	0.0313
C	7.2	1	0.0379	0.0143	0.0239	0.0599	0.0279	0.0483	0.0161	0.0362
C	7.2	2	0.0326	0.0087	0.0207	0.0638	0.0286	0.0355	0.0222	0.0364
C	7.2	3	0.0599	0.0057	0.0476	0.0758	0.0458	0.0192	0.0371	0.0454
C	7.2	4	0.0387	0.0114	0.0291	0.0784	0.0345	0.0237	0.0299	0.0314
C	7.2	5	0.0348	0.0098	0.0252	0.081	0.039	0.0308	0.0325	0.0358
C	9	1	0.0358	0.0082	0.0291	0.0757	0.0341	0.0273	0.0321	0.0397
C	9	2	0.0301	0.005	0.02	0.0756	0.0206	0.0225	0.0149	0.0411
C	9	3	0.0443	0.0078	0.0346	0.0668	0.0343	0.0373	0.0204	0.0327
C	9	4	0.0355	0.0102	0.0327	0.0749	0.0365	0.0328	0.0296	0.0323
C	9	5	0.0295	0.0117	0.0215	0.0856	0.0344	0.0375	0.0278	0.0322
C	10.8	1	0.0149	0.0063	0.0067	0.0743	0.0135	0.0294	0.0136	0.0399
C	10.8	2	0.0285	0.0052	0.0186	0.0737	0.0178	0.0356	0.0099	0.04
C	10.8	3	0.0233	0.0061	0.0179	0.0742	0.0244	0.0286	0.0155	0.0348
C	10.8	4	0.0298	0.0105	0.0194	0.087	0.0213	0.0259	0.0162	0.0468
C	10.8	5	0.0317	0.0064	0.0259	0.0774	0.0304	0.0321	0.0208	0.0447
D	5.4	1	0.0582	0.0173	0.0418	0.0545	0.0454	0.0511	0.0395	0.0237
D	5.4	2	0.0662	0.0102	0.0654	0.0584	0.0613	0.0585	0.0456	0.0513
D	5.4	3	0.0766	0.0061	0.0676	0.0603	0.0672	0.0404	0.0677	0.0485
D	5.4	4	0.0653	0.0141	0.0513	0.0661	0.0452	0.0559	0.0325	0.056
D	5.4	5	0.054	0.0126	0.0364	0.0738	0.0396	0.0388	0.0421	0.05
D	7.2	1	0.0534	0.0082	0.0427	0.0718	0.0411	0.0371	0.0254	0.0366
D	7.2	2	0.0438	0.0085	0.0411	0.0695	0.0427	0.0454	0.0285	0.0551
D	7.2	3	0.0509	0.0117	0.0429	0.0717	0.0383	0.0422	0.0295	0.0485
D	7.2	4	0.0435	0.0127	0.0393	0.0739	0.0378	0.0333	0.0364	0.0482
D	7.2	5	0.0493	0.0068	0.0431	0.0732	0.0475	0.0358	0.039	0.0508
D	9	1	0.039	0.0063	0.0294	0.0624	0.0363	0.0423	0.0343	0.0425
D	9	2	0.0334	0.011	0.0301	0.071	0.0284	0.0382	0.0189	0.0535
D	9	3	0.0379	0.0085	0.0334	0.0664	0.0335	0.0492	0.0242	0.0606
D	9	4	0.0326	0.0056	0.0311	0.0717	0.0296	0.041	0.0316	0.0485
D	9	5	0.0461	0.0089	0.0407	0.0708	0.041	0.0513	0.0209	0.049
D	10.8	1	0.0267	0.0094	0.0203	0.0546	0.0201	0.0347	0.0109	0.0476
D	10.8	2	0.0321	0.0052	0.0311	0.0673	0.0292	0.0426	0.0213	0.051
D	10.8	3	0.0362	0.0079	0.0264	0.0608	0.0255	0.035	0.017	0.061
D	10.8	4	0.0233	0.0049	0.0179	0.0754	0.021	0.0415	0.0193	0.0544
D	10.8	5	0.0415	0.0079	0.0316	0.0725	0.03	0.0383	0.0165	0.0642
E	5.4	1	0.0106	0.0119	0.0028	0.0767	0.0045	0.0372	0.0032	0.0447
E	5.4	2	0.0198	0.0096	0.005	0.0731	0.0039	0.0392	0.0046	0.0463
E	5.4	3	0.009	0.0136	0.0021	0.0785	0.0035	0.0386	0.0038	0.0509
E	5.4	4	0.012	0.0138	0.0048	0.0818	0.0025	0.0401	0.0062	0.0438
E	5.4	5	0.0132	0.012	0.0051	0.0856	0.0043	0.035	0.0056	0.0441
E	7.2	1	0.0067	0.01	0.0059	0.0799	0.0033	0.0326	0.0031	0.0448
E	7.2	2	0.0225	0.009	0.0073	0.0839	0.0057	0.036	0.0058	0.0535
E	7.2	3	0.0097	0.0088	0.0095	0.0865	0.0042	0.0331	0.0054	0.0467
E	7.2	4	0.0095	0.0101	0.0087	0.08	0.0033	0.0382	0.0061	0.0453

Table 2. *Cont.*

Electrode Configuration	Voltage (kV)	d (mm)	HD_2	HD_3	HD_4	HD_5	HD_6	HD_7	HD_8	HD_9
E	7.2	5	0.0108	0.0089	0.0102	0.0817	0.0043	0.032	0.0066	0.0434
E	9	1	0.0241	0.0106	0.0083	0.0812	0.004	0.0339	0.0042	0.0417
E	9	2	0.0188	0.0115	0.0094	0.0737	0.0027	0.0366	0.0088	0.0398
E	9	3	0.0106	0.0083	0.0049	0.0803	0.0071	0.0386	0.0026	0.044
E	9	4	0.0101	0.0092	0.0036	0.0728	0.0066	0.0328	0.0048	0.0458
E	9	5	0.0113	0.0092	0.0055	0.0742	0.0064	0.038	0.0043	0.0426
E	10.8	1	0.0263	0.0074	0.0044	0.0787	0.0055	0.0254	0.0037	0.0489
E	10.8	2	0.0211	0.0109	0.0042	0.0804	0.0074	0.0353	0.0049	0.0413
E	10.8	3	0.0107	0.0137	0.006	0.077	0.0055	0.0371	0.0046	0.0467
E	10.8	4	0.0119	0.0123	0.0063	0.0796	0.0048	0.0373	0.0061	0.0452
E	10.8	5	0.2206	0.1052	22.253	0.0824	0.1706	0.2423	2.882	0.4432
F	5.4	1	0.0102	0.0093	0.0044	0.0604	0.0067	0.0357	0.0074	0.0675
F	5.4	2	0.0145	0.0082	0.0049	0.0728	0.0069	0.0474	0.0087	0.0609
F	5.4	3	0.0151	0.0076	0.0063	0.0704	0.0076	0.0503	0.0089	0.0611
F	5.4	4	0.0102	0.005	0.0088	0.0723	0.0063	0.0451	0.0051	0.0621
F	5.4	5	0.0109	0.0097	0.0056	0.069	0.0056	0.0484	0.0071	0.0718
F	7.2	1	0.011	0.0081	0.0074	0.0742	0.0067	0.0448	0.0095	0.058
F	7.2	2	0.0106	0.0075	0.0084	0.0717	0.0074	0.0476	0.0066	0.0611
F	7.2	3	0.0106	0.0058	0.0084	0.0731	0.0072	0.0471	0.0073	0.0611
F	7.2	4	0.01	0.0058	0.0057	0.0651	0.0081	0.0373	0.0044	0.0641
F	7.2	5	0.0079	0.0062	0.009	0.0692	0.006	0.0418	0.0066	0.0708
F	9	1	0.009	0.0084	0.0034	0.065	0.0084	0.0495	0.0072	0.0621
F	9	2	0.0121	0.0067	0.0055	0.0726	0.0082	0.0484	0.0065	0.0595
F	9	3	0.0112	0.0056	0.006	0.0648	0.0071	0.034	0.0057	0.061
F	9	4	0.0132	0.0086	0.0059	0.0601	0.0062	0.0369	0.0058	0.0632
F	9	5	0.01	0.0076	0.008	0.0634	0.0078	0.0368	0.007	0.0683
F	10.8	1	0.0139	0.008	0.0073	0.071	0.0059	0.0453	0.0081	0.0619
F	10.8	2	0.0136	0.0083	0.0057	0.0694	0.0066	0.0458	0.0071	0.0603
F	10.8	3	0.0035	0.0181	0.0049	0.0819	0.0053	0.0529	0.0039	0.0714
F	10.8	4	0.0056	0.0194	0.0036	0.0752	0.0083	0.0472	0.0025	0.0622
F	10.8	5	0.2866	0.2173	20.5486	0.1514	0.1719	0.2391	4.1126	0.959

To explain why harmonic distortions were specifically calculated instead of just measuring leakage currents, Figure 3 was quoted to show the example of the recorded current signal for the selected case—that is, for a value of voltage equal to 7.2 kV and electrode configuration A. The distinctive courses from 1 to 5 represent the five gap distances considered. Figure 3a presents the original waveforms obtained during the measurements, while Figure 3b, in order to better visualize the similarities, shows the magnified fragments of the courses from Figure 3a.

As can be seen in Figure 3, no significant differences are observed between the collected leakage current signals. Hence, it is preferable to analyze the harmonic distortions, as this gives the possibility of performing meaningful comparisons between the cases considered. In detail, when the FFT analysis is performed (which is presented later in the paper), the effect of the applied voltage level, the electrode diameter, and the gaps between the electrodes on the partial discharge characteristics in the analyzed electrode system can be revealed significantly.

In general, it is important to emphasize herein that there is not a linearity between the electrode gap and grounded electrode diameter increase with the harmonic component change. In particular, specific changes in some harmonic distortions are remarkable. As is well-known, the capacitive reactance increases when the gap between electrodes is greater. However, within this experimental study, it is reported that the capacitive reactance, which was expected to increase with the increase in the electrode gap and the grounded electrode diameter, decreases at some voltage levels for some harmonic distortion levels, even if every parameter is kept the same.

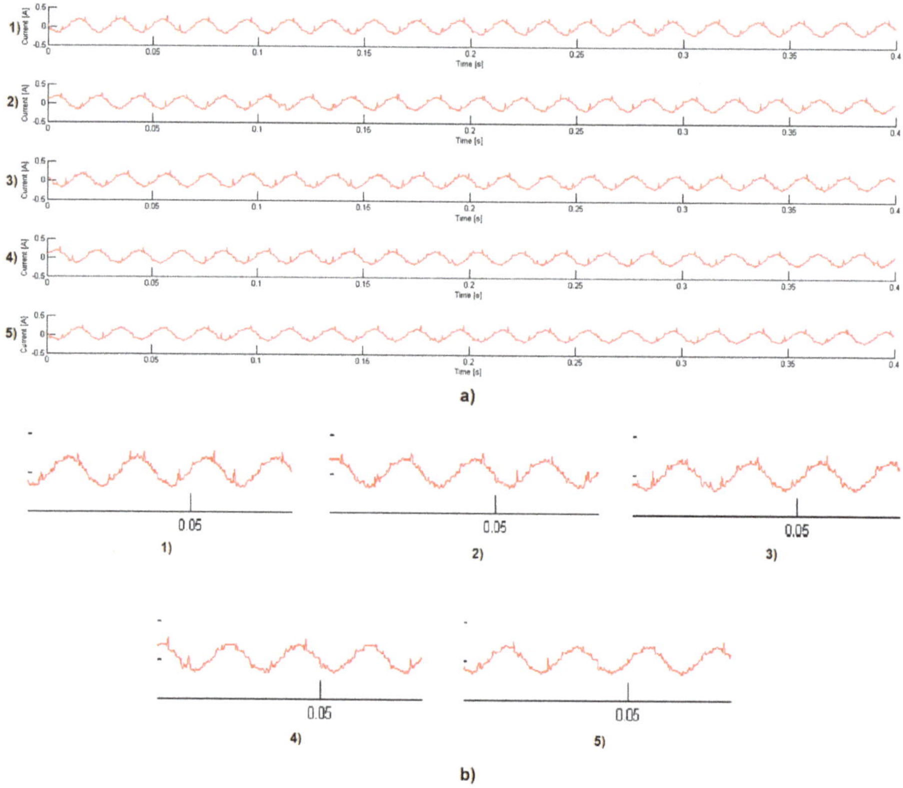

Figure 3. Current signal at electrode configuration A under a 7.2 kV voltage level and for the selected gap distances: (**a**) original waveforms; (**b**) magnified fragments of the waveforms; (**1**) 1, (**2**) 2, (**3**) 3, (**4**) 4, (**5**) 5 mm.

The current signals were more often distorted in the 1 and 2 mm gap distances, as seen in Figure 4 when comparing waveforms (1) and (2) with the other waveforms from this Figure. This is especially visible when observing the magnified fragment presented in Figure 4b. However, for a meaningful comparison of the PD behavior between the cases of the 1 and 2 mm gap distances, the proposed harmonic analysis method is needed. Additionally, for other electrode gaps, it is clearly seen in Figure 4 that it is so hard to analyze the PD behavior just by using the current signal. Hence, the proposed method of analyzing the FFT modified signals may give more information about the PD behavior in the analyzed electrode system immersed in oil.

The discharges seen in the oils cause the oil to locally warm up and eventually the oil to expand. This phenomenon is visible as the oil movement [11,20,24,34]. Such behavior of the oil was also observed during the studies performed herein, with a significant fluctuation in the oil during intensive discharges. In the following parts of the study, this movement made by the oil after the discharge appeared was considered in the analyses conducted. In some cases, the oil movement might eliminate the deteriorations in signals collected, especially when the distance increased to 5 mm. In such cases, no comparable harmonic distortion behavior was observed, and thus the harmonic components were evaluated as a possible source of information on PD activity, which is presented in the further part of the study.

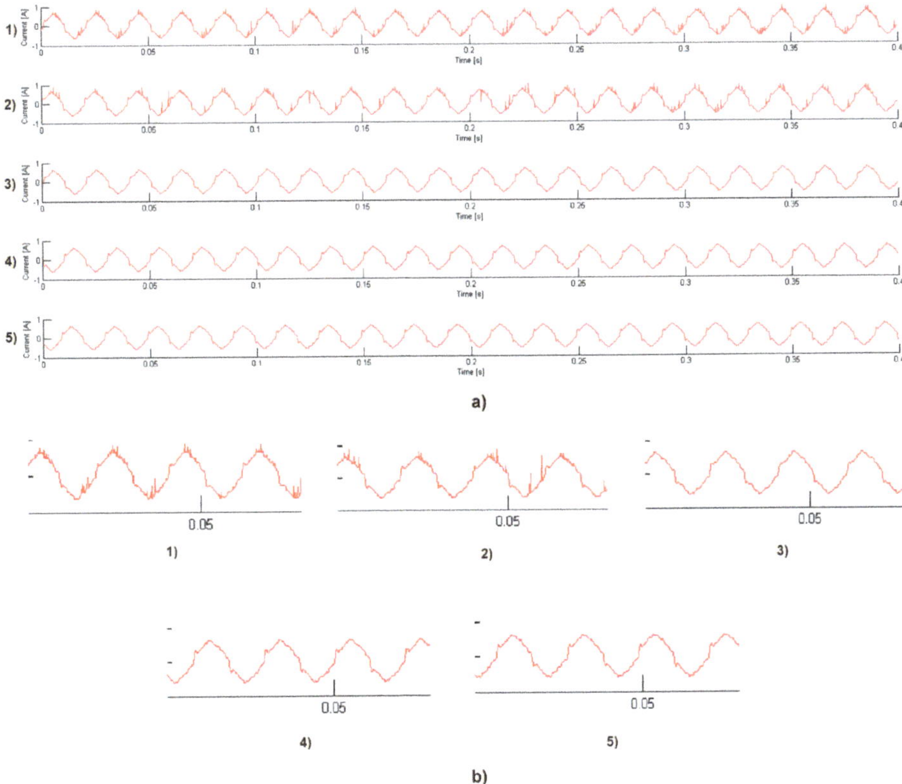

Figure 4. Current signal at electrode configuration E under a 10.8 kV voltage level and for the selected gap distances: (**a**) original waveforms; (**b**) magnified fragments of the waveforms; (**1**) 1, (**2**) 2, (**3**) 3, (**4**) 4, (**5**) 5 mm.

First, in Figure 5 in the form of a 3D chart, the data concerning configuration A are presented.

As can be seen from Figure 5 and simultaneously by analyzing the data from Table 2, it is first conspicuous that the value of the 5th harmonic for all electrode gaps and all voltage levels is the greatest. Due to the fact that the stimulated electrons at low voltage levels could not flow more easily into the grounded electrode, the potential capacitive effects around the electrodes at low voltage (5.4 kV) levels put forward the second harmonic (Penning effect) for this experiment. It is, however, well-known that the Penning effect is noted in the reactions between molecules and neutral atoms, and this is defined as chemi-ionization, especially occurring in gaseous insulators. Herein, this effect is referred to indicate that the 2nd harmonics arise from structures similar to chemical ionization [33,37,38].

It was observed also that the value of the 7th harmonic is greater at gaps of 4 mm between the electrodes, where harmonic distortion was observed intensely. The harmonic values at gaps of 1 mm between electrodes are expected to be higher due to the small electrode gap between the electrodes, which influences the value of electric field stress around the HV point. Since the closest gap between the electrodes is comparing the same value of testing voltage, a higher electrical field stress is created and more intense PDs occur with the higher generation of harmonics in the current signal. However, since the partial discharge is continuous at this gap, there are no sudden signal distortions. At the gaps of 3 and 4 mm, partial discharges from the leakage current signal are also noticed, and they are damped by the oil movement. Nevertheless, when partial discharge is initiated, since as the same time it can initiate harmonic distortions in the leakage current signal at a distance greater than the gap of

1 mm between electrodes, the harmonic distortion values here are high. In addition, it was observed that even the harmonic values were found to be higher.

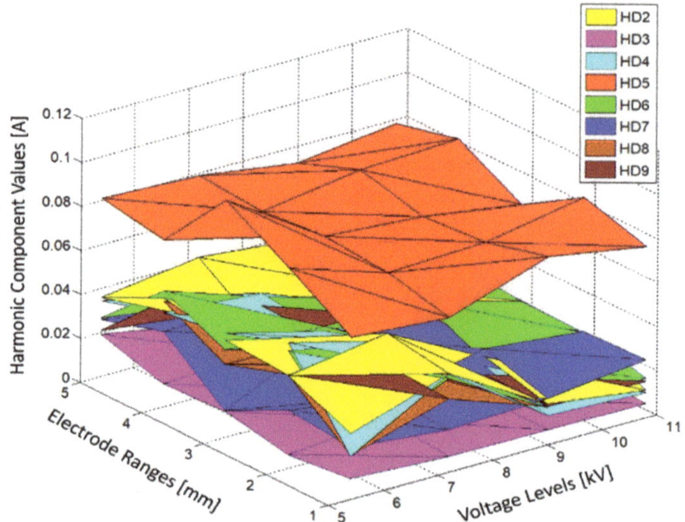

Figure 5. A 3D chart of the HD values for electrode configuration A and all voltage levels.

The results, in a similar way to those concerning the A configuration, are presented in Figure 6 for configuration B. When Table 2 and Figure 6 are examined at the 9 kV voltage level, it is seen that the harmonic values of the 2nd, 3rd, 4th, and 5th are higher at the gap of 2 mm compared to the harmonic values found for other electrode ranges. When the harmonic values are examined for all ranges, it is seen that the 5th harmonic values are 2–4 times higher than the other harmonic values.

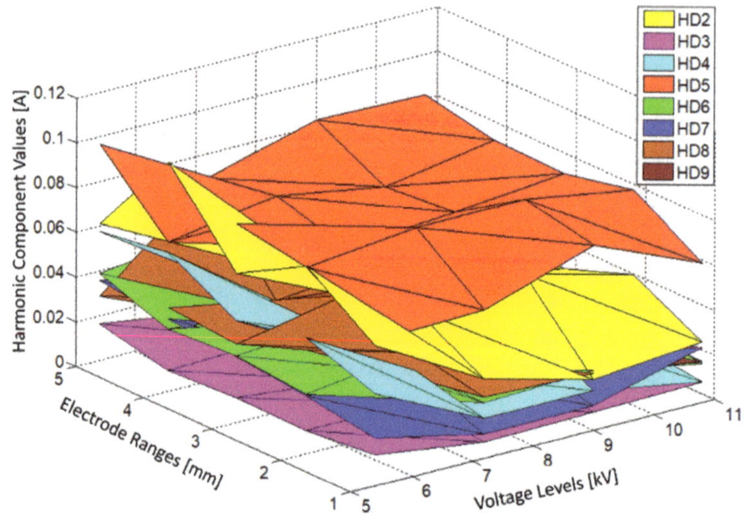

Figure 6. A 3D chart of the HD values for electrode configuration B and all voltage levels.

The steel (medical) needle has a lower electrical conductivity than the copper needle. Therefore, partial discharge started at a higher voltage level for the configurations based on this type of HV electrode. This is because, in the case of gaseous or liquid insulation (mineral oil herein), PDs appear as a result of a local electric field stress increase, especially around the HV electrode. In such a case, the developing discharge needs energy for propagation, and this energy is ensured by fast electron movements (in the form of an electron avalanche) to the grounded electrode. In the initial phase of the PDs, they can continue their development through the electron movements. Additionally, when the number of electrons which are "knocked" from the HV electrode increase, the PD formation becomes faster. The conductivity of the electrode is thus important from the point of view of the number and accumulation of these movement electrons (the number of electrons in the avalanche). The more conductive a material is, more "knocked" electrons may take part in the discharge process [15,33]. Since the partial discharges at the 5.4 and 7.2 kV voltage levels have a low energy, no current signal is given in this section. The partial discharges began at 9 kV and have a high amplitude, especially at the gap of 2 mm. The current signal distortions at this range were seen at more points than for other ranges.

The results concerning configuration C are presented in Figure 7. In this electrode configuration, the diameter of the grounded electrode is smaller than that previously described. Thus, when a high voltage is applied, the distance it takes the free electrons to reach the sphere is shortened. The ionization path between the electrodes was shortened and partial discharge started earlier. In Figure 7, as the distance between the electrodes increases, the values of the harmonic components generally are less due to the oil movement as the ionization path increases.

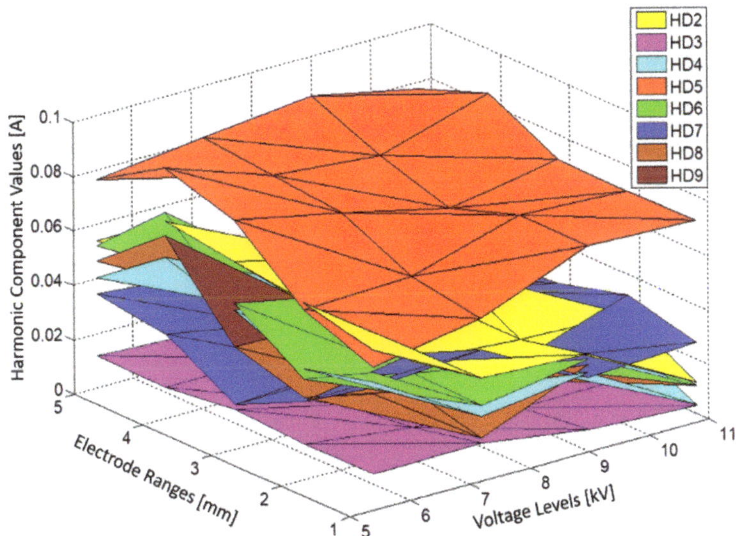

Figure 7. A 3D chart of the HD values for electrode configuration C and all voltage levels.

Analyzing jointly Table 2 and Figure 7 for all the electrode ranges, we can state that the 2nd and 5th harmonics predominantly have higher values than the others. In turn, the 6th harmonic value is higher than 2nd when the gap distance is 4 mm. When the 5th and 2nd harmonics are compared with each other, it is seen that value of the 2nd harmonic is higher at the 1 mm and 2 mm gaps. However, at the gaps of 3, 4, and 5 mm, the value of the 5th harmonic is higher. Even harmonics are known to occur when the oil movement decreases. Since the distance at which oil can move is increased at 3 mm and at higher distances, the 2nd harmonic value is less than the 5th harmonic.

Figure 8 shows the HD values for electrode configuration D when the high-voltage electrode is constituted of steel.

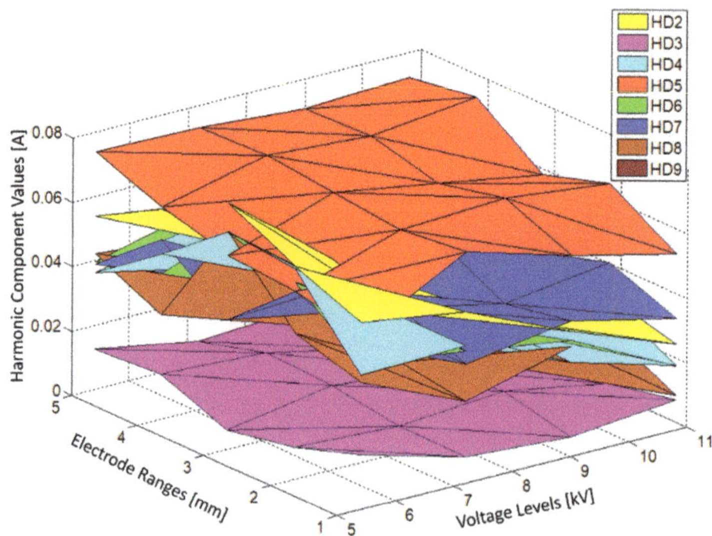

Figure 8. A 3D chart of the HD values for electrode configuration D and all voltage levels.

As can be seen in Figure 8, the harmonic distortion behavior in electrode configuration D is similar to the behavior in electrode configuration C. However, more ripples were formed at the 5.4 kV voltage level at the 2 and 3 mm distances. It can be obviously seen in Table 2 that the 2nd harmonic values at the gap of 1, 2, and 3 mm between electrodes are higher than the other harmonic values. The 5th harmonic value is the highest at the gaps of 4 and 5 mm. In addition, it is observed that the 3rd harmonic value decreases for the gaps 1, 2, and 3 mm and differs from the 5th harmonic value, but then increases again for the gaps of 4 and 5 mm.

For the E and F electrode configurations, the results are presented in Figures 9 and 10, respectively. The grounded electrode has, in these cases, a diameter of 2 cm. The harmonic component surfaces at electrode configuration E are shown in Figure 9.

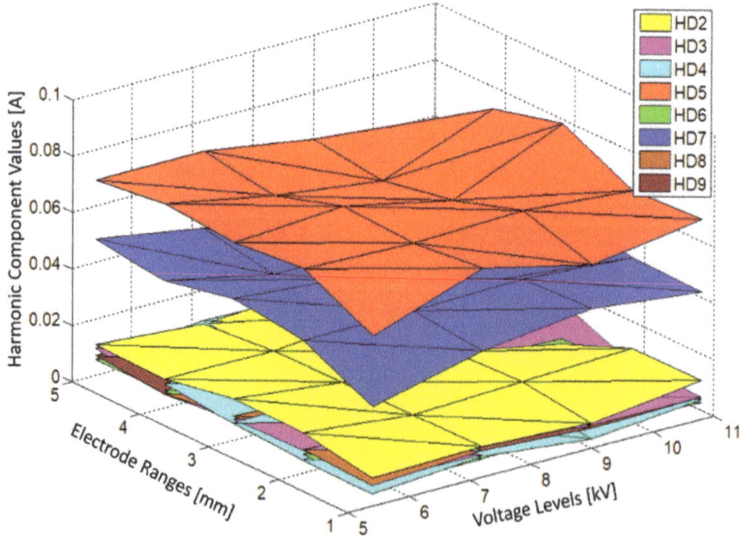

Figure 9. A 3D chart of the HD values for electrode configuration E and all voltage levels.

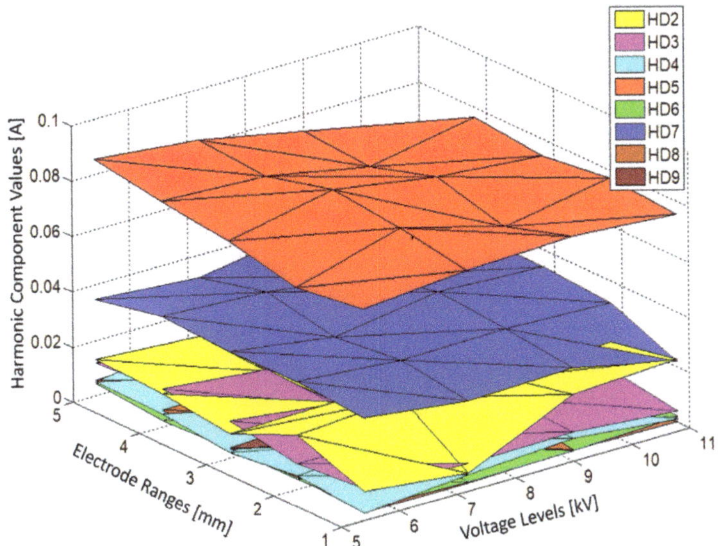

Figure 10. A 3D chart of the HD values for electrode configuration F and all voltage levels.

In this electrode configuration, characteristic points were formed during harmonic distortion at different electrode ranges at a voltage level of 9 kV. In Figure 9, the values of the 5th and 7th harmonics are significantly higher than the others and the 2nd harmonic is slightly higher. At the gap of 1 mm value the 5th harmonic is the highest, thereby at this distance the partial discharge and arc energy are reach the distorted value for the insulation strength. However, at the gap of 2 mm, the value of the 3rd harmonic was identified as higher than the counterparts in the other ranges.

The steel (medical) pointed-needle high-voltage electrode has a lower conductivity than the copper one used in the F-type electrode configuration. Therefore, in the case of configuration F the distortion in the signal is less than that of the E-type electrode configuration, which can be noticed from Figure 10. As shown also in Figure 10 and in Table 2, the 5th and 7th harmonic values are generally higher than the other harmonic values predominantly. At the gap of 2 mm, the value of the 5th harmonic component of the leakage current was the highest among all the harmonics. Conversely, at the gap of 1 mm the 4th harmonic value is lowest. The capacitive effect of the bubbles in the oil also caused the formation of even harmonics.

5. Conclusions

In the presented studies, a measurement method based on collecting the leakage currents through the grounded electrode under a non-uniform electric field was applied in order to assess the PD behavior in the designed electrode system immersed in mineral oil. To do this, the collected leakage currents were differentiated with FFT analysis and the harmonic distortion level was obtained.

From the results obtained, the following findings may be stated:

(1) By using the technique proposed in the paper, it was possible to measure the value of the leakage current corresponding with the partial discharges of corona type in oil at the different metal points, creating high-voltage electrode and different electric field distributions based on a non-invasive measurement technique.

(2) In all electrode configurations and gap distances, even harmonics components were measured.

(3) The amplitude of the 5th harmonic was noticed to be dominant in almost all cases, giving an unequivocal indication of which of the harmonic components can be used to identify the corona PDs in oil.

(4) The 2nd and 7th harmonics were also found to be clearly identified within the registered range of leakage current. They are characteristic in a slightly lower range for the configurations E and F, wherein the 7th harmonic has been identified with discharges for a small sphere creating the grounded electrode.

(5) It was observed that the calculated harmonic distortion values for all voltage levels increase when the sphere diameter decreases and the voltage level increases. This may be a result of the increase in the energy of the discharges formed.

(6) For the lowest voltage level, the harmonics (except the 5th and 9th harmonics) were measured to be higher in the case of the steel HV electrode, regardless of the gap distance between the electrodes. The reason for this may be attributed to the material (its conductivity) as the factor determining this finding.

(7) The studies allowed us to notice that the 7th harmonic is more visible when the E and F configurations were analyzed. This means that appearance of this harmonic may be accompanied by a higher level of electrical field uniformity.

(8) It has been observed that the 5th harmonics are the most recognized from others when the electric field uniformity reaches its maximum, as in the case of electrode configuration A.

(9) As the conductivity of the steel (medical) needle was less than that of the copper one, it was found that the harmonics were higher due to the conductivity disadvantage of the steel compared to the copper, and thus the electrode tip deformed faster.

(10) The measurements performed opened the way to continue studies of other electrode configurations, creating the systems generating surface-type PDs or PDs in internal cavities. Such studies are being planned by the authors to be performed in the near future. In the studies planned, also other dielectric liquids such as synthetic and natural esters will be considered as possible sources of changes in the registered quantities. An important part of the future studies will be also the verification of the correlation between the PDs' behavior and harmonics using the methods based on indexes of similarity, such as Pendry, van Hove, Integrated against Error Log Frequency (IELF), or Frequency Selective Validation (FSV) [39].

Author Contributions: Conceptualization, A.E.Y.; methodology, A.E.Y.; software, A.A.; validation, A.E.Y. and P.R.; formal analysis, F.A. and P.R.; investigation, F.A.; resources, A.A.; data curation, A.A.; writing—original draft preparation, A.A., F.A., and P.R.; writing—review and editing, F.A. and P.R.; visualization, F.A.; supervision, A.E.Y.; project administration, A.E.Y.; funding acquisition, A.A. All authors have read and agreed to the published version of the manuscript.

Funding: This work was supported by the Istanbul University-Cerrahpaşa Scientific Research Projects Executive Secretariat with the Project number of 33689.

Conflicts of Interest: The authors declare no conflict of interest.

References

1. Li, S.; Yu, S.; Feng, Y. Progress in and prospects for electrical insulating materials. *High Volt.* **2016**, *1*, 122–129. [CrossRef]

2. Fofana, I. 50 Years in the development of insulating liquids. *IEEE Electr. Insul. Mag.* **2013**, *29*, 13–25. [CrossRef]

3. Lu, W.; Liu, Q. Effect of cellulose particles on streamer initiation and propagation in dielectric liquids. In Proceedings of the IEEE 11th International Conference on the Properties and Applications of Dielectric Materials (ICPADM), Sydney, Australia, 19–22 July 2015; pp. 939–942. [CrossRef]

4. Zhang, J.; Li, J.; Huang, D.; Zhang, X.; Liang, S.; Li, X. Influence of Nonmetallic Particles on the Breakdown Strength of Vegetable Insulating Oil. In Proceedings of the 2015 IEEE Conference on Electrical Insulation and Dielectric Phenomena (CEIDP), Ann Arbor, MI, USA, 18–21 October 2015; pp. 609–612. [CrossRef]

5. Tao, Z.; Yunpeng, L.; Fangcheng, L.; Ruihong, D. Study on dynamics of the bubble in transformer oil under non-uniform electric field. *IET Sci. Meas. Technol.* **2016**, *10*, 498–504. [CrossRef]

6. Berg, G.; Lundgaard, L.E.; Becidan, M.; Sigrnond, R.S. Instability of electrically stressed water droplets in oil. In Proceedings of the IEEE 14th International Conference on Dielectric Liquids (ICDL), Graz, Austria, 12–12 July 2002; pp. 220–224. [CrossRef]

7. Yilmaz, H.; Guler, S. The effect of electrode shape, gap and moisture on dielectric breakdown of transformer oil. In Proceedings of the 12th International Conference on Conduction and Breakdown in Dielectric Liquids (ICDL), Roma, Italy, 15–19 July 1996; pp. 354–357. [CrossRef]

8. Uzunoğlu, C.P. A Comparative study of empirical and variational mode decomposition on high voltage discharges. *Electrica* **2018**, *18*, 72–77. [CrossRef]

9. Piotrowski, T.; Rozga, P.; Kozak, R. Analysis of excessive hydrogen generation in transformers in service. *IEEE Trans. Dielectr. Electr. Insul.* **2015**, *22*, 3600–3607. [CrossRef]

10. Dombek, G.; Nadolny, Z. Liquid kind, temperature, moisture, and ageing as an operating parameters conditioning reliability of transformer cooling system. *Maint. Reliab.* **2016**, *18*, 413–417. [CrossRef]

11. Rozga, P.; Stanek, M.; Rapp, K. Lightning properties of selected insulating synthetic esters and mineral oil in point-to-sphere electrode system. *IEEE Trans. Dielectr. Electr. Insul.* **2018**, *25*, 1699–1705. [CrossRef]

12. Khaled, U.; Beroual, A. AC dielectric strength of mineral oil-based Fe_3O_4 and Al_2O_3 nanofluids. *Energies* **2018**, *11*, 3505. [CrossRef]

13. *High-Voltage Test Techniques—Partial Discharge Measurements*; IEC Standard 60270; International Electrotechnical Commission: Geneva, Switzerland, 2000.

14. Ersoy, A. Investigating Boron Contributions and Electrical Properties of Polymeric Insulators Used in Electrical Insulation Systems. Ph.D. Dissertation, Institute of Science and Technology, Istanbul University, Istanbul, Turkey, 2007.

15. Bartnikas, R. Partial discharges. Their mechanism, detection and measurement. *IEEE Trans. Dielectr. Electr. Insul.* **2002**, *9*, 763–808. [CrossRef]

16. Kunicki, M.; Cichon, A.; Nagi, L. Statistics based method for partial discharge identification in oil paper insulation systems. *Electr. Power Syst. Res.* **2018**, *163*, 559–571. [CrossRef]

17. Zhao, J.; An, Z.; Lv, B.; Wu, Z.; Zhang, Q. Characteristics of the partial discharge in the development of conductive particle-initiated flashover of a GIS Insulator. *Energies* **2020**, *13*, 2481. [CrossRef]

18. Abu-Siada, A. Improved consistent interpretation approach of fault type within power transformers using dissolved gas analysis and gene expression programming. *Energies* **2019**, *12*, 730. [CrossRef]

19. Sikorski, W. Active dielectric window: A new concept of combined acoustic emission and electromagnetic partial discharge detector for power transformers. *Energies* **2019**, *12*, 115. [CrossRef]

20. Kunicki, M.; Cichon, A.; Borucki, S. Measurements on partial discharge in on-site operating power transformer: A case study. *IET Gener. Transm. Distrib.* **2018**, *12*, 2487–2495. [CrossRef]

21. Walczak, K.; Sikorski, W.; Gil, W. Multi-module system for partial discharge monitoring using AE, HF and UHF methods. *Prz. Elektrotech.* **2016**, *92*, 5–9. [CrossRef]

22. Liang, R.; Wu, S.; Chi, P.; Peng, N.; Li, Y. Optimal placement of UHF sensors for accurate localization of partial discharge source in GIS. *Energies* **2019**, *12*, 1173. [CrossRef]

23. Florkowski, M.; Krzesniak, D.; Kuniewski, M.; Zydron, P. Partial Discharge Imaging Correlated with Phase-Resolved Patterns in Non-Uniform Electric Fields with Various Dielectric Barrier Materials. *Energies* **2020**, *13*, 2676. [CrossRef]

24. Koziol, M. Energy distribution of optical radiation emitted by electrical discharges in insulating liquids. *Energies* **2020**, *13*, 2172. [CrossRef]

25. *Partial Discharges in Transformer Insulation*; CIGRE Task Force 15.01.04; International Council on Large Electric Systems (CIGRE): Paris, France, 2000.

26. Chen, W.; Chen, X.; Peng, S.; Li, J. Canonical correlation between partial discharges and gas formation in transformer oil paper insulation. *Energies* **2012**, *5*, 1081–1097. [CrossRef]

27. Kreuger, F.H.; Gulski, E.; Krivda, A. Classification of partial discharges. *IEEE Trans. Electr. Insul.* **1993**, *28*, 917–931. [CrossRef]

28. Gulski, E.; Kreuger, F.H. Computer-aided recognition of discharge sources. *IEEE Trans. Electr. Insul.* **1992**, *27*, 82–92. [CrossRef]

29. Cavallini, A.; Contin, A.; Montanari, G.C.; Puletti, F. Advanced PD inference in on-field measurements. I. Noise rejection. *IEEE Trans. Dielectr. Electr. Insul.* **2003**, *10*, 216–224. [CrossRef]

30. Cavallini, A.; Conti, M.; Contin, A.; Montanari, G.C. Advanced PD inference in on-field measurements. II. Identification of defects in solid insulation systems. *IEEE Trans. Dielectr. Electr. Insul.* **2003**, *10*, 528–538. [CrossRef]

31. Contin, A.; Pastore, S. Classification and separation of partial discharge signals by means of their auto-correlation function evaluation. *IEEE Trans. Dielectr. Electr. Insul.* **2009**, *16*, 1609–1622. [CrossRef]

32. Alvarez, F.; Ortego, J.; Garnacho, F.; Sanchez-Uran, M.A. A clustering technique for partial discharge and noise sources identification in power cables by means of waveform parameters. *IEEE Trans. Dielectr. Electr. Insul.* **2016**, *23*, 469–481. [CrossRef]

33. Kufel, E.; Zaengl, W.S.; Kufel, J. *High Voltage Engineering—Fundamentals*, 2nd ed.; Elsevier: Oxford, UK, 2000.

34. *Fluids for Electrotechnical Applications—Unused Mineral Insulating Oils for Transformers and Switchgear*; IEC Standard 60296; International Electrotechnical Commission: Geneva, Switzerland, 2003.

35. Patrissi, S.; Pompili, M.; Yamashita, H.; Forster, E.O. A study of the effect of electrical breakdown in dielectric liquids on the needle point structure. In Proceedings of the IEEE 11th International Conference on Conduction and Breakdown in Dielectric Liquids, Baden-Dättwil, Switzerland, 19–23 July 1993; pp. 376–382. [CrossRef]

36. Lesaint, O. Prebreakdown phenomena in liquids: Propagation modes and basic physical properties. *J. Phys. D Appl. Phys.* **2016**, *49*, 14401–14422. [CrossRef]

37. Falcinelli, S.; Candori, P.; Bettoni, M.; Pirani, F.; Vecchiocattivi, F. Penning ionization electron spectroscopy of hydrogen sulfide by metastable helium and neon atoms. *J. Phys. Chem. A* **2014**, *118*, 6501–6506. [CrossRef] [PubMed]

38. Korenev, S. The internal Penning Ion Source of negative hydrogen ions for isochronous cyclotrons. In Proceedings of the IEEE Pulsed Power Conference (PPC), Austin, TX, USA, 31 May–4 June 2015; pp. 1–6. [CrossRef]

39. Bongiorno, J.; Mariscotti, A. Evaluation of performances of indexes used for validation of simulation models based on real cases. *Int. J. Math. Models Methods Appl. Sci.* **2015**, *9*, 29–43.

MDPI

St. Alban-Anlage 66

4052 Basel

Switzerland

Tel. +41 61 683 77 34

Fax +41 61 302 89 18

www.mdpi.com

Energies Editorial Office

E-mail: energies@mdpi.com

www.mdpi.com/journal/energies